53 Springer Series in Solid-State Sciences
Edited by Hans-Joachim Queisser

Springer Series in Solid-State Sciences

Editors: M. Cardona P. Fulde H.-J. Queisser

Volume 40 **Semiconductor Physics** - An Introduction By K. Seeger

Volume 41 **The LMTO Method** Muffin-Tin Orbitals and Electronic Structure
By H.L. Skriver

Volume 42 **Crystal Optics with Spatial Dispersion, and Excitons**
By V.M. Agranovich and V.L. Ginzburg

Volume 43 **Resonant Nonlinear Interactions of Light with Matter**
By V.S. Butylkin, A.E. Kaplan, Yu.G. Khronopulo, and E.I. Yakubovich

Volume 44 **Elastic Media with Microstructure II** Three-Dimensional Models
By I.A. Kunin

Volume 45 **Electronic Properties of Doped Semiconductors**
By B.I. Shklovskii and A.L. Efros

Volume 46 **Topological Disorder in Condensed Matter**
Editors: F. Yonezawa and T. Ninomiya

Volume 47 **Statics and Dynamics of Nonlinear Systems**
Editors: G. Benedek, H. Bilz, and R. Zeyher

Volume 48 **Magnetic Phase Transitions**
Editors: M. Ausloos and R.J. Elliott

Volume 49 **Organic Molecular Aggregates,** Electronic Excitation and Interaction Processes
Editors: P. Reineker, H. Haken, and H.C. Wolf

Volume 50 **Multiple Diffraction of X-Rays in Crystals**
By Shih-Lin Chang

Volume 51 **Phonon Scattering in Condensed Matter**
Editor: W. Eisenmenger, K. Laßmann, and S. Döttinger

Volume 52 **Magnetic Superconductors and Their Related Problems**
Editors: T. Matsubara and A. Kotani

Volume 53 **Two-Dimensional Systems, Heterostructures, and Superlattices**
Editors: G. Bauer, F. Kuchar, and H. Heinrich

Volumes 1 - 39 are listed on the back inside cover

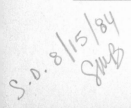

Two-Dimensional Systems, Heterostructures, and Superlattices

Proceedings of the International Winter School
Mauterndorf, Austria, February 26 – March 2, 1984

Editors:
G. Bauer F. Kuchar H. Heinrich

With 231 Figures

Springer-Verlag
Berlin Heidelberg New York Tokyo 1984

Professor Dr. Günther Bauer
Institut für Physik, Montanuniversität Leoben, Franz-Josef-Straße 18
A-8700 Leoben, Austria

Professor Dr. Friedemar Kuchar
Institut für Festkörperphysik, Universität Wien, A-1090 Wien, and
Ludwig Boltzmann Institut für Festkörperphysik, A-1060 Wien, Austria

Professor Dr. Helmut Heinrich
Institut für Experimentalphysik, Universität Linz, A-4040 Linz, Austria

Series Editors:

Professor Dr. Manuel Cardona
Professor Dr. Peter Fulde
Professor Dr. Hans-Joachim Queisser

Max-Planck-Institut für Festkörperforschung, Heisenbergstrasse 1
D-7000 Stuttgart 80, Fed. Rep. of Germany

ISBN 3-540-13584-7 Springer-Verlag Berlin Heidelberg New York Tokyo
ISBN 0-387-13584-7 Springer-Verlag New York Heidelberg Berlin Tokyo

Offset printing: Beltz Offsetdruck, 6944 Hemsbach/Bergstr. Bookbinding: J. Schäffer OHG, 6718 Grünstadt
2153/3130-543210

Preface

This volume contains the proceedings of the International Winter School on "Heterostructures and Two-Dimensional Electronic Systems in Semiconductors". The school took place in Mauterndorf, Austria, from February 26 - March 2, 1984, and was the third one in a series of winter schools on "New Developments in Solid-State Physics", organized by the Austrian Physical Society, Solid State Physics Division.

The school was attended by about 150 scientists from 12 countries, including the United States of America, Japan and Poland. Most of the research groups of Western Europe working in this field participated and demonstrated the still increasing interest in the topics. These topics cover the wide area ranging from fundamental phenomena in solid-state physics, like the quantum Hall effect, to new semiconductor devices, all based on two-dimensional electronic systems.

We hope that the spirit of this school, the combination of basic and applied physics (including skiing) will stimulate further progress in this field of research.

This conference was sponsored by the

Austrian Physical Society,
Austrian Federal Ministry for Science and Research,
European Research Office of the US Army,
European Office of Aerospace Research and Development,
Federal Province of Salzburg, and
Österreichische Forschungsgemeinschaft

Further financial support came from the following companies: Balzers, IBM Austria, Messer-Griesheim, Oxford Instruments, Klaus Schäfer & Co., Siemens Österreich and Varian.

March, 1984 *G. Bauer · F. Kuchar · H. Heinrich*

Contents

Part I Physics of Heterostructures and Inversion Layers

Part II Growth and Devices

Part III Multi Quantum Wells and Superlattices

Part IV Doping Superlattices

Part V Quantum Hall Effect

Part I

Physics of Heterostructures and Inversion Layers

Electric and Magnetic Quantisation of Two-Dimensional Systems - Elementary Theory

Wlodek Zawadzki

Institute of Physics, Polish Academy of Sciences, 02668 Warsaw, Poland

We describe electric and magnetic quantisation of two-dimensional systems for standard and narrow-gap semiconductors, using extensively the semiclassical quantisation procedure. The density of electron states is discussed.

1. Electric Quantisation

If one considers a movable electron gas in the presence of an external potential, the spatial electronic distribution is described in general by the Schrödinger equation for the wavefunction (containing an effective potential) and the Poisson equation for the potential (containing the wavefunction through the charge density). These have to be solved together in a selfconsistent way. However, in order to simplify the problem we assume that the effective potential is fixed and consider only most important idealized cases.

1a. Quantisation. In the effective mass approximation for a spherical energy band the Schrödinger equation reads

$$\left(\frac{p^2}{2m^*} + U\right)\Psi = E\Psi . \tag{1}$$

If the potential is one-dimensional: $U = U(z)$, one looks for solutions in the form

$$\Psi = A \exp(ik_x x + ik_y y)\phi(z) \tag{2}$$

which reduces (1) to

$$\left(\frac{1}{2m^*}p_z^2 + U\right)\phi = \varepsilon\phi \tag{3}$$

in which

$$\varepsilon = E - \frac{\hbar^2}{2m^*}(k_x^2 + k_y^2) . \tag{4}$$

In order to proceed further one has to specify the potential. We consider four different types of effective potentials, which define quantum wells for various structures: 1) Infinite square well (heterostructures of different semiconductors); 2) Parabolic well (n-i-p-i periodically doped crystals |1|); 3) Triangular asymmetric well (metal-insulator-semiconductor devices);

4) Symmetric triangular well (bi-crystals |2|). Figure 1 shows schematically the potentials in question. Finite square wells are more difficult to treat and they are not considered here (cf. |3|).

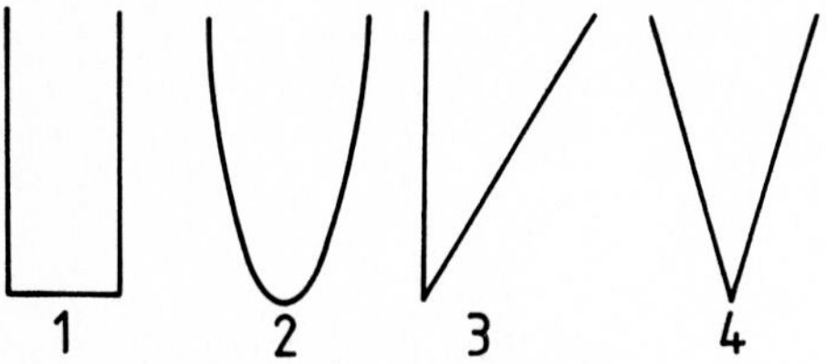

Fig.1 Four potential wells discussed in the text: 1) infinite square; 2) parabolic; 3) triangular asymmetric; 4) triangular symmetric. (Schematically)

It is not difficult to compute exact energies of the eigenvalue problem (3) for the above potentials. However, for reasons of simplicity and in view of more advanced applications, we use when possible the semiclassical Wentzel - Kramers - Brillouin approximation. In its lowest order it reduces to the famous Bohr-Sommerfeld quantisation condition.

$$\int_{z_1}^{z_2} p_z dz = \hbar\pi(r + c_1 + c_2) \qquad (5)$$

where p_z is to be treated as a non-operator and determined from (3), z_1 and z_2 are turning points of the classical motion (at which $p_z = 0$), c_1 and c_2 are phases to be determined, $r=0,1,2,...$ is the number of zeros of the wavefunction. The first correction to condition (5) involves d^2U/dz^2, cf. |4,5,6|, it is then clear that the WKB approximation is particularly good for linear potentials.

A. Infinite Square Well. The potential is

$$\begin{aligned} U(z) &= 0 \quad \text{for } |z| < a \\ U(z) &= \infty \quad \text{for } |z| > a. \end{aligned} \qquad (6)$$

Exact eigenenergies for this problem are quoted in all textbooks on quantum mechanics. In order to use (5) we observe that, if the walls are infinitely high, the wavefunction must vanish at the turning points $z = \mp a$, hence the phases are $c_1 = c_2 = 1/2$. Thus we have

$$(2m^*\varepsilon)^{1/2} \int_{-a}^{+a} dz = \hbar\pi(r+1) \qquad (7)$$

which gives

$$\varepsilon_r = \frac{\hbar^2\pi^2(r+1)^2}{8m^*a^2} \qquad r = 0,1,2,.... \qquad (8)$$

Thus the semiclassical quantisation gives here the exact result!

The exact wavefunctions for the problem are well known and easy to handle.

B. Parabolic Well. We take the potential in the form

$$U(z) = \frac{m^*\omega^2}{2} z^2 . \tag{9}$$

The turning points are $z_{1,2} = \mp(2\varepsilon/m^*\omega^2)^{1/2}$ and the phases are $c_1 = c_2 = 1/4$ because at the turning points the potential is not far from linear. Thus we have

$$(2m^*)^{1/2} \int_{z_1}^{z_2} (\varepsilon - \frac{m^*\omega^2}{2} z^2)^{1/2} dz = \hbar\pi(r + \frac{1}{2}) \tag{10}$$

which gives after simple integration

$$\varepsilon = \hbar\omega(r + \frac{1}{2}) \qquad r = 0,1,2,\ldots . \tag{11}$$

Again the exact result! The wavefunctions for the problem (harmonic oscillator) are well known and easy to handle.

C. Asymmetric Triangular Well. The potential is

$$\begin{aligned} U(z) &= eFz \quad &\text{for } z \geqslant 0 \\ U(z) &= \infty \quad &\text{for } z < 0 . \end{aligned} \tag{12}$$

F is an effective electric field. In this case $c_1 = 1/2$ since at the left turning point $z_1 = 0$ the wave function must vanish, and $c_2 = 1/4$ since at the right turning point $z_2 = \varepsilon/eF$ the potential is linear (cf. |7|). Equation (5) becomes

$$(2m^*)^{1/2} \int_0^{z_2} (\varepsilon - eFz)^{1/2} dz = \hbar\pi(r + \frac{3}{4}) \qquad r = 0,1,2,\ldots . \tag{13}$$

An elementary integration gives semiclassical energies

$$\varepsilon_r = (\frac{\hbar^2 e^2 F^2}{2m^*})^{1/3} [\frac{3}{2}\pi(r + \frac{3}{4})]^{2/3} . \tag{14}$$

Formula (14) approximates exact eigenvalues up to a fraction of a percent. Exact wavefunctions for the problem are given by appropriate parts of the Airy function, but in practical calculations it is easier to deal with variational functions of proper symmetry (cf. |8|).

D. Symmetric Triangular Well. The potential is

$$U(z) = eF|z| . \tag{15}$$

The turning points occur at $z_{1,2} = \mp \varepsilon/eF$. The phases are $c_1 = c_2 = 1/4$, since at the turning points the potential is linear. An integration similar to that in (13) gives

$$\varepsilon_r = (\frac{\hbar^2 e^2 F^2}{2m^*})^{1/3} [\frac{3}{4}\pi(r + \frac{1}{2})]^{2/3} . \tag{16}$$

Equation (16) approximates exact eigenenergies up to a fraction of a percent. Exact wavefunctions for the problem are given by symmetrised or antisymmetrised parts of the Airy function, but in practical calculations it is easier to deal with variational functions of proper symmetry.

1b. Density of States

As follows from (4), in all above cases the complete energy is

$$E_{rk_xk_y} = \varepsilon_r + \frac{\hbar^2}{2m^*}(k_x^2 + k_y^2) \qquad (17)$$

so that we deal with a series of two-dimensional electric subbands. It is of interest to know the corresponding density of electron states. To this end one should sum over all states belonging to the same energy. Introducing polar coordinates: $k_\parallel = (k_x^2 + k_y^2)^{1/2}$ and ϕ, and taking into account the spin degeneracy, we get for one subband

$$\rho_r(E)dE = \frac{2\cdot 2\pi}{(2\pi)^2} k_\parallel \frac{dk_\parallel}{dE} dE = \frac{m^*}{\pi\hbar^2} \ . \qquad (18)$$

Thus each subband contributes the same constant density of states! Clearly, this contribution begins at the subband edge ε_r, cf. (17), so that the total density of states is

$$\rho(E) = \frac{m^*}{\pi\hbar^2}\sum_r \theta\,(E - \varepsilon_r) \qquad (19)$$

where θ is the step function: $\theta(x) = 0$ for $x < 0$, $\theta(x) = 1$ for $x \geqslant 0$. This density of states is shown schematically in Fig.2. It is clear that the system is truly two-dimensional when one deals with only one electric subband. If more subbands come into play, the situation loses gradually its two-dimensional features.

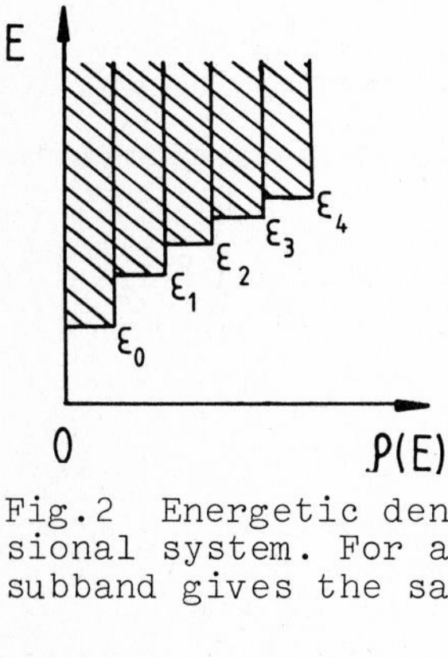

Fig.2 Energetic density of electron states for a two-dimensional system. For a simple parabolic energy band each electric subband gives the same constant contribution. (Schematically)

1c. Optical Transitions

It is of interest to know what optical transitions are possible in the system. The electron-photon interaction within the model described by (1) is

$$H_R = \frac{eA_0}{m^*c} (\vec{a}\cdot\vec{p}) \qquad (20)$$

where A_0 is the amplitude and $\vec{a}$ is the polarization of the vector potential of radiation. Let us consider optical transitions between the lowest and the first excited electric subband. There are two possibilities of interest: light polarization transverse or parallel to the surface. For the transverse polarization a_z the matrix element of optical transition is, cf. Eq. (2),

$$(\Psi_1|H_R^z|\Psi_0) = \frac{eA_0}{m^*c} \delta_{k'_\parallel k_\parallel} a_z (\phi_1|p_z|\phi_0) . \qquad (21)$$

The above integral over z does not vanish, so that the optical transitions for the transverse polarization are possible. On the other hand, for the parallel light polarization (say a_x) there is,

$$(\Psi_1|H_R^x|\Psi_0) = \frac{eA_0}{m^*c} a_x \hbar k_x \; \delta_{k'_\parallel k_\parallel} (\phi_1|\phi_0) = 0 \qquad (22)$$

because ϕ_1 and ϕ_0 are orthogonal, belonging to different energy eigenvalues. Thus, in the one-band effective mass approximation the inter-subband optical transitions are not possible for the light polarization parallel to the interface. Clearly, this result does not depend on the form of the effective potential U(z).

2. Electric and Magnetic Quantisation

We assume now that, in addition to electric field in the z direction, there exists also an external magnetic field $\vec{H}$ parallel to it. For parallel fields the situation is particularly simple, because electric field affects only the electron motion in the z direction, while magnetic field affects only the motion in the x-y plane.

2a. Quantisation. In the effective mass approximation for a simple parabolic band the Schrödinger equation (excluding spin) reads

$$\left[\frac{1}{2m^*} (\vec{p} + \frac{e}{c}\vec{A})^2 + U(z)\right]\Psi = E\Psi . \qquad (23)$$

$\vec{A}$ is the vector potential of magnetic field. For $\vec{H} = (0,0,H)$ the vector potential depends only on x and y variables, so that (23) can be rewritten as

$$[H_1(x,y) + H_2(z)]\,\Psi = E\Psi . \qquad (24)$$

Looking for solutions in the form

$$\Psi(x,y,z) = g(x,y)\,\phi(z) \qquad (25)$$

we separate the variables in the usual way. The eigenvalue problem (23) becomes equivalent to two separate equations

$$(\frac{1}{2m^*} p_z^2 + U)\phi = \varepsilon\phi \qquad (26)$$

$$\frac{1}{2m^*}[(p_x + \frac{e}{c}A_x)^2 + (p_y + \frac{e}{c}A_y)^2]g = \varepsilon_\parallel g \qquad (27)$$

$$E = \varepsilon + \varepsilon_{\parallel} . \tag{28}$$

This separation corresponds to the above-mentioned property of electron motion in parallel fields.

Equation (26) has been discussed above. In order to solve (27) one has to specify $\vec{A}$. We take the asymmetric Landau gauge $\vec{A} = (-Hy,0,0)$, which gives $\mathrm{curl}\vec{A} = (0,0,H)$, as it should. In this case the variables in (27) can be further separated by looking for solutions in the form

$$g(x,y) = A \exp(ik_x x) f(y) . \tag{29}$$

It is well known that with the above choice of gauge, (27) becomes similar to the eigenvalue problem for the harmonic oscillator with a characteristic frequency $\omega_c = eH/m^*e$, called the cyclotron frequency |9,10|. The energies are

$$\varepsilon_{\parallel} = \hbar\omega_c(n + \tfrac{1}{2}) \qquad n = 0,1,2,... \tag{30}$$

n is called Landau quantum number and the energies (30) are called Landau levels. The eigenfunctions are

$$f(y) = \Phi_n(\frac{y-y_o}{L}) \tag{31}$$

where Φ_n are harmonic oscillator functions, $y_o = k_x/L^2$, and $L = (\hbar c/eH)^{1/2}$ is called magnetic (Landau) radius. y_o has the meaning of the center of oscillations and L characterises an extent of the cyclotron orbit. It can be seen that the wavefunctions of magnetic motion do not depend on the electron mass. On the other hand, the energy (30) does not depend on the quantum number k_x, which results in a high degeneracy of each Landau level. Due to the choice of the asymmetric gauge for $\vec{A}$, the wavefunctions (29) and (31) do not reflect spiral symmetry of the magnetic motion. If one chooses a symmetric (Dingle) gauge: $\vec{A} = (-Hy/2, + Hx/2,0)$, the spiral symmetry is preserved but the wavefunctions are more complicated |9,11|. Clearly, all observable quantities (energies, transition probabilities, selection rules, etc.) do not depend on the choice of gauge.

If the electron spin is included, the Schrödinger equations (23) and (27) should be completed by adding the Pauli term $\mu_B g^* \vec{H}\cdot\vec{\sigma}$, where μ_B is the Bohr magneton, g^* is the effective spin-splitting (Landé) factor and $\vec{\sigma}$ is Pauli spin operator. This gives the magnetic energies

$$\varepsilon_{\parallel} = \hbar\omega_c(n + \tfrac{1}{2}) \pm \tfrac{1}{2} g^* \mu_B H . \tag{32}$$

The corresponding wavefunctions have spin-up and spin-down components.

As follows from (28) and (32) the total energies are

$$E_{rns} = \varepsilon_r + \hbar\omega_c(n + \tfrac{1}{2}) + \tfrac{1}{2} g^* \mu_B Hs \quad r=0,1,2,... \; n=0,1,2,... \tag{33}$$

ε_r are the electric levels discussed above, $s = \pm 1$ is the spin variable. Thus the energies are now completely quantised. In the presence of a magnetic field each electric subband develops into a series of Landau and spin levels counted from the subband edge.

2b. Density of States

In order to calculate the density of states due to the Landau quantisation one has to count the number of states belonging to the same energy (33). We assume that the sample has dimensions L_x and L_y. Then the allowed k_x values are $(2\pi/L_x)\cdot l$, where l is an integer. Thus there are $L_x/2\pi$ allowed values of k_x within the unit length of k_x in k space. The center of electron oscillations $y_0 = k_x L^2$ should be within the sample, which gives the condition $0 < k_x L^2 < L_y$, hence $0 < k_x < L_y/L^2$. Thus, there are $L_x L_y/2\pi L^2$ allowed values of k_x in the sample, belonging to each Landau and spin level (32). Since the energies are discrete, the density of states is

$$\rho(E) = \frac{L_x L_y}{2\pi L^2} \sum_{rns} \delta\,(E-E_{rns}). \tag{34}$$

$\delta(x)$ denotes the Dirac delta function. The density of states (34) is characterised by the delta-like singularities, in contrast to free electrons in a magnetic field, for which the density of states is a sum of square-root singularities $(E-E_{ns})^{-1/2}$.

However, in real systems the density of states is not singular. Due to scattering, inhomogeneities, etc., the singularities are damped and broadened. This feature is particularly important in two-dimensional systems, because of the strongly singular initial form (34). There are various ways of broadening the delta functions. In practice a compromise is often made between physical considerations and mathematical convenience. There exist four important models for more realistic peaks of the density of states: rectangular, elliptic, Gaussian, Lorentzian. The e are shown schematically in Fig.3. It appears that the Gaussian shape is the most realistic, cf. discussion in |12|. An example of its use can be found in the description of magnetic and thermal properties of two-dimensional electron gas |13|. In our elementary considerations we leave aside the problems of mobility edges, localised states, etc.

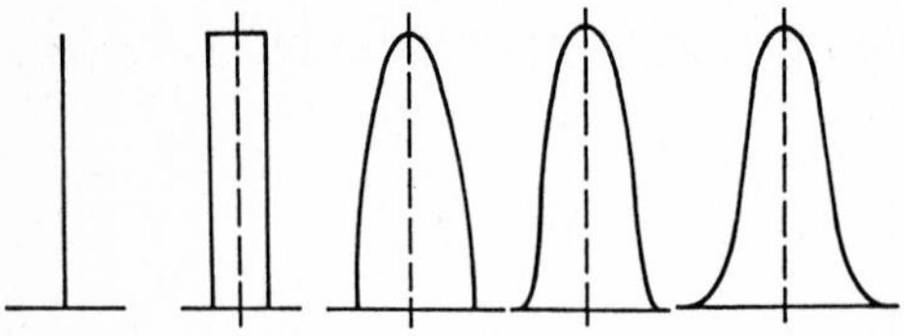

Fig.3 Models for the density of states in two-dimensional systems in a magnetic field: a) delta function, b) rectangular, c) elliptic, d) Gaussian, e) Lorentzian

3a.Electric and Magnetic Quantisation. Narrow-gap Semiconductors

Narrow-gap semiconductors are characterised by a nonparabolic dispersion relation E(k). We are concerned here with nonparabolic but spherical conduction bands, which occur in III-V compounds: InSb, InAs, GaSb, GaAs and their mixed crystals. The main feature of the band structure in question is that one may not neglect the proximity of the valence band when considering the conduction band effects. In two-dimensional systems non-parabolicity effects are particularly strong, because one deals often with electric subband energies comparable to the gap value ε_g. We do not go into calculations of the band model, as it has been done before, cf. |8,14|. If, as assumed above, the potential is U = U(z) and the magnetic field $\vec{H} = (0,0,H)$, the effective eigenenergy problem for the conduction band in the two-band model reads

$$\left[\frac{1}{2m_o^*} p_z^2 - \frac{1}{\varepsilon_g}(E-\varepsilon_\parallel - U)(\varepsilon_g+E+\varepsilon_\parallel - U)\right]\phi(z) = 0 \tag{35}$$

where for H=0

$$\varepsilon_\parallel = -\frac{\varepsilon_g}{2} + \left[\left(\frac{\varepsilon_g}{2}\right)^2 + \varepsilon_g \frac{\hbar^2 k_\parallel^2}{2m_o^*}\right]^{1/2} \tag{36}$$

and for $H \neq 0$

$$\varepsilon_\parallel = -\frac{\varepsilon_g}{2} + \left[\left(\frac{\varepsilon_g}{2}\right)^2 + \varepsilon_g D_{ns}\right]^{1/2} \tag{37}$$

with

$$D_{ns} = \hbar\omega_c\left(n + \frac{1}{2}\right) + \frac{1}{2} g_o^* \mu_B H s. \tag{38}$$

m_o^* and g_o^* are the effective mass and the effective spin-splitting factor at the bottom of the conduction band (in absence of fields), respectively. The energy E is counted from the conduction band edge (in absence of fields). It can be seen that when the potential U and the energies in question are much smaller than the gap ε_g, Eqs.(35) (36) (37) reduce to the one--band approach considered above.

Equation (35) can be conveniently used for the semiclassical quantisation. We consider briefly four important cases treated in Section 1.

A. Infinite Square Well. The Bohr-Sommerfeld condition reads now

$$\left(\frac{2m_o^*}{\varepsilon_g}\right)^{1/2} (E-\varepsilon_\parallel)^{1/2} (\varepsilon_g+E+\varepsilon_\parallel)^{1/2} \int_{-a}^{+a} dz = \hbar\pi(r+1) \tag{39}$$

which leads to

$$\frac{(E-\varepsilon_\parallel)(\varepsilon_g+E+\varepsilon_\parallel)}{\varepsilon_g} = \frac{\hbar^2\pi^2(r+1)^2}{8m_o^* a^2} \qquad r = 0,1,2,\dots . \tag{40}$$

This presents a quadratic equation for the energies E_r.

B. Parabolic Well. The phases are $c_1 = c_2 = 1/4$, and the turning points for the conduction band are, as before: $z_{1,2} = \mp|2(E-\varepsilon_\parallel)/m_o^*\omega^2|^{1/2}$. The BS condition is of the form

$$\left(\frac{2m_o^*}{\varepsilon_g}\right)^{1/2} \int_{z_1}^{z_2} (A-c^2z^2)^{1/2} (B-c^2z^2)^{1/2} dz = \hbar\pi(r + \tfrac{1}{2}) \; . \qquad (41)$$

This presents an elliptic integral, which is to be carried out numerically.

C. Asymmetric Triangular Well. The phases are $c_1 = 1/2$ and $c_2 = 1/4$, and the turning points for the conduction band are, as before: $z_1 = 0$ and $z_2 = (E-\varepsilon_\parallel)/eF$. The BS condition is

$$\left(\frac{2m_o^*}{\varepsilon_g}\right)^{1/2} \int_{o}^{z_2} (E-\varepsilon_\parallel - eFz)^{1/2} (\varepsilon_g+E+\varepsilon_\parallel -eFz)^{1/2} dz = \hbar\pi(r+\tfrac{3}{4}) \, . \qquad (42)$$

The integral can be carried out analytically and the resulting semiclassical energies are given by the relation (see |8|)

$$(a+b)a^{1/2}b^{1/2}+(b-a)^2\ln\left|\frac{b^{1/2}-a^{1/2}}{(b-a)^{1/2}}\right| = \left(\frac{\varepsilon_g}{2m_o^*}\right)^{1/2} 4eF\hbar\pi(r+\tfrac{3}{4}) \qquad (43)$$

where

$$a = E-\varepsilon_\parallel \qquad \text{and} \qquad b = \varepsilon_g + E + \varepsilon_\parallel \; . \qquad (44)$$

D. Symmetric Triangular Well. The phases are $c_1 = c_2 = 1/4$, and the turning points for the conduction band are, as before: $z_{1,2} = \mp(E-\varepsilon_\parallel)/eF$. The integration similar to that in (42) leads to the following relation for the eigenenergies,

$$(a+b)a^{1/2}b^{1/2}+(b-a)^2\ln\left|\frac{b^{1/2}-a^{1/2}}{(b-a)^{1/2}}\right| = \left(\frac{\varepsilon_g}{2m_o^*}\right)^{1/2} 2eF\hbar\pi(r+\tfrac{1}{2}) \qquad (45)$$

where a and b are defined in (44).

Thus, as long as the electron motion is confined within one semiconductor material whose band structure is known, the semiclassical quantisation of energies does not present serious difficulties. The narrow-gap band structure introduces the following important modifications, as compared to the case of a standard band:

1. Spacing of electric subbands is modified.
2. Relation $E_r(k_\parallel)$ in a subband is nonparabolic. In the presence of a magnetic field this corresponds to an uneven spacing of the Landau and spin levels.
3. Inter-subband optical transitions become possible for light polarisation parallel to the interface, see |8,15|.

3b. Density of States. In a nonparabolic energy band the effective mass depends in general on the energy, so that expression (19) has to be modified. Defining an effective mass by the relation

$$\frac{1}{m_r^*(k_\parallel)} = \frac{1}{\hbar^2 k_\parallel} \frac{dE_r(k_\parallel)}{dk_\parallel} \qquad (46)$$

and using the first equality in (18), we can write

$$\rho(E) = \frac{1}{\pi\,\hbar^2} \sum_r m_r^*(E)\; \theta(E-\varepsilon_r) \qquad (47)$$

in which $m_r^*(E) = m_r^*(k_{||})$ and the relation $E(k_{||})$ for a given subband are to be determined from relations (40), (41), (43) or (44) for the system in question.

Acknowledgments. I am grateful to Dr. Jacek Kossut for informative discussion. I appreciate expert help of Miss Maryla Borejko in the preparation of the typescript.

References

1. P.Ruden and G.H.Döhler, Phys. Rev. B27, 3538 (1983)
2. S.Uchida, G.Landwehr, and E.Bangert, Solid State Commun. 45, 869 (1983)
3. G.Bastard, Phys. Rev. B25, 7584 (1982)
4. J.L.Dunham, Phys. Rev. 41, 713 (1932)
5. J.B.Krieger, M.L.Lewis, and C.Rosenzweig, J.Chem.Phys. 47, 2942 (1967)
6. P.N.Argyres, Physics 2, 131 (1965)
7. A.B.Migdal, Qualitative Methods in Quantum Theory, Moscow (Nauka: 1975, in Russian)
8. W.Zawadzki, J. Phys. C: Sol. St. Phys. 16, 229 (1983)
9. L.D.Landau and E.M.Lifshitz, Quantum Mechanics, 3d Edition, Moscow (Mir: 1974, in French)
10. D.I.Blokhintzev, Quantum Mechanics, 3d Edition, Moscow 1961 (in Russian)
11. R.B.Dingle, Proc.Roy.Soc. A211, 500 (1952)
12. T.Ando, A.B.Fowler, and F.Stern, Rev.Mod.Phys. 54, 537 (1982)
13. W.Zawadzki, this volume
14. W.Zawadzki, in Lecture Notes in Physics, Vol. 133, Ed. W.Zawadzki (Springer: 1980) p. 85
15. H.Reisinger, PhD. Thesis, Technical University of Munich 1983 (unpublished)

The Effects of a Quantizing Magnetic Field

P.J. Stiles*

Cavendish Laboratory, Madingley Road, Cambridge CB3 OHE, United Kingdom

A discussion is given of some of the effects due to the presence of a quantizing magnetic field. This includes the interplay of single particle effects such as spin, valley and Landau levels with the manybody effects. The results of tilted field, the thermoelectric, de Haas-van Alphen and capacitance experiments, the beat effect and the quantized resistance are discussed.

1 Introduction

This volume discusses much of the background and the latest results in a field of study some twenty years old. An important milestone for people interested in learning more of this field was the work "Electronic properties of two-dimensional systems" by Ando, Fowler and Stern [1], published recently. Therefore we will not attempt to cover the background material in those other presentations. They should be consulted for further information and a full reference list. We will always be speaking of two-dimensional electronic systems in the conduction band of Si formed on a 100 surface unless noted otherwise. Further one also assumes $kT < \hbar\omega_C < E_F$ where these symbols have their usual meaning.

We have made this choice for a number of reasons. It is easy to collect material on many effects. The effective mass of the carriers is essentially constant because of the parabolic bands. The spin orbit coupling is very weak. This latter fact allows one to consider a decoupled spin and orbit system. However such a choice as Si 100 means we will not discuss the effect of non-parabolicity or spin orbit coupling. Further it limits the maximum value of $\omega_C\tau$ studied to about 50 (τ is the H=0 conductivity τ) whereas values for heterostructures of over 1000 exist.

2 Microscopic Effects

In the absence of a magnetic field, our two-dimensional system has a density of states that is independent of energy for those energies above the ground state level. This is rigorously true for a non-fluctuating potential, manybody interactions and non-parabolicity. For the case of a fluctuating potential, the density of states is altered appreciably. There are two ways in which the density of states is affected. Firstly, the lifetime of the states is altered due to the scattering by the random potential. Secondly there is a spatial inhomogeneity in any particular sample. This inhomogeneity can have two causes. If we have a gaussian distribution of the position of coulomb centres, the mean potential in any size region will vary by having a net number different from similar regions elsewhere in the crystal. The second cause can be that there is an additional gaussian

* Permanent address: Physics Dept., Brown Univ., Providence, RI 02912, USA

distribution, i.e. on a longer scale in a space which is not given by the initial distribution. In either case the density of states that we speak of has an energy dependence, most markedly near the ground state level.

The study of the behaviour of this two-dimensional electron gas for H=0 and for $\omega_c\tau < 0.5$ where localized effects modify the simple electron behaviour has recently been discussed [2]. We do not treat this region here.

The two major effects of the quantizing magnetic field are a modification of the energy spectrum and the degeneracy. The energy of the states due to motion in the xy plane is modified by the presence of a magnetic field in the following way [3] (in the absence of disorder)

$$E_{nN} = \hbar\omega_c \left(N + \frac{1}{2}\right) + \frac{\hbar^2}{2m_t \ell_y^4} \left[(z^2)_{nn} - (z_{nn})^2\right] + E_n \tag{1}$$

where n is the electrically quantized level corresponding to motion in the z direction, $\omega_c = eH_z/m_t c$, $\ell_y^2 = c\hbar/eH_y$ and m_t is the isotropic effective mass in the xy plane. The degeneracy for each state is given by eB/h. We must include the energy of each of the two spin states and each of the two valley states (that is the band structure effect for the Si 100 surface). The energy relative to the magnetically shifted ground state is given by

$$E_N = \hbar\omega_c \left(N + \frac{1}{2}\right) \pm \frac{1}{2}g^*\beta H_T \pm E_v . \tag{2}$$

This is a single-particle energy. The manybody effects modify the energy through dependence of g^* and E_v on n_s, the surface electron density.

When we consider modification of the density of states due to the effect of disorder, we do not as yet have a definitive answer. What we probe are macroscopic parameters, those which are averages of the microscopic density of states. We do not know the distribution of the disorder. We do have fairly convincing evidence that some states are localized while others are extended. The quantum Hall effect is a good example of the latter (see this volume). Here we must be careful to include in our definition of localized states those whose spatial extent may be much smaller than the magnetic length ℓ_y as well as states with extent orders of magnitude larger than ℓ_y. In keeping with the discussion of the effect of disorder without a magnetic field, we must remember that there are many possible length scales of disorder, and that of the real probability of macroscopic disorder dominating some situations.

3 The Tilted Field Effect

We recognize from the previous treatment that the separation of Landau levels depends on the component of the magnetic field perpendicular to the surface while the spin splitting depends on the total field (for Si, spin orbit coupling ~ 0). At some angle of tilt, the separation of spin levels in a Landau level can be made equal to those between spins on neighbouring Landau levels. From such an experiment Fang and Stiles [4] noted that the g^* was enhanced and the enhancement depended on n_s, the surface electron density. They did a detailed measurement of the temperature dependence and found that of the density chosen, $g^* = 2.7$ and $m_t^* = (0.185 \pm 1.6\%)m_0$, where m_0 is the free electron mass. The mass value is 3% less than that measured in bulk whereas the "enhanced" mass appears to be 10% higher than the bulk. Why this is so is unresolved.

The effective mass measured by Smith and Stiles [5] for H perpendicular to the surface showed a small enhancement (~ 10%) that depended on n_s. However when they rotated and increased the field to keep the normal component of the magnetic field constant, the properties of the effective mass changed drastically [6]. The n_s dependence remained, but the scale factor changed. It appears to become a very heavy mass when one had the spin splitting about half of the Landau splitting. This behaviour is also not understood at present. The tilted field experiments indicate that the macroscopic behaviour is not governed by: 1, the single-particle picture; or 2, a simple application of manybody theory.

4 The beat effect

Previously we discussed the effect of having the magnetic field not perpendicular to the 100 surface. Here we discuss the case where the magnetic field is perpendicular to the surface, but the 100 axis has been rotated a small angle from the normal to the surface (a different sample). Ando [7] about the time that experimental evidence was being considered, described what the valley splitting would be for the case listed above. Theories existed for the valley splitting as a function of n_s for H=0, with ΔE_v being nearly linearly dependent on n_s, ($\Delta E_v = a n_s$ where a is about 1 mev/5 x $10^{12}/cm^2$). To calculate the splitting of the two valley states, one needs the matrix element between states from each valley of the surface potential. In the presence of quantizing magnetic field, these are harmonic osicllator states. As the centre of the orbits project to different values of k parallel to the surface, the matrix element of the surface potential oscillates as [7]

$$\Delta E_v(n_s,H) \sim \Delta E_v(n_s,0)\,|J_{NN}(2k_o \ell \sin\theta)| \qquad (3)$$

where θ is the angle between the 100 axis and the surface normal, k_o is the conduction band minimum k as measured from Γ and ℓ is our previously defined ℓ_y.

Let us consider the behaviour of the Shubnikov–de Haas (SdH) effect for low magnetic fields. The qualitative behaviour of the valley splitting for two magnetic fields is illustrated in Fig. 1a. Here we show the case for $\theta = 0.3°$. We have included the values of 0.5 $\hbar\omega_c$ for the two cases. At those points where 0.5 $\hbar\omega_c = \Delta E_v(n_s,H)$ the fundamental period disappears, we see a "beat", the phase of the oscillations for 0.5 $\hbar\omega_c > \Delta E_v$ and 0.5 $\hbar\omega_c < \Delta E_v$ differ by 180°. Such behaviour [8] was used to plot out the behaviour of ΔE_v in the H,n_s plane. The loci of the points for $\hbar\omega_c = \Delta E_v$

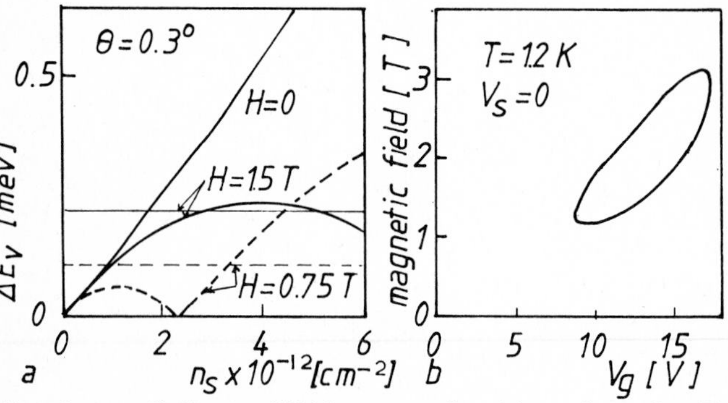

Fig. 1: a, Valley splitting as a function of n_s for three magnetic fields; b, Loci of points where 0.5 $\hbar\omega_c$

is shown in 1b. From this the value of the valley splitting for low magnetic fields was determined for the first time. Ando's theory proved essentially correct, except that the argument of the Bessel function contains not k_o but $(0.15/0.85)\ k_o$. This was due to the overlap matrix element of the projection through the X point rather than through the Γ point being the larger.

The justification for this statement involves recognition that the spin splitting is smaller than the valley splitting in the region of interest and is presently ignored, and that at low fields, $\omega_c\tau$ is sufficiently low that manybody enhancement is essentially zero.

5 Magnetocapacitance

There has been much recent effort towards trying to obtain the density of states via the measurement of the magnetocapacitance. This effort is also discussed in AFS.

Stern's [9] modelling of the problem involved treating the charging of the linear capacitor from both ends as that of the lossy stripline problem. The problem then reduces to the solution of the one-dimensional diffusion equation (1DDS) and the application of boundary conditions. The complex capacitance C_{eff} is given by

$$C_{eff} = C_1[1 + (C^*/C_d)(2/a)\tan(a/2)]$$

where $C_1 = C^*C_d/C_t$, $C_t = C^*+C_d$, $C^* = C_oC_i/(C_o+C_i)$, $C_o = C_{oxide}$, $C_i = C_{inv}$, $a = (i\omega RC_t)^{1/2}$ and R is the source drain resistance.

If this were an adequate description of the behaviour, then one could measure the real and imaginary parts of the capacitance and from the ratio and one of the absolute values determine the capacitance and the resistivity of the inversion layer. The change in C_{inv} could then be correlated with a model for the density of states (which would have to include a frequency dependence). The behaviour in the absence of the magnetic field and for small magnetic fields is shown in Fig. 2 [10]. We show also in the Fig. the behaviour expected from the 1DDE. We see that the behaviour is quite similar, but that there is not agreement. The lack of agreement under these conditions is most striking in the region where we expect a high percentage of the states to be localized.

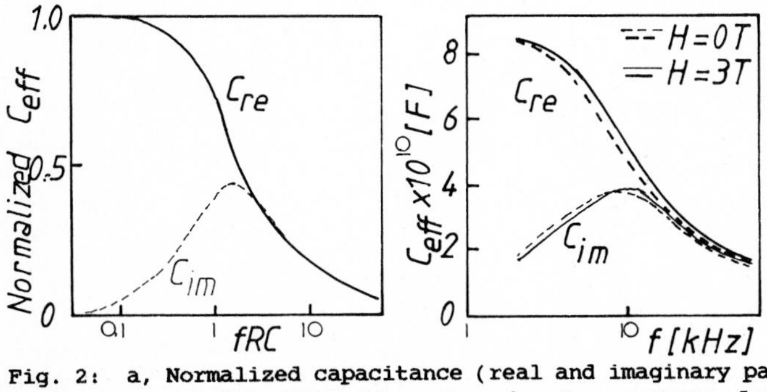

Fig. 2: a, Normalized capacitance (real and imaginary parts) of a MOSFET as a function of fRC (1DDE theory); b, C_{eff} versus frequency for a real MOSFET

We now focus on the case for a high quantizing magnetic field. When we are in a region in which the Fermi energy is between the centres of the Landau levels a region of localized states, we would expect that the 1DDE would give poor results. This is shown in Fig. 3. There are two causes for the behaviour deviating from that expected. The first is that we should compare the results with the solution for a 2DDE not a 1DDE. The second reason is that although the two-terminal resistance is quantized (see the following section), the conductance involved with charging the capacitor is that via localized states, not via the Hall term. This is obvious as there are no conducting states empty near the Fermi level or we would have loss. When the real and imagninary parts are plotted vs R_{SD}, we see the strange results in Fig. 4. This is obviously not the correct procedure. The case where the one-dimensional model should work, the Corbino disk structure, gives results similar to those in Fig. 3. However, as noted before, the results do not fit well at all. We conclude that there is reason to believe that proper modelling and fitting experimental results will indeed give vital information on the density of states.

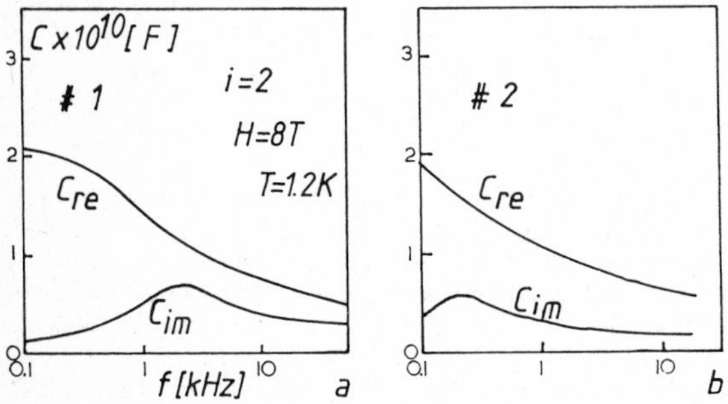

Fig. 3: Real and imaginary parts of C_{eff} as a function of frequency for a sample with: a, #1, μ=1.2 m^2/V-s; b, #2, μ=0.8 m^2/V-s

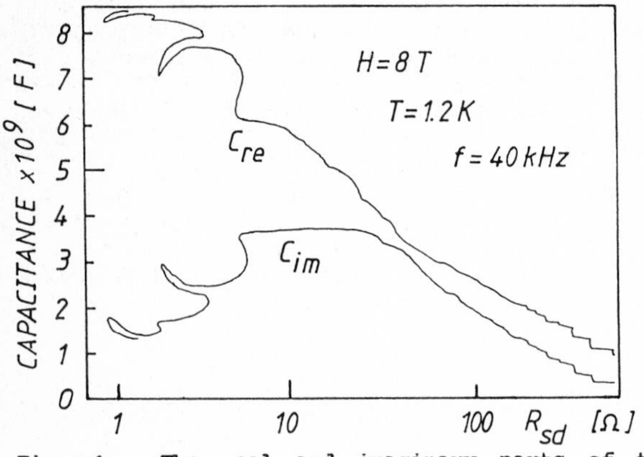

Fig. 4: The real and imaginary parts of the capacitance versus the source-drain resistance

6 The Quantized Resistance

Under the conditions that the quantized Hall effect is observed, there exists another curious phenomena known as the quantized resistance effect.

This latter refers to the fact that if we measure the two-terminal resistance of the sample with any pair of contacts, the resistance is always the same independent of the sample geometry provided that the two terminals are on the periphery of the sample [11]. It has been shown to be the same value as that of the quantized Hall resistance to the accuracy of the measurements, $1/10^5$. One would expect that behaviour only if the contact resistance $< 10^{-5}\ h/e^2$, as it is expected to be. Further if under the above conditions one interconnects other contacts on the periphery with a low resistance [12], one obtains quantized resistances that are integer ratios of the quantized Hall resistance. These ratios are given by $R_q' = (N/M) \times R_q$ where N and M are integers equal to or less than one-half of the number of contacts. Similar behaviour has been obtained when one interconnects more than one such quantized Hall effect region [12]. Here one has the option to connect regions where the number of filled Landau levels i and i' are different. A simple analysis in terms of current and potentials solves this set of cases also.

Of perhaps more interest is the case where the two layers of properly prepared quantized regions for two values of i are contiguous. Here it is hard to achieve experimentally as one is not certain that one has not set up a "normal" region between them. However it appears that there are some peculiar results on the best samples obtained to date. The result , that only certain Landau levels communicate with others, has some of the mystery of the quantized resistance in a lossless medium [13].

Consider the fact that as σ is zero, that there can be no loss as $J.E=0$. However we have seen that for our two-terminal resistor, $I \times V = 0$, with the resistance being determined by the lossless regions. How can such be, and where is the energy going? Clearly we cannot have I^2R losses in the contacts or the measured two-terminal resistance would not be quantized as it would depend on the choice of contacts. Further, the conductivity of the contacts is much higher than the σ_{xy} of the quantized two-dimensional system. This argues that the current must enter and leave at the corners of the contacts. What is the scale of the region where the perimeter of the 2DEG is not an equal potential? It cannot be less than the cyclotron radius. The electric field at the corner can be high, but the integral of the field, the voltage can be made orders of magnitude less than the corresponding energy between Landau levels. It is thought that the current is carried by the extended states in the middle of the Landau levels. These are $\sim \hbar\omega_c$ below the Fermi energy. How do the carriers rise up to the Fermi energy of the contact?

There appears to be no concensus on the above. Further experiments will have to be performed. It is expected that the understanding of the quantized resistance effect will shed more light on the unique state that exists when the two-dimensional system is in the quantized Hall effect regime.

7 The de Haas-van Alphen Effect

The dHvA effect has recently been studied by two groups, one working on multilayer heterostructures [14], and the other on Si MOSFET [15] inversion layers. The former experiments, performed in the hope that one could obtain information on the density of the states, made quantitative measurements and observed for a fixed density as a function of field a signal many times smaller than predicted. It appears that the results were limited by the macroscopic inhomogeneities discussed before. The experiments performed on Si inversion layers did not do quantitative measurements. They, however, while fixing the magnetic field, did the experiments as a function of density by modulating the density. This latter approach of modulating the density requires the passage of charge into the inversion layer, with its accompanying eddy currents. Fortunately, the eddy currents have an ω^2 amplitude dependence while the dHvA has a linear one. This allows a distinction between the two. Striking spike structure was observed between Landau levels as was expected. However without quantitative measurements, no attempt was made to determine the density of states from the experimental results.

8 Thermoelectric Effect

A study of the magnetoconductivity tensor allows one to gather information about the two independent components of the magnetoconductivity tensor. The thermoelectric effect studied in the same sample allows two other parameters to be obtained, which are effectively the energy derivatives of the two components of the magnetoconductivity tensor. Studies in the two dimensional systems have been studied by a number of investigators [16,17]. In addition, a number of theoretical treatments are available [18,19,20,21]. Results to date have been focused in two main areas, those which approach the ideal limit, and hence yield little about the material properties, results far from the ideal limit, which appear to be dominated by the breadth of the density of states.

Of particular interest are those with the dependence on disorder. The thermoelectric effect with its four parameters is very sensitive to the properties of the density of states, certainly more so than the magnetoconductivity tensor. Preliminary results for heterostructures and Si MOSFETs indicate that the approach is likely to be very fruitful. Recent theoretical results including disorder are encouraging [22].

9 Conclusions

The recognition of the quantum Hall effect has brought intense interest to the two-dimensional systems. While the quantum Hall effect is understood, the details of the density of states and how the latter relates to the quantum Hall effect has not been answered.

References

1 T. Ando, A.B. Fowler and F. Stern: Rev. Mod. Phys. 54, 437 (1982)
2 M.J. Uren, R.A. Davies, M. Kaveh and M. Pepper: J. Phys. C: Solid State Phys. 14, 5737 (1981)
3 F. Stern and W.E. Howard: Phys. Rev. 163, 816 (1967)
4 F.F. Fang and P.J. Stiles: Phys. Rev. 174, 823 (1968)
5 J.L. Smith and P.J. Stiles: Phys. Rev. Lett. 29, 102 (1972)
6 J.L. Smith and P.J. Stiles: Low Temperature Physics - LT13, edited by K.D. Timmerhaus, W.J. O'Sullivan and E.F. Hammel (Plenum, New York) 4, 32 (1974)
7 T. Ando: Phys. Rev. B19, 3089 (1979)
8 P.J. Stiles, Teresa Cole and Amir A. Lakhani: to be published
9 F. Stern: IBM Internal report RC 3758 (1972)
10 L.C. Zhao, D.A. Syphers, B.B. Goldberg and P.J. Stiles: to appear in Solid State Comm
11 F.F. Fang and P.J. Stiles: Phys. Rev. B27, 6487 (1983)
12 D.A. Syphers, F.F. Fang and P.J. Stiles: to be published in the proceedings of EP2DSV
13 F.F. Fang and P.J. Stiles: accepted for publication in Phys. Rev. B. (1984)

14 T. Haavasoja, H.J. Stormer, D.J. Bishop, V. Narayanamurti, A.C. Gossard and W. Wiegmann: to be published in the proceedings of EP2DSV and references therein.
15 F.F. Fang and P.J. Stiles: Phys. Rev. B28, 6992 (1983)
16 H. Obloh, K. v. Klitzing and K. Ploog: to be published in the proceedings of EP2DSV and references therein
17 R.P. Smith, H. Closs and P.J. Stiles: to be published in the proceedings of EP2DSV and references therein.
18 S.M. Girvin and M. Jonson: J. Phys. C15, L1147 (1982)
19 H. Oji: preprint
20 P. Streda and H. Oji: Internal report IC/83/94 Miramare-Trieste
21 W. Zawadski and R. Larsnig: to be published in the proceedings of EP2DSV and references therein
22 M. Jonson and S.M. Girvin: preprint

Subband Physics for the Narrow-Gap Semiconductor $Hg_{1-x}Cd_xTe$

F. Koch

Physik-Department, Technische Universität München
D-8046 Garching, Fed. Rep. of Germany

A survey of the work on the surface bands for that exemplary narrow-gap semiconductor $Hg_{1-x}Cd_xTe$.

I. Introduction

The compound $Hg_{1-x}Cd_xTe$ is the classic very narrow-gap semiconductor. Not only is E_g small and comparable to typical subband energies, but it can easily be varied. For stoichiometric compositions $0.17 \lesssim x \lesssim 0.30$, the gap rises from 0 to $\sim$ 250 meV at low temperatures. It is readily tuned by temperature and pressure. The material is obtainable with a high degree of crystalline perfection and good homogeneity. Its use as an infrared detector material has spurred a $Hg_{1-x}Cd_xTe$ technology for the production of high-mobility single crystals with specially prepared surface layers.

$Hg_{1-x}Cd_xTe$ is ideally suited for narrow-gap subband physics, because the relevant physical parameters are within easy experimental reach. Thus we are able to illustrate all the essential effects from investigations on this material. This is the vantage point in our review of the experiments on the subband structure and surface transport on $Hg_{1-x}Cd_xTe$.

The paper begins with some remarks on why surface charge layers are of importance for the detectors and how they are achieved. There follows a discussion of the theory of subbands for narrow-gap semiconductors. The experimental part begins with the investigation of the subband structure and then turns to an account of the surface transport phenomena.

II. The why and how of charge layers on $Hg_{1-x}Cd_xTe$

An important step in the preparation of infrared detectors, whether of the photoconductive- or photovoltaic-type, from single crystals of this material is "surface passivation". A sensitive detector requires the photoexcited minority carriers to have long lifetimes. The untreated surface with its unsatisfied chemical valences, many irregularities, and defect structures provides electronically active centers which trap carriers and allow them to recombine. Surface recombination is in many cases the dominant lifetime killer. The remedy is a chemical treatment to form an electronically inert and stable compound in a surface layer. If the process of interface formation is accompanied by band bending, it serves the purpose all the better. The surface electric field will act to keep the photoexcited carriers apart. It is evident that surface passivation is closely related to the requirements for obtaining a gate-voltage tunable surface charge layer.

In standard $Hg_{1-x}Cd_xTe$ detector technology the surface is made passive by anodic oxidation in an aqueous electrolyte. In this process approximately 0.1 µm of native oxide is grown onto the surface of the crystal with the passage of an electrical current. In a second step a ZnS layer is sputter-deposited to provide encapsulation. Thus processed the crystal provides an ideal starting point for the subband experiments. We evaporate an optically transparent Ni-Cr layer to form a metal-insulator-semiconductor (MIS) sandwich structure. The semiconductor substrate is n-type with $n \sim 10^{15}$ cm^{-3}. The single crystals are unoriented, of order 10 µm thick, and are provided with ohmic contacts.

For purposes of comparison we have in addition made samples without the anodic oxidation technology. Such "naked" surfaces are chemically polished in a bromine-methanol solution. They are coated with an insulating lacquer.

III. The subband structure of $Hg_{1-x}Cd_xTe$

The quantized, z-directed motion of carriers bound in the surface potential well is properly described in terms of wavefunctions and energies as they are calculated in various approximations. The simplest of these, which incorporates nonparabolicity of the electron bands, is that introduced by Uemura and coworkers. Explicit results for $Hg_{1-x}Cd_xTe$ are contained in ref. /1/. In essence the approach works with electrons described by a $\vec{k}\cdot\vec{p}$ Hamiltonian and being subject to a self-consistent surface potential V(z). The nonparabolicity is that induced by the $\vec{k}\cdot\vec{p}$ mixing of the Bloch states. The potential is introduced "a posteriori" as an external influence on the electrons in nonparabolic volume bands. The subbands found by Takada are nonparabolic in $k_{\parallel}$. Wavefunctions and binding energies depend

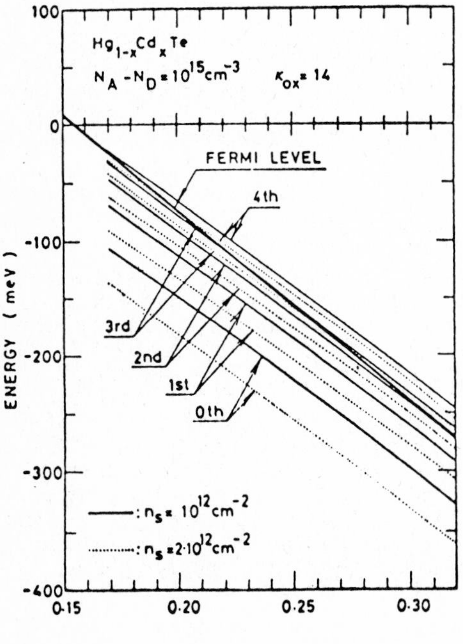

Fig. 1:

Energies of the subbands on $Hg_{1-x}Cd_xTe$ for different compositions x according to ref. /1/. Energies are for $k_{\parallel} = 0$ and for a small depletion charge

on $k_{\parallel}$. For InSb the calculation gives good energy values. Fig. 1 is for $Hg_{1-x}Cd_xTe$ and shows the $k_{\parallel} = 0$ subband energies for two values of N_s for a range of x values. It is remarkable how insensitive the energy separations such as $E_1 - E_0 \equiv E_{01}$ are with changing x.

The simple treatment of nonparabolicity will fail when E_g is small compared to E_{01}. In this case the potential V(z) will act to mix the bands. It can therefore not be included as an afterthought in the subband calculation. A lucid treatment of the problem can be found in Zawadzki's recent paper /2/. This derivation introduces V(z) at the outset in the $\vec{k}\cdot\vec{p}$ matrix. The 4-band, spin-less treatment is not applicable immediately for $Hg_{1-x}Cd_xTe$, but the reduced one-band Hamiltonian in that paper clearly shows the term responsible for band mixing by V(z). Reisinger /3/ has used this formulation to evaluate numerically subbands and matrix elements for parallel-excitation subband spectroscopy on InAs surfaces.

The fact that measured subband energies for $Hg_{1-x}Cd_xTe$ disagree "violently" with the Takada result in Fig. 1 was noted in ref. /4/. This has been taken as evidence for band mixing by V(z). A rather ingenious perturbation-theory treatment of the band mixing which makes use of Takada's wavefunctions and potential has confirmed this supposition. The derivation by Brenig and Kasai /5/ correctly gives the magnitude of band mixing found in the experiments.

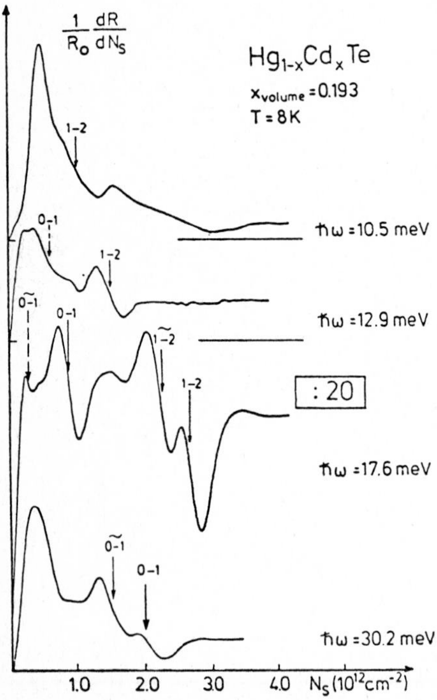

Fig. 2: Subband spectroscopy (Ref.6)

IV. Subband spectroscopy

The resonant excitation of the subband electrons at infrared frequencies serves to determine the energy separations of the bound states on $Hg_{1-x}Cd_xTe$. As previously found for other nonparabolic systems, the $Hg_{1-x}Cd_xTe$ subbands can be excited in both the parallel and perpendicular polarization modes. For the reflection derivative data in Fig. 2, the laser beam is incident at 45° to the detector-type sample. Transitions marked with the ∿ sign are the depolarization-shifted, perpendicularly excited modes. The other transitions are excited by parallel radiation and give the genuine subband energy separation /6/.

The comparison of experiments, such as those in Fig. 2, with the Takada calculation cited in the previous section does not fare well. For example, the $0 \rightarrow 1$ resonance for $\hbar\omega_{12} = 30.2$ meV occurs at $N_s \sim 2 \times 10^{12}$ cm^{-2}. According to the calculation the splitting should be ∿ 50 meV. The discrepancy is more than a reasonable uncertainty in the calculation and the experiments can account for. The data in Fig. 2 contains another interesting feature. The 17.6 meV spectrum is unusually large in amplitude (note division by factor 20) and the depolarization splitting at this frequency is very pronounced. These characteristics have been identified as the result of coupling the electronic polarization of the subband excitation with that of the optical phonons /6/.

In order to explore the role of the small E_g in the subband energies, a series of experiments on samples with different x values in the range 0.17 to 0.20 was initiated. It was found that samples with the same x value but with the "naked", unprocessed surfaces did not reproduce the data in Fig. 2. Instead, the energy values for the x = 0.20 sample agreed reasonably with ref. /1/ and showed a consistent deviation for small x and E_g. The table below is from ref. /8/. It is intended to show the sensitive dependence of subband energy on x.

Table 1: Subband energies for surface density $N_s = 1 \times 10^{12} cm^{-2}$

x	E_g(meV)	E_{01}(meV)	E_{12}(meV)
0.200	60	46	16
0.181	28	30	14
0.171	10	16	$\sim$ 9.5

The comparison of subband spectra led us to realize that the anodically oxidized surfaces were somehow different. The measured energies indicated a lower x value for the surface layer. The data in Fig. 2 for the x = 0.193 sample will fit the scheme of things implied by the table, if the surface composition were $x_{surface} = 0.175$. We shall return later to this point in the discussion of gap-grading in a later section.

The essential feature of the data in Table 1 is the strong E_g dependence of the subband energies. The Takada calculations without band-mixing have energies E_{01} or E_{12} that increase very slowly with decreasing x. At x = 0.2 the calculated energies are 37 and 18 meV, where the experiments give 46 and 16 meV. In view of experimental uncertainties this must be considered good agreement. For the cases where $E_g < E_{01}$ the calculation obviously fails. The recent perturbation treatment /5/ of the surface-potential-induced band-band coupling confirms both the sign and the magnitude of the effect demonstrated in Table 1.

After this short account of the status quo of subband spectroscopy we turn to phenomena associated with the motion of carriers in the surface layer.

V. The surface conductivity

Since the samples are n-type and degenerate with $n \sim 10^{15}$ cm^{-3}, the bulk source-drain contacts register the surface channel conduction as an additional conductivity $\Delta\sigma$. The latter rises with N_s starting at the flat-band voltage. For detector-type samples flat band lies at negative V_g.

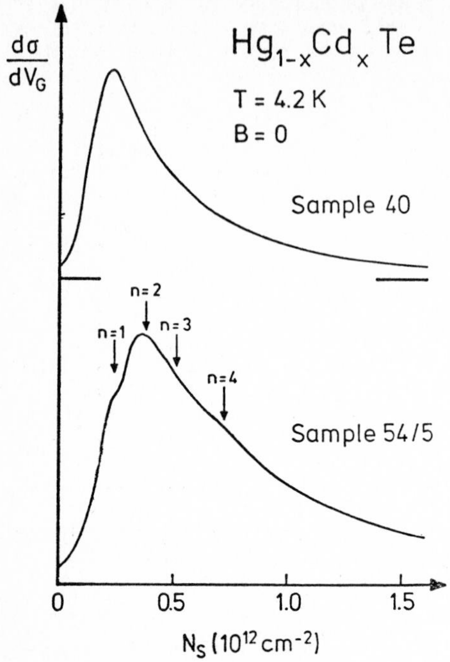

Fig. 3: Field-effect mobility curves from different samples (Ref. /7/)

In the absence of an applied voltage there is accumulation with $N_S \sim 1.5 \times 10^{12}\ cm^{-2}$. The derivative of $\Delta\sigma$ with N_S, the so-called field-effect mobility, reveals a distinct maximum which resembles similar curves in Si. The changes $\Delta\sigma$ have the magnitude expected from the number of carriers induced.

Like other narrow-gap materials, $Hg_{1-x}Cd_xTe$ has such a small density of 2-D states that already modest N_S values lead to several partially filled subbands. In filling the subband ladder, each time a new band is reached, the density of states at the Fermi level has a step-like increase. If the scattering mechanisms are sensitive to such density-of-state jumps, one expects structure in the mobility. In analogy with the better known conductivity structures that occur because of magnetic quantization, one may refer to this effect as the "electric" Shubnikov-de Haas effect. The observability obviously depends on the nature of the interface scattering mechanism. As demonstrated in Fig. 3 we observe the structures only in certain of the samples. The effect is best seen for those crystals where the mobility is not peaked sharply. In Fig. 3 are marked structures that relate to the onset of occupancy of nth subband. As marked, these agree reasonably with occupancies N_S determined from conventional Shubnikov-de Haas experiments.

VI. The Shubnikov-de Haas oscillations

The quantization of the electronic motion in the surface layer by a $B_\perp$ field leads to the usual pronounced oscillations of the surface conductivity. They can be observed either in a sweep of N_S at fixed $B_\perp$, or in a $B_\perp$ sweep at fixed N_S. In spite of the fact that there is strong nonparabolicity and the Landau levels are not spaced equidistantly, the degeneracy per spin-split level remains at $\Delta N_S = eB_\perp/h$. It is difficult to interpret N_S sweep data on $Hg_{1-x}Cd_xTe$ because of the large number of occupied surface bands. In practice it is easier to vary $B_\perp$. Fig. 4 is an example of such data for a detector-type sample. The spin-split oscillation series for each of three partially occupied subbands are observed in this figure. Oscillations persist down to ~ 0.1 T.

The $1/B_\perp$ periodicity of the oscillations gives the partial occupancies N_S^i of the subbands. For all the detector samples whose nominal $x \sim 0.2$,

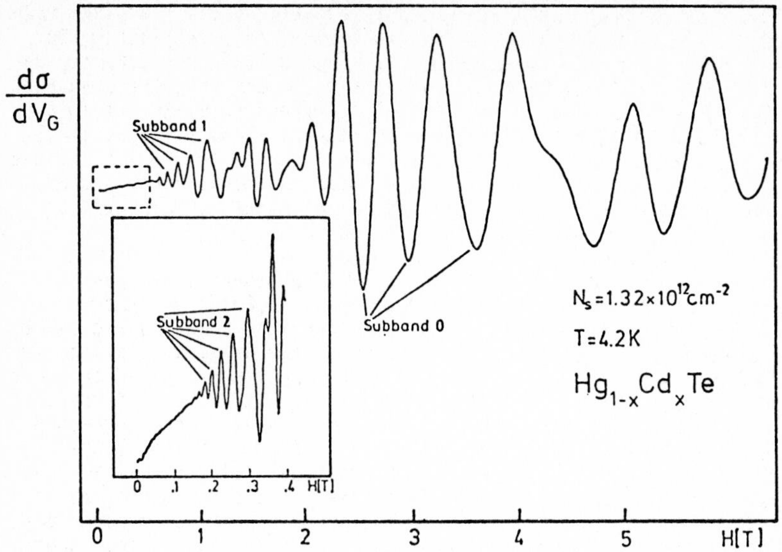

Fig. 4: Shubnikov-de Haas oscillations for a detector-type sample with x ∿ 0.20 (ref. /7/)

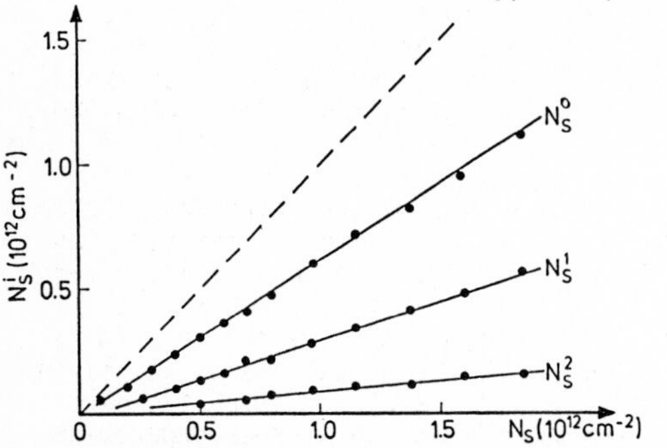

Fig. 5: Partial occupancies N_S^i of the subbands for samples with x ∿ 0.2

as well as others with various x values between 0.18 and 0.21, we measure within the experimental uncertainty the same N_S^i. The results in Fig. 5 are typical. For densities $N_S > 1 \times 10^{12}$ the lines are sufficiently straight and can be extrapolated without serious error to $N_S \rightarrow 0$. In this sense, the partial occupancies can be quoted as 0.62, 0.27, and 0.08 of N_S, each with an uncertainty of $\pm$ 0.02. For higher x near 0.3 the number rises somewhat. We find in agreement with earlier work /8/ occupancies 0.70, 0.23 for the first two subbands. There is only a weak dependence of the occupancies on x. They are quite insensitive to the band-mixing effect that entered so prominently in the discussion of the subband energies.

VII. Conductivity structures in a $B_{\parallel}$. Magneto-electric surface bands

A new development in subband physics is the discovery of regular structures in the surface conductivity in a $B_{\parallel}$, when N_S is varied. The effect has been noted in InSb /9/ and InAs /10/. It exists very prominently also in $Hg_{1-x}Cd_xTe$ /7/.

While the detailed analysis of the surface levels in the crossed electric and magnetic fields of the above materials is still lacking, one can easily describe the qualitative features. It is known that the strongly bound electric subbands of such a material as Si in a weak $B_{\parallel}$ is adequately described by perturbation theory. The shifts and changes of energies for a typical $B_{\parallel}$ field are modest. The states remain essentially "electric" subbands.

For the narrow-gap materials with their large binding lengths and small m_c^* a situation is easily achieved for which $\hbar\omega_c > E_{01}$. In this case the Landau quantization may be considered as dominant. The electric field provides a perturbation which causes the Landau energies to be position-dependent. Such a tilting of the levels, as it is shown in Fig. 6c) schematically, determines their relative occupancies in the surface charge layer. They are emptied and filled in a sweep of N_s, with each touching of E_F registered as a structure in the conductivity. In contrast with the subbands which are purely electric states, and the Landau levels which relate only to the magnetic field, these new states in the crossed fields may be named magneto-electric surface bands.

In a strong $B_{\parallel}$ the $d\sigma/dV_g$ vs. N_s curve shows a sequence of structures that indicate the onset of filling of a given magneto-electric surface band. In Fig. 7 these onset points are labeled and identified in terms of the spin-split (-, +) pairs of surface bands. The spin-splitting serves as

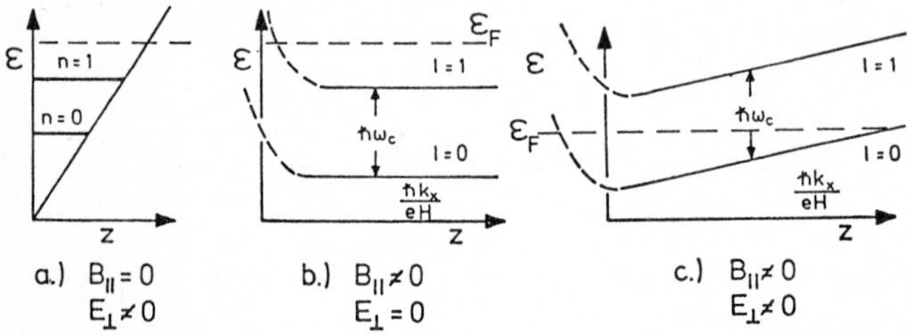

Fig. 6: Electric, magnetic, and magneto-electric surface bands

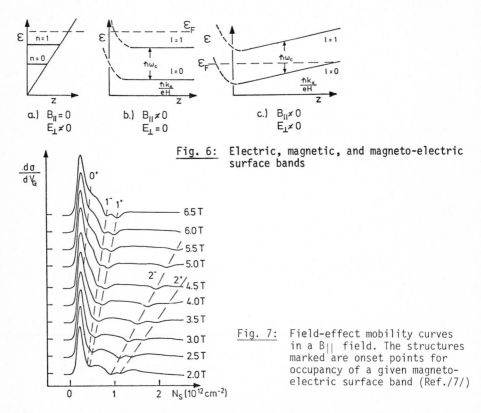

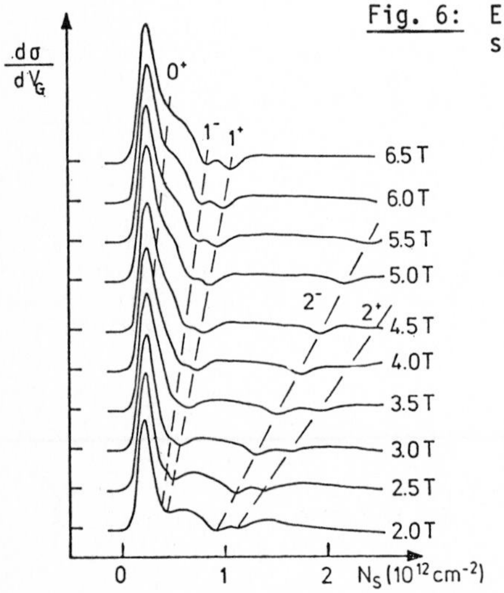

Fig. 7: Field-effect mobility curves in a $B_{||}$ field. The structures marked are onset points for occupancy of a given magneto-electric surface band (Ref./7/)

an energy marker allowing us to assign an energy scale for the data. The plot of N_s vs. $B_\parallel$ for each of the signatures in $d\sigma/dV_g$ is a perfectly straight line. The set of structures makes up a fan chart whose regularities, in particular the straight-line behavior, is not understood at present. At weak $B_\parallel$ there is some curvature. At $B_\parallel = 0$ the structures must go over into the features identified in Fig. 3. Comparing the two figures makes it clear that in the presence of $B_\parallel$ the "electric" Shubnikov-de Haas effect is enhanced. The reason is likely to be the singularity in the density of states associated with magneto-electric subbands. For these bands the momentum in the direction of $B_\parallel$ is a free variable. The density of states is one-dimensional and has the usual singularity. It remains to give a quantitative account of the $B_\parallel$-conductivity experiments.

VIII. Cyclotron resonance in perpendicular and parallel fields

Surface cyclotron resonance experiments are usually done in a perpendicular-field geometry. The resonance represents transitions between neighboring Landau levels in a given subband. For nonparabolic bands the transition energies not only depend on the subband in question, but also on the specific Landau level pair $\ell \to \ell + 1$ between which the transitions are taking place. One observes distinctly the discontinuities of m_c^* when with density changes a new transition becomes possible. Fig. 8, taken from ref. /11/, gives a sampling of data obtained on a detector-type specimen with $x = 0.20$ and using circular polarization in a reflection experiment. In the analysis of such data contributions from four subbands can be identified at the highest densities. At the low-field end of the spectrum is recorded the volume resonance. Note that the m_c^* values for carriers in the surface bands are as much as a factor of 7 higher than the bulk electron mass.

The cyclotron resonance experiments, because they sample transitions at the Fermi energy, are generally quite insensitive to the x value /11/.

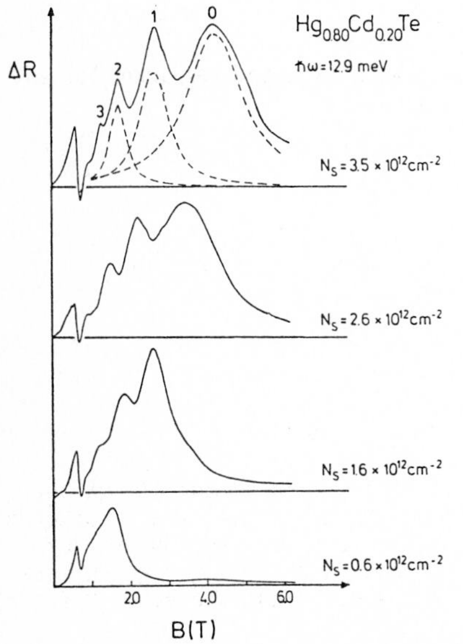

Fig. 8: Cyclotron resonance spectra at different surface carrier densities (Ref. /11/)

For even the lowest density in Fig. 8 the surface electrons have more than double the volume mass. In order to search for the possibly lower E_g in the surface region, which should lead to a surface mass lower than the volume value, it is necessary to work at extremely low N_s. Only in the limit $N_s \rightarrow 0$ can one expect to confirm the reduced surface E_g value. We have in ref. /12/ succeeded in finding the effect for the anodically oxidized samples. Fig. 9 is obtained in transmission. The resonance shows as a dip in the transmitted intensity. For $N_s = 4 \times 10^{10} cm^{-2}$ the surface resonance observed in the usual N_s-chopping mode appears at fields below the volume cyclotron resonance. The latter is measured by chopping the laser source. The volume signal is folded into the upper trace, appearing as an antiresonance. The observation qualitatively confirms the gap-grading hypothesis.

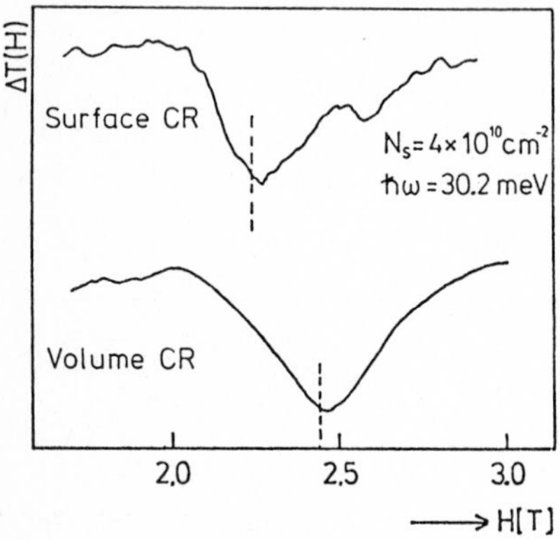

Fig. 9: Camparison of the surface cyclotron resonance at very low N_s with the volume resonance. The resonance position is marked according to a lineshape fit (Ref. /12/)

The search for polaron effects in the surface cyclotron resonance on $Hg_{1-x}Cd_xTe$ so far has proved negative /13/. The effect at the typical densities N_s is certainly reduced over that for the volume signal. It remains to search carefully at the lower densities such as those in Fig. 9.

At high $B_\perp$, a situation can be realized where for typical N_s the Landau level energies and even the spin energies are much greater than subband splittings such as E_{01}. In this case the electrons will magnetically freeze out to the 0^- Landau level. With rising N_s successive subbands n = 0, 1, 2, etc. will be filled with 0^- electrons. This is the reverse of the situation usually encountered in surface cyclotron resonance.

Preliminary work has been carried out on cyclotron resonance in a parallel magnetic field /14/. In this case transitions are observed between the magneto-electric levels illustrated in Fig. 6. The surface resonance in this Voigt-geometry mode is wholly different from either the volume resonance or the $B_\perp$ surface resonance. The signal is not measurably shifted although the surface plasma frequency would lead one to expect a substantial shift. Moreover, it is quite small in amplitude and asymmetrically broadened. These effects can be understood on the basis of an inhomogeneous broadening resulting from the fact that the neighboring magneto-electric levels need not really be equidistant. The effect has recently been discussed also for InSb /15/.

IX. Spin resonance, magnetophonon resonance, and other loose ends

The account of subband physics for $Hg_{1-x}Cd_xTe$ would not be complete without mention of spin resonance. The fact that an electric-dipole-excited spin resonance has been found for the bulk material has prompted our search for a corresponding surface signal. The curves in Fig. 10 show preliminary results. The resonance appears as a small signal on the high-field side of the cyclotron line. It is N_s-dependent, as the change with chopping voltage in the figure shows. The reversal of the line is similar to that found for spin resonance in PbTe /16/. More work is under way. The present results are cited in the way of a progress report.

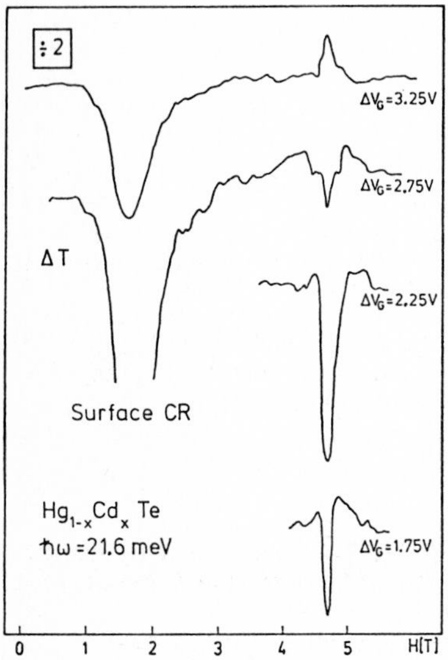

Fig. 10: Surface spin resonance is found at fields above the cyclotron resonance. ΔVg is the voltage above flat-band (1 Volt corresponds to 5×10^{10} cm^{-2}) (Ref. /12/)

In spite of the fact that polaron effects did not show up in surface cyclotron resonance, we have searched for the magnetophonon effect. At 77 K and fields between 0.8 and 1.0 T the resonance has recently been found. The experiments were done at fixed $B_\perp$ sweeping N_s. The strong N_s-dependence of m_c^* that was seen in the data of Fig. 8 can be used to advantage here for tuning the surface resonance through the $\hbar\omega_{LO}$ frequency of 17.2 meV. Three resonances from the occupied bands were identified. Additional work is in progress.

The recent progress in studying the Landau levels in the extreme quantum limit, where fractional occupancies of the lowest level are achieved in high-mobility samples, provides a stimulus for related work on $Hg_{1-x}Cd_xTe$. Such experiments are being planned.

$Hg_{1-x}Cd_xTe$ heterojunction interfaces are currently being prepared and will no doubt provide a further field of investigation. At some future time transport effects in a periodic structure based on $Hg_{1-x}Cd_xTe$ would seem worthwhile. Another loose end could be a study of electrons in deliberately gap-tailored material.

X. The case for gap-grading

We consider in this final chapter the case for a different E_g in the surface region. Several pieces of evidence have surfaced in the experiments that point to a reduced E_g at the interface on the anodically oxidized detector samples. The situation is sketched schematically and without proper regard for the modification of the potential by the charges in Fig. 11.

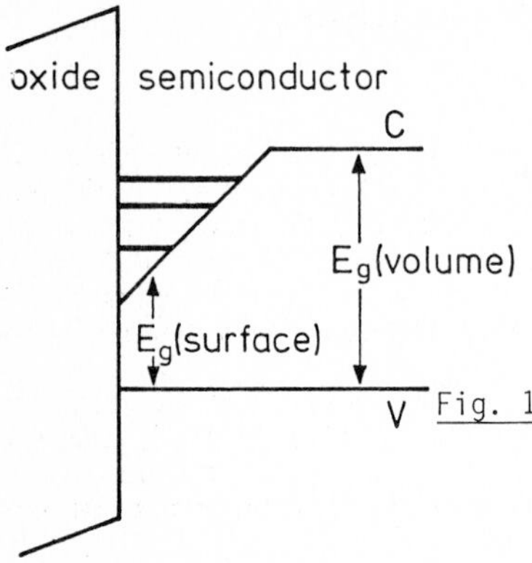

Fig. 11: Gap-grading model at the oxide-semiconductor interface. Charges accumulated in the surface will modify the potential well

A first bit of evidence, albeit circumstantial, is that passivated samples always have an accumulation layer ($N_s \sim 1.5 \times 10^{12}$ cm^{-2}). This would be the case if a natural potential well such as provided by gap-grading existed at the surface. Accumulated charge could come from both the volume of the semiconductor or from defect states of the insulator. There are other ways to explain the band bending, for example oxide charge or interface states, but gap-grading provides a natural explanation.

The second hint came from the subband spectroscopy results that we discussed in Sec. IV. Not only did the measured energies E_{01} and E_{12} on the detector-sample disagree with those obtained for "naked-surface" samples of the same x, but the values of these energies could be explained by band-band coupling /5/. This in turn occurs only if the gap were smaller than the bulk value.

Additional support of the gap-grading hypothesis came from the cyclotron resonance in the $N_s \to 0$ limit. This is a difficult experiment and requires cautious evaluation, but the qualitative evidence is admissible.

The final link in our chain of reasoning is an experiment that suggests itself after an examination of the Fig. 11. The flat-band V_g for electrons and holes must be different. The flat-band voltage for electrons is the point where $N_s = 0$. It is easily determined from the conductivity experiments (cf. Fig. 3) with good accuracy. A reduced gap would require that when $N_s = 0$ for electrons that the surface potential be attractive for the minority carriers, the holes. This in turn should lead to a marked reduction of the photoconductive response as holes make it to the surface and recombine at defects. In measurements of the photoresponse we have confirmed this expectation. A distinct decrease of the photosignal allows us to identify flat-band for holes. It lies at a V_g which is far less negative than the $N_s = 0$ voltage.

The sum total of the evidence for gap-grading hypothesis is substantial. We believe that the surface channel electrons occupy a region at the interface between the oxide and the semiconductor where the E_g is reduced. There are two possible reasons for such a reduction. One of these, a changed stoichiometry of the surface layer with reduced Cd content, has already been implied in the subband spectroscopy discussion. The bulk concentration x = 0.20 would have to be reduced to something like $x \sim 0.175$ to explain these results quantitatively. An alternative explanation that merits consideration involves interface strain. In this case the oxide strains the $Hg_{1-x}Cd_xTe$ surface layer in such a way as to reduce E_g.

XI. Concluding remarks

We have given a review of past and current work on the subband physics of a classic narrow-gap semiconductor. Some of the work mentioned is of the nature of a progress report. There is a chance that it will need to be revised and rethought in future publications. By trying to be as complete as possible, the lecture has tried to exemplify the wide range of physical investigation on the surface charge layer of a narrow-gap semiconductor.

I have cited heavily from theses and "to be published" work of many students and coworkers of the group here in Munich. Credits go to J. Scholz, E. Schindlbeck, Wen-Qin Zhao, C. Mazurê, and R. Wollrab. We have enjoyed the active collaboration on the materials side with J. Ziegler and H. Maier of the Telefunken Electronic research group.

References:

/1/ Y. Takada, K. Arai, and Y. Uemura, in "Physics of Narrow-Gap Semiconductors", p. 101 Proceedings (Linz, Austria 1981), Lecture Notes in Physics 152, Springer-Verlag
/2/ W. Zawadzki, J. Phys. C. 16, 229 (1983)
/3/ H. Reisinger, Ph.D. Thesis, Techn. Univ. München (1983) and to be published
/4/ J. Scholz, F. Koch, J. Ziegler, and H. Maier, Surface Sci. (to be published)
/5/ W. Brenig and H. Kasai, Z. Phys. B, Condensed Matter 54, 191 (1984)
/6/ J. Scholz, F. Koch, H. Maier, and J. Ziegler, Solid State Commun. 45, 39 (1983)
/7/ Wen-Qin Zhao, F. Koch, J. Ziegler, and H. Maier, Phys. Rev. (submitted)
/8/ J.P. Dufour, R. Machet, J.P. Viton, J.C. Thuillier, and F. Baznet, in "Insulating Films on Semiconductors", Proc. Erlangen 1981, ed. M. Schulz and J. Pensl, Springer-Verlag, p. 259
/9/ H.P. Grassl, Diplom Thesis, Techn. Univ. München (1975)
/10/ R.E. Doezema, M. Nealon, and S. Whitmore, Phys. Rev. Letters 45, 1593 (1980)
/11/ J. Scholz, Ph.D. Thesis, Techn. Univ. München (1983), and to be published
/12/ E. Schindlbeck, Diplom Thesis, Techn. Univ. München (1983)
/13/ J. Scholz, F. Koch, H. Maier, and J. Ziegler, Solid State Commun. 46, 665 (1983)
/14/ Wen-Qin Zhao, C. Mazurê, F. Koch, J. Ziegler, and H. Maier, Surface Sci. (to be published)
/15/ See for example, J.P. Kotthaus, this volume
/16/ H. Schaber and R.E. Doezema, Phys. Rev. B 20, 5257 (1979)

Infrared Spectroscopy of Electronic Excitations in MOS Structures

Jörg P. Kotthaus

Institut für Angewandte Physik der Universität Hamburg, Jungiustraße 11
D-2000 Hamburg 36, Fed. Rep. of Germany

In metal-oxide-semiconductor capacitors voltage-tunable quasi-two-dimensional electron systems can be realized by a gate voltage induced band bending at the semiconductor-oxide interface. Here recent experimental results serve to illustrate how infrared spectroscopy of electronic excitations can be used to obtain important information on the electronic properties of such two-dimensional electron systems.

1. Introduction

In metal-oxide-semiconductor (MOS) structures electrons can be confined at the semiconductor-oxide interface in a narrow potential well formed by the oxide barrier and the conduction band edge E_c in the semiconductor which is bend downward towards the interface by the application of a suitable gate voltage V_g (see Fig.1). In this potential well the motion of electrons perpendicular to the interface is quantized in discrete eigenstates E_i (i=0,1,...), whereas the motion parallel to the interface is free-electron-like [1]. The energy spectrum of the spatially confined electrons thus consists of a set of two-dimensional subbands with energy dispersion

$$E_i(k_x,k_y)=\hbar^2k_x^2/(2m_x)+\hbar^2k_y^2/(2m_y)+E_i \ . \qquad (1)$$

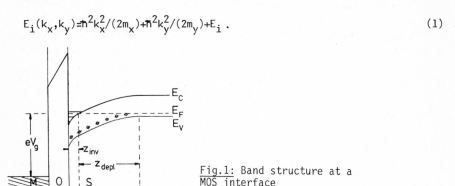

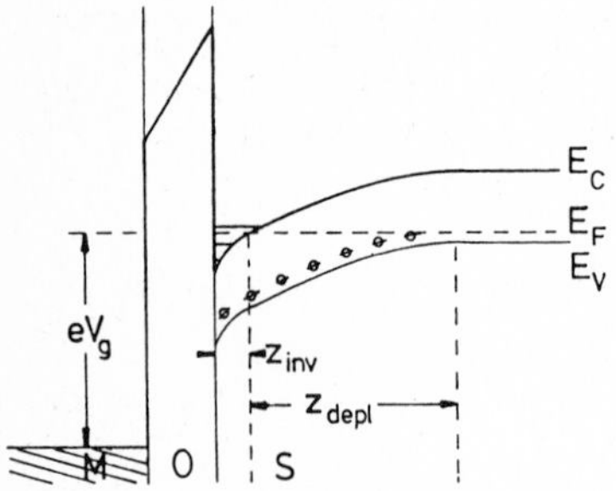

Fig.1: Band structure at a MOS interface

Here m_x and m_y are the components of the effective mass tensor for motion parallel to the interface. In systems with parabolic bands the subband density of states is

$$D(E_i)=g_v g_s (m_x m_y)^{1/2}/2\pi\hbar^2 \quad , \tag{2}$$

where g_v and g_s denote valley and spin degeneracy, respectively.

In MOS structures the electronic properties of this quasi-two-dimensional electron system (2DES) such as areal electron density N_s, Fermi energy E_F, and subband energies E_i can be easily controlled over typically two orders of magnitude by variation of the gate voltage V_g. Therefore MOS systems are particularly attractive for studies of fundamental electronic excitations in a 2DES. As illustrated in the following infrared spectoscopy of such fundamental excitations can be used both to understand better the excitation spectrum of a 2DES as well as to extract information on bandstructure and fundamental interactions in a 2DES.

2. Experimental Techniques

To observe the electronic excitations in a 2DES one usually measures the gate voltage induced relative change of the transmission coefficient T of infrared radiation of frequency ν incident normally onto the surface of a MOS capacitor, i.e.:

$$\Delta T/T=(T(V_g)-T(V_t))/T(V_t) \quad , \tag{3}$$

where V_t is the conductivity threshold. In the small signal limit $-\Delta T/T$ is directly proportional to the real part of the dynamical conductivity $\sigma_{xx}(\omega,N_s)$ of the 2DES. Infrared radiation is provided either monochromatically by an infrared laser or as a continuous spectrum by a conventional lamp source and, in the latter case, is frequency analysed by rapid-scan Fourier-transform spectroscopy [2]. The transmitted radiation is usually detected by a broadband cryogenic bolometer. Amplitude modulation techniques and phase-sensitive detection are employed to enhance sensivity. To minimize free carrrier absorption in the bulk and to reduce phonon scattering in the 2DES samples are cooled to cryogenic temperatures. As samples MOS capacitors are fabricated on various semiconductors (e.g. Si, InSb, ...) with a large diameter ($\sim$3 mm) semitransparent gate. The sample thickness is made non-uniform under the gate area to avoid interference effects.

3. Intersubband Spectroscopy in Parallel Excitation

The energy spectrum of the oscillatory motion of inversion electrons perpendicular to the interface can be studied most accurately by intersubband resonance experiments [1]. On surfaces where there exists coupling between the electronic motion perpendicular and parallel to the surface, caused,e.g. by bandanisotropy or non-parabolicity, i.e. where the dynamical conductivity tensor has off-diagonal components, intersubband resonance transitions may be excited also by infrared radiation polarized parallel to the surface [3]. We have made use of this parallel excitation mechanism to carry out extensive frequency domain studies of the intersubband resonance in electron and hole space charge layers on Si [4]. Figure 2 shows typical experimental spectra for electron space charge layers on p-Si (110) at fixed N_s and for various depletion charges N_{depl} which are established in a substrate bias technique [2]. For polarization along the $[\bar{1}10]$ direction the spectrum consists of a monotonically decreasing Drude-type intraband contribution and prominent intersubband resonances, whereas for [001] polarization only Drude absorption is observed as expected [3]. A comparison of the observed dependence of the lineshape on N_{depl} with theoretical calculations [3] strongly suggests a valley degeneracy of $g_v=4$, though the period of Shubnikov-de-Haas oscillations measured on samples from the same batch shows a twofold Landau-level degeneracy. Further insight into the problem of valley degeneracies can be gained by plasmon resonance experiments [5].

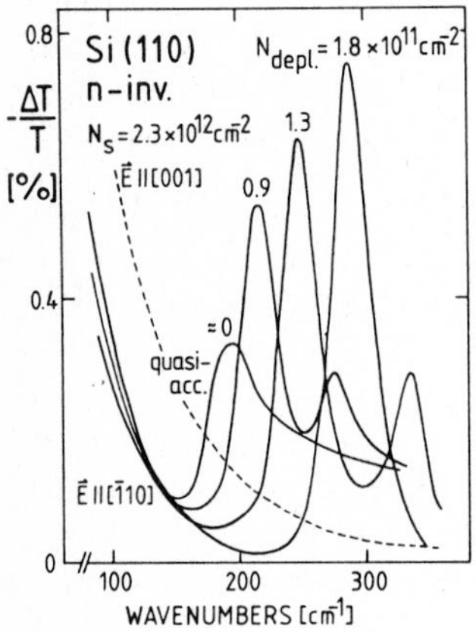

Fig.2: Intersubband resonances in an electron space charge layer on Si (110) (from Ref. 4)

Though intersubband resonance for electrons on surfaces of Si has been studied rather extensively both experimentally and theoretically, the agreement between experimental and theoretical resonance positions is only good on Si (100) [4,6,7]. Rather surprising is the observation [4] that for electron accumulation layers the resonance energies are nearly identical for all high symmetry surfaces of Si. This seems to contradict simple models which predict the resonance energies to scale like $(m_z)^{-1/3}$. Intersubband spectroscopy also gives good insight into the rather complex subbandstructure of hole space charge layers on Si [4] and should now allow a more quantitative understanding of these systems.

4. Plasmon Resonance Spectroscopy

Infrared excitation of collective modes in a 2DES, i.e. plasmon resonance spectroscopy, has been increasingly used in recent years to study intrasubband dispersion in electron and hole space charge layers on Si [1,5,8]. Quite different from the 3D case the plasmon dispersion in a 2DES depends strongly on wavevector q, roughly as:

$$\omega_p^2 = N_s e^2 q/(2\varepsilon m_p) \quad , \tag{4}$$

where ε is the average dielectric constant at the interface and m_p the optical mass for motion parallel to the interface. In a 2DES plasmon frequencies in the infrared correspond to wavevectors $q \gg \omega_p/c$. Using grating couplers with typically submicrometer periodicities a on top of the gate electrode, one can couple infrared radiation to the plasmon modes at discrete wavevectors $q=2\pi n/a$ ($n=1,2,..$) and observe plasmon resonances in the relative change in transmission [1]. Typical spectra for electron accumulation layers on the anisotropic (110) surface are depicted in Fig.3. The anisotropic optical mass m_p extracted from such experiments is shown in Fig.4. At $N_s \geqq 2 \times 10^{12}$ cm^{-2} m_p is essentially N_s-independent as expected for a parabolic subband but has a smaller anisotropy than theo-

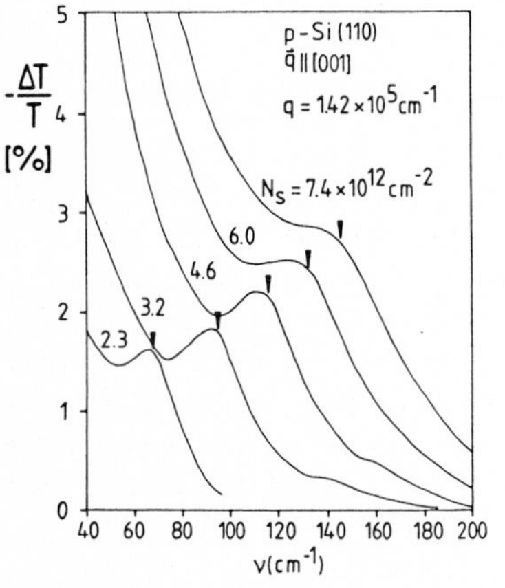

Fig.3: Plasmon resonances in an electron accumulation layer on Si (110) (from Ref. 5)

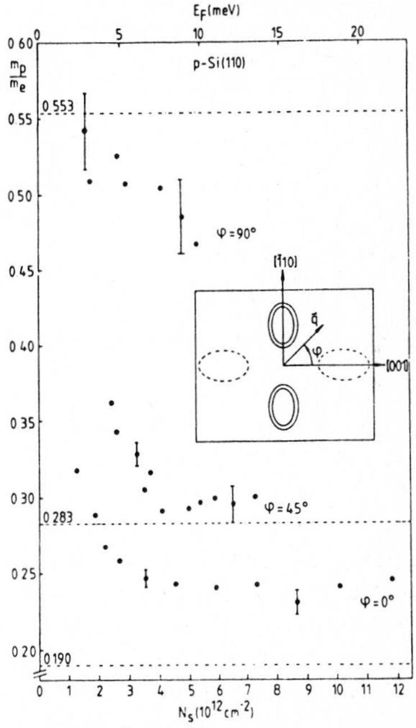

Fig.4: Optical masses m_p extracted from plasmon resonance experiments for electrons on Si (110). The dashed lines give the theoretical values [1] (from Ref. 5)

retically predicted. The observed anisotropy is consistent with a valley degeneracy of $g_v = 4$. At densities below 2×10^{12} cm^{-2} the plasmon mass increases with decreasing N_s. Similar behavior is observed on Si (100) and (111) and the onset of this rise in m_p appears to be correlated with the absolute value of the Fermi energy [5]. The origin of this increase of m_p is not yet understood and could be localization or electron-electron interaction.

Plasmon resonance spectroscopy has also been successfully used to measure optical masses in hole space charge layers on Si. In comparison to cyclotron resonance experiments plasmon resonance studies have the advantage of not needing a magnetic field and allowing a mass determination on rather low mobility samples, since the resonance condition $\omega_p \tau > 1$ can be realized by increasing q. In addition, plasmon resonance experiments give information on anisotropy. Plasmon spectroscopy may also be used to investigate 2DES in which non-equivalent subbands are occupied such as Si (100) under uniaxial compression [9] or 2DES with spatially periodic charge modulation. Therefore one may expect that such experiments are extended to other interesting 2DES in the near future.

5. Spectroscopy in High Magnetic Fields

The energy spectrum of a 2DES can be strongly modified by the additional application of a high magnetic field. For a magnetic field applied perpendicular to the surface the energy spectrum consists of discrete Landau levels, one ladder for each subband. Infrared-induced transitions between partially occupied adjacent Landau levels, i.e. cyclotron resonance (CR), has first been used to study spectroscopically subband masses and now has been applied to 2DES on various semiconductor MOS and heterostructures [1,10]. CR studies can be used not only to extract bandstructure information but may also be employed to study characteristic interactions in a 2DES. One example are recent frequency-dependent CR studies in electron inversion layers on InSb [11]. There one is able with moderate magnetic fields to sweep the cyclotron frequency $\omega_c = eB/m^*$ through the optical phonon frequencies. As in bulk InSb one observes resonant polaron coupling of the CR to the LO phonons. The coupling effect may be expressed as an anomaly in the frequency-dependent CR mass around $\hbar\omega_{LO}$ (see Fig.5). On InSb this anomaly is much stronger in the 2DES than in the bulk. Recent calculations [12] show that this does not correspond to a change of the Fröhlich coupling constant at the interface but rather reflects the modification of the wavefunction and the matrix elements in the transition from a 3DES to a 2DES.

If one applies a magnetic field B parallel to the surface of a MOS structure (x direction) the subband dispersion remains two-dimensional and may be written as:

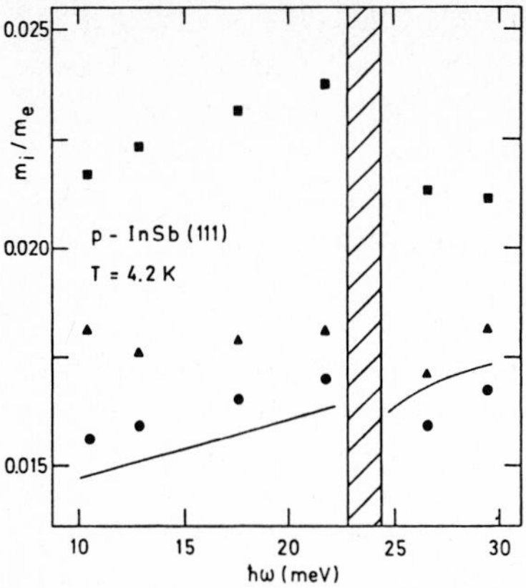

Fig.5: Frequency dependence of the cyclotron masses in electron inversion layers on InSb at $N_s = 2\times10^{11}$ cm^{-2} in the ground subband (circles) and at $N_s = 1\times10^{12}$ cm^{-2} in the ground subband (squares) and first subband (triangles). The solid line represents bulk values in InSb, the hatched region indicates the reststrahlen band (from Ref. 11)

$$E_i(k_x,k_y)=\hbar^2k_x^2/(2m_x)+E(k_y,B,F_s)+E_i(B,F_s) \tag{5}$$

where $F_s = eN_s/\varepsilon_s$ is the surface electric field in the semiconductor. Because of the competition between the Lorentz and the Coulomb force one obtains so-called hybrid subbands [13,15] with their dispersion critically dependent on the ratio of the Landau radius l to the electric binding length L which are, respectively:

$$l=(\hbar/eB)^{1/2} \quad \text{and} \quad L=(\hbar^2/2m^*eF_s)^{1/3}\,. \tag{6}$$

Particularly interesting is the regime $(l/L)\approx1$, where the electronic motion in real space exhibits a transition from 2D to 3D behavior. This transition regime can be spectroscopically studied,e.g. on InSb,where the low m^* yields relatively large L values. Fig. 6 shows the band structure of such hybrid subbands in the triangular well approximation at fixed ratio l/L as a function of the normalized center coordinate of cyclotron motion $z_0/l= l\cdot k_y -(l/L)^3/2$ with respect to the interface [13,14]. Also shown are the classical electron orbits in the x-y plane omitting a drift term. In a situation as in Fig.6 the CR observed in Voigt geometry ($q\perp B$, where q is the wavevector of the incident radiation) behaves rather 3D-like as shown in Fig.7. The resonance is narrow and does not shift with increasing N_s. In contrast to the real 3D case, however, the resonance is not depolarisation shifted. The nonresonant 2D electron states screen the polarization that the 3D cyclotron motion wants to establish [14]. At lower resonance magnetic fields, i.e. $l/L\gg1$, the spectra are qualitatively different (see Fig.8). On a relatively large Drude-like, field-independent background one observes CR-like structures that shift with increasing N_s to higher magnetic fields, thus reflecting nonparabolicity [13]. Such spectroscopic studies show that a magnetic field parallel to the interface may be used to study a transition from 3D to 2D behavior and to create rather exotic subband systems.

Fig. 6

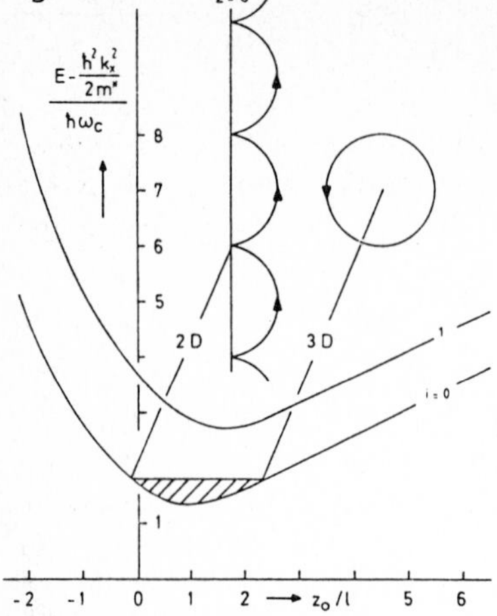

Fig. 7

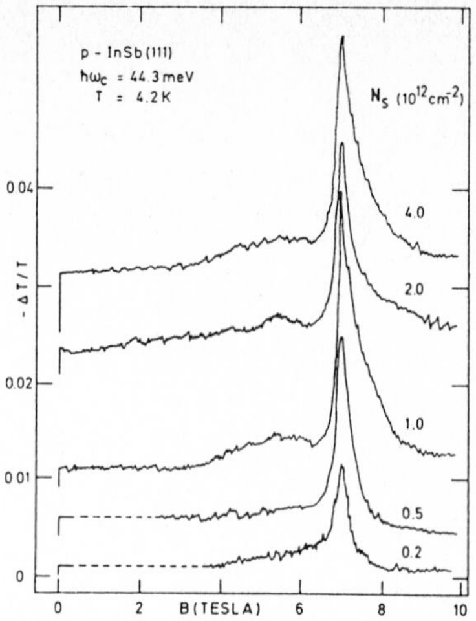

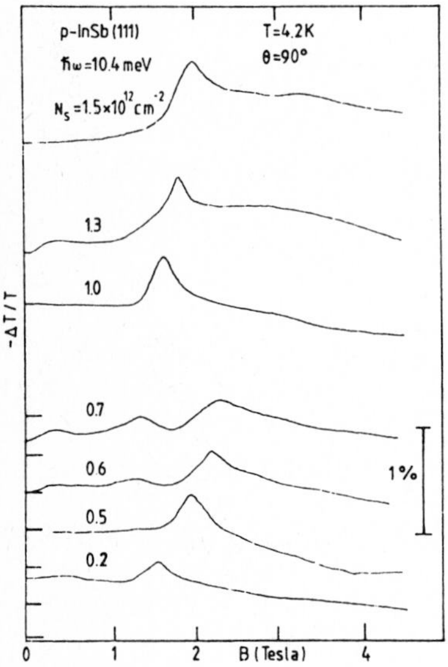

Fig.6: Surface band structure of inversion electrons on InSb in a parallel magnetic field for the lowest two subbands at $N_S = 2.5 \times 10^{11}$ cm^{-2}, $B = 7.0$ Tesla, $1/L \simeq 1$. Occupied states are indicated by the hatched region (from Ref. 14)

Fig.7: Hybrid resonances in an inversion layer on InSb in a parallel magnetic field at high resonance fields $B \simeq 7T$ (from Ref. 14)

Fig.8: Hybrid resonances in an inversion layer on InSb in a parallel magnetic field at low resonance fields $B \simeq 2T$ (from Ref. 13)

6. Conclusions

I have tried to demonstrate that infrared spectoscopy can be used in a variety of ways to study the energy spectrum and fundamental interactions in a 2DES. The same spectroscopic techniques can be used on semiconductor heterostructures [15,16] or on 2DES with superimposed 1D periodic potentials [1].

Acknowledgements

The experiments presented here have been a result of a collaborative effort and I wish to thank my colleagues as identified in the reference section. I also want thank W. Beinvogl of the Siemens Forschungslabor for continuously providing oxidized Si wafers used in part of this work. Finally I wish to acknowledge the financial support of the Deutsche Forschungsgemeinschaft and the Stiftung Volkswagenwerk.

References

1. For a recent review see: T. Ando, A. B. Fowler, and F. Stern: Rev. Mod. Phys. 54, 437 (1982)
2. For details see: E. Batke and D. Heitmann: Infrared Physics, in press
3. T. Ando, T. Eda, and M. Nakayama: Solid State Commun. 23, 751 (1977)
4. A. D. Wieck, E. Batke, D. Heitmann, and J. P. Kotthaus: submitted for publication to Phys. Rev. B
5. E. Batke and D. Heitmann: Solid State Commun. 47, 819 (1983) and references quoted therein
6. T. Ando: Z. Phys. B26, 263 (1977)
7. S. Das Sarma and B. Vinter: Phys. Rev. B28, 3639 (1983)
8. E. Batke, D. Heitmann, A. D. Wieck, and J. P. Kotthaus: Solid State Commun. 46, 269 (1983)
9. Th. Englert, D. C. Tsui, and R. A. Logan: Solid State Commun. 39, 483 (1981)
10. see,e.g.,J. P. Kotthaus: J. Magn. Magn. Materials 11, 20 (1979)
11. M. Horst, U. Merkt, and J. P. Kotthaus: Phys. Rev. Lett. 50, 754 (1983).
12. R. Lassnig and W. Zawadzki: Surface Sci.,in press
13. J. H. Crasemann, U. Merkt, and J. P. Kotthaus: Phys. Rev. B28 2271 (1983)
14. M .Horst, U .Merkt, and J. P. Kotthaus: Solid State Commun., in press
15. J. C. Maan: this volume
16. see,e.g.,E. Gornik: this volume

Self-Consistent Calculations of Subbands in p-Type Semiconductor Inversion Layers

G. Landwehr and E. Bangert

Physikalisches Institut der Universität Würzburg
D-8700 Würzburg, Fed. Rep. of Germany

After a review of previous calculations of electric subbands in p-type silicon inversion layers new results for a p-type GaAlAs-GaAs hetero-structure are reported. Due to the electric field at the interface the spin degeneracy is lifted, leading to two heavy hole masses for finite wave vector. The calculated effective masses are compared with experimental results. In addition, self-consistent subband calculations for p-type germanium inversion layers adjacent to grain boundaries in artificially grown bicrystals are reported.

1. Introduction

Recently, the interest in p-type inversion layers has been revived. New intersubband optical absorption experiments in p-type silicon space charge layers have been performed by KOTTHAUS and coworkers /1/, and Raman scattering in this system has been studied by ABSTREITER et al. /2/, giving information about subband energies. Moreover, it has turned out that silicon p-channel MOSFETs have properties at high source drain fields which make them attractive for high speed integrated circuits.

Previous calculations of electric subbands in p-type silicon inversion layers of (100), (110) and (111) orientation predicted a lifting of the Kramers degeneracy at finite wave vectors /3,4/. Such a splitting was not clearly observed experimentally up to now. However, recent investigations of the Shubnikov-de Haas effect in p-type GaAlAs-GaAs hetero-structures by STÖRMER et al. /5/ indicated the presence of two heavy holes with effective masses of 0.38 m_0 and 0.60 m_0. No comparison with theory could be made, because no subband calculations were available. For this reason, we performed self-consistent calculations in the Hartree approximation with parameters relevant for the performed experiments.

There is another two-dimensional p-type system in which the effects of boundary quantization should show up. These are inversion layers adjacent to the grain boundary in germanium bicrystals with n-type bulk doping. For bicrystals tilted from a common (100) plane, the grain boundary is stabilized by a regular array of edge dislocations. Because these have acceptor character, a p-type inversion layer with free hole concentrations between $10^{12}/cm^2$ and $10^{13}/cm^2$ arises /6/. Recently, it has been demonstrated that this system is especially suitable for the study of many-body effects which are expected in two-dimensional disordered systems /7/. The observation of Shubnikov-de Haas oscillations in a particular bicrystal in which the period of the oscillations depended only on the normal component of the magnetic field /8/ initiated us to perform subband calculations for the bicrystal system. As expected, it turned out that the band structure is substantially modified due to the high electric field existing in the grain boundary.

In the following, we shall first briefly review previous subband calculations for p-type silicon inversion layers. Some results will be given which were not published previously. Subsequently, we shall report results of self-consistent subband calculations for p-type inversion layers in (100) GaAs. The boundary conditions were chosen to be representative for a modulation doped hetero-structure of GaAlAs-GaAs. Finally, we shall present our data for p-type germanium inversion layers /8/. In all cases the dependence of the effective mass on the carrier concentration as well as the constant energy contours were calculated.

2. Self-consistent calculations for p-type silicon inversion layers

The strong electric field which is present at the Si-SiO_2 interface of a MOSFET (Metal Oxide Semiconductor Field Effect Transistor) confines the charge carriers to a thin layer with a thickness of 100 Å or less. As a consequence of this boundary quantization arises. The motion parallel to the interface is hardly affected, resulting in quasi two-dimensional behaviour. The first self-consistent subband calculations for n-type silicon inversion layers were performed by Stern and Howard /9/. They used the effective mass approximation which turned out to be appropriate. For the (100) orientation the motion of the electrons parallel and perpendicular to the surface is completely decoupled so that intrasubband properties like effective masses do not depend on the subband index or the surface carrier concentration. However, the situation is quite different for p-type silicon inversion layers. This is so because in the degenerate valence band the motion parallel and perpendicular to the surface is strongly coupled, with the coupling depending on the surface electric field. It is necessary to take into account all three valence bands, because the lower one is depressed by only 44 meV with respect to the upper two. Consequently, the calculations for p-type inversion layers are considerably more complicated than the counterpart for n-type layers. Practical considerations dictate to restrict the calculations to zero temperature, which should not be a serious drawback because most experimental data are obtained at helium temperatures. Because many experiments are performed in high magnetic fields it would be desirable to incorporate a magnetic field in the calculations. This, however, would make them almost intractable. Fortunately, there are indications that even a high magnetic field does not modify the band structure substantially although this could not be principally excluded. It turned out that the effective masses determined from the temperature dependence of Shubnikov-de Haas oscillations /10/ did not depend on the magnetic field. Also, hole masses in GaAlAs-GaAs hetero-junctions derived from cyclotron resonance experiments were independent of the magnetic field /5/.

The first detailed calculations for the subband structure in p-type silicon inversion layers were independently performed by BANGERT, von KLITZING and LANDWEHR /3/ and by OHKAWA and UEMURA /4/. The work has been reviewed by LANDWEHR /11/ and by ANDO, FOWLER and STERN /12/. Both groups solved a one-dimensional one-electron Schrödinger equation in the effective mass approximation:

$$\left[H_0\left(k_x, k_y, \frac{1}{i}\frac{\partial}{\partial z}\right) + V(z)\right] \psi_{i,k_x,k_y}(z) = E_i(k_x, k_y)\, \psi_{i,k_x,k_y}(z) \qquad (1)$$

k_x and k_y are the wave vectors associated with the motion in the plane of the inversion layer and z is the perpendicular direction. In order to obtain the z-dependent part of the envelope function $\psi_{i,k_x,k_y}(z)$ the well-known Luttinger and Kohn 6x6 matrix /13/ is inserted:

$$H_0(k_x,k_y,k_z)=\begin{pmatrix} H_1 & R+i\Delta & S & 0 & 0 & \Delta \\ R-i\Delta & H_2 & T & 0 & 0 & i\Delta \\ S & T & H_3 & -\Delta & -i\Delta & 0 \\ 0 & 0 & -\Delta & H_1 & R-i\Delta & S \\ 0 & 0 & i\Delta & R+i\Delta & H_2 & T \\ \Delta & -i\Delta & 0 & S & T & H_3 \end{pmatrix}$$

$$H_1 = 1/2\,L(k_z-k_x)^2 + 1/2\,M(k_x^2+k_y^2+k_z^2+2k_xk_z) \qquad R = 1/2\,N(k_z^2-k_x^2)$$
$$H_2 = 1/2\,L(k_z+k_x)^2 + 1/2\,M(k_x^2+k_y^2+k_z^2-2k_xk_z) \qquad S = 1/\sqrt{2}\,Nk_y(k_z-k_x)$$
$$H_3 = Lk_y^2 + M(k_x^2+k_z^2) \qquad T = 1/\sqrt{2}\,Nk_y(k_z+k_x).$$

For the calculation k_z is replaced by $1/i(\partial/\partial z)$. The band parameters L, M and N are taken from experimental data.

The Hartree potential V(z) is obtained by solving Poisson's equation

$$V''(z) = \frac{e}{\varepsilon\varepsilon_0}\,\rho(z) = \frac{e}{\varepsilon\varepsilon_0}\left[\sum_{\text{occ.states}} e|\psi_{i,k_x,k_y}(z)|^2 + \rho_{depl}\right]. \quad (2)$$

The charge density $\rho(z)$ consists of the free carrier charge and that arising from impurities located in the depletion layer. It is necessary to solve equations (1) and (2) self-consistently. One starts with an approximate potential and performs so many iterations until the desired accuracy is obtained. BANGERT and coworkers /3/ expanded the envelope wave function into a linear combination of six Airy functions whereas OHKAWA and UEMURA /4/ employed a more complicated set of trial functions. The boundary conditions were chosen that the envelope function vanished at the interface and deep inside the bulk. The matrix Hamiltonian has to be diagonalized for a sufficiently large number of points in the k_x-k_y plane.

A typical result for the potential energy as a function of the distance from the interface is shown in Fig. 1 for the (100) orientation for a free hole concentration of $4.5 \cdot 10^{12}/cm^2$. The difference between donor- and acceptor concentration N_D - N_A was assumed to be $10^{15}/cm^3$. The band bending is substantial and amounts to more than 200 meV. Four subbands can be distinguished which have been characterized as heavy and light hole bands. An identification is possible by inspecting the shape of the wave functions. It turns out that the upper two bands above the Fermi energy E_F are the ground states of the heavy and light hole with a shape of the wave function like curve a in Fig. 2. The wave functions b and c with one and two nodes can be attributed to the first and second excited state of the heavy hole subband, respectively.

The energy versus wave vector relation is highly nonparabolic. This can be recognized immediately in Fig. 3. It is obvious that for finite wave vectors the spin degeneracy is lifted. The reason for this is the breaking of the inversion symmetry by the high surface electric field. The amount of the spin splitting depends on the band parameters chosen, the splitting predicted by OHKAWA and UEMURA /4/ is larger than that indicated in Fig. 3. The calculations show that it is not possible to interpret a splitting of Shubnikov-de Haas peaks in terms of a g factor. The dispersion in the light hole subbands is especially large, leading to a positive hole mass in the upper branches.

The constant energy contours are strongly warped as one can see in Fig. 4. The E = constant lines have been plotted for carrier concentrations of

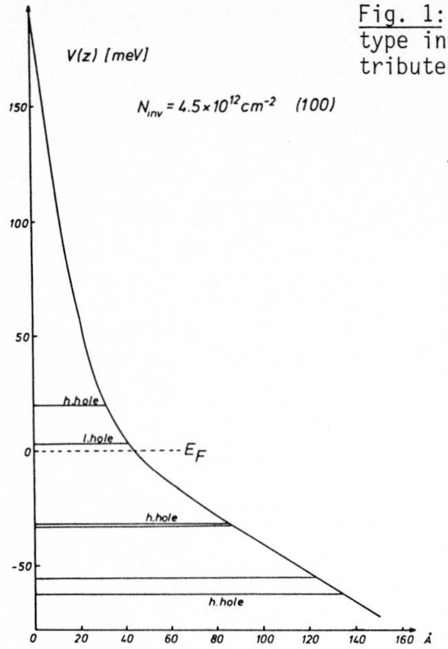

Fig. 1: Potential energy of holes in a (100) p-type inversion layer. Electric subbands are attributed to heavy and light holes

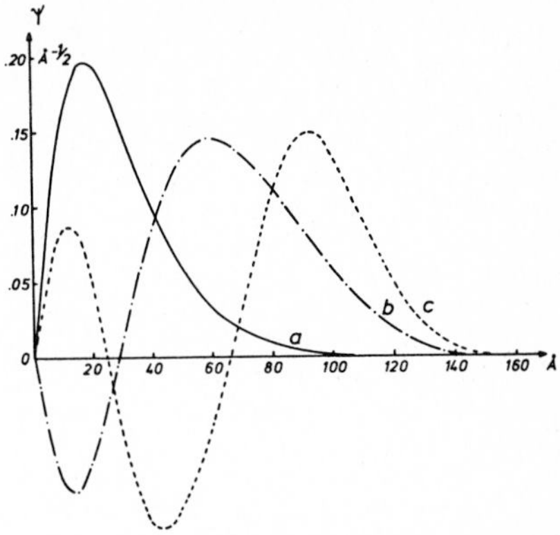

Fig. 2: Envelope functions for 3 different heavy hole subbands for a total hole concentration of $4.5\times10^{12}/cm^2$ (k = 0, (100) orientation) a) ground state, b) first excited state, c) second excited state

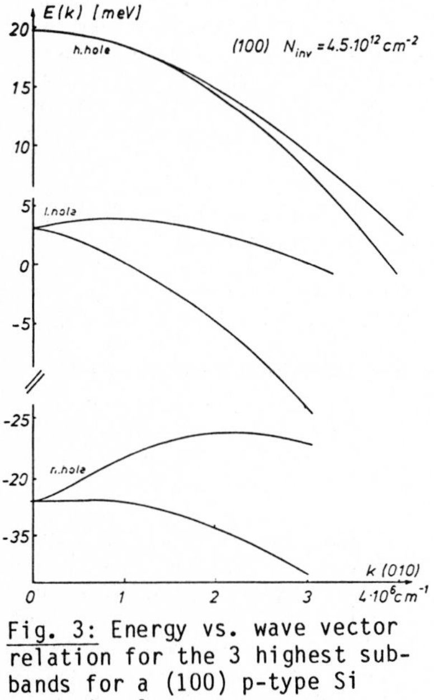

Fig. 3: Energy vs. wave vector relation for the 3 highest subbands for a (100) p-type Si inversion layer

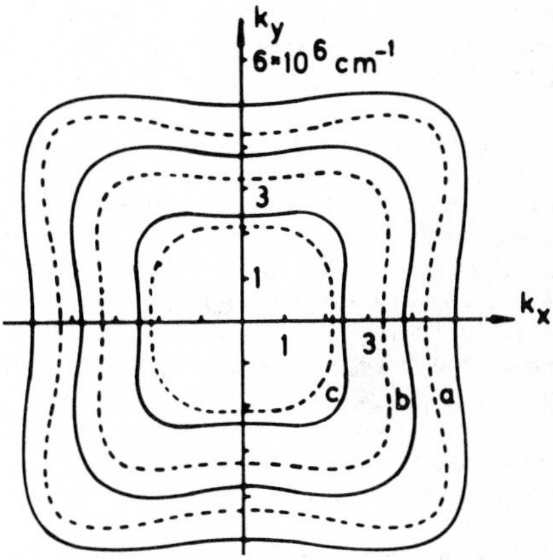

Fig. 4: Constant energy contours for a (100) silicon inversion layer at 3 surface hole concentrations a) N = $6.10^{12}/cm^2$, b) = $3.10^{12}/cm^2$, c) $1.10^{12}/cm^2$. The lifting of the spin degeneracy is again evident

$1 \cdot 10^{12}/cm^2$, $3 \cdot 10^{12}/cm^2$ and $6 \cdot 10^{12}/cm^2$. The lifting of the Kramers degeneracy results in two lines for each carrier concentration.

The subband energies depend strongly on the surface carrier concentration. This is shown in Fig. 5 where the subband energy has been plotted as a function of the surface carrier concentration. According to the calculations the light hole subband occupation should start at a carrier concentration of about $2.5 \cdot 10^{12}/cm^2$. It should be mentioned that no light holes have been observed in Shubnikov-de Haas experiments on (100) silicon MOSFETs even at hole concentration as high as $10^{13}/cm^2$. For the (110) orientation the population of the light hole subband has been predicted /3/ for p about $4 \cdot 10^{12}/cm^2$. This has been roughly confirmed by the experiment /14/. The agreement must be considered as fortuitous, however, because the predictions of theory for both the (100) and (111) orientations could not be verified by the experiment /10/. The arrows in Fig. 5 indicate possible optical dipole transitions with the electric vector E polarized perpendicular to the interface. The dipole matrix element was calculated between different subbands. A typical magnitude was 50 $Å^2$ for the allowed transitions. Values smaller than 10^{-9} $Å^2$ were interpreted as forbidden transitions. For (110) planes appreciable mixing of the valence band branches occurs so that no forbidden transitions are predicted.

From the constant energy contours the cyclotron mass can be calculated making use of the semiclassical relation

$$m_c^i = \hbar^2/2\pi \quad (dA_i(E)/dE) . \tag{3}$$

$A_i(E)$ is the area in the k_x,k_y plane enclosed by the contour of constant energy. Whether this approach is justified is somewhat doubtful, because the Onsager quantization scheme, which is reasonable for large quantum numbers, might break down. There is, however, empirical support for our

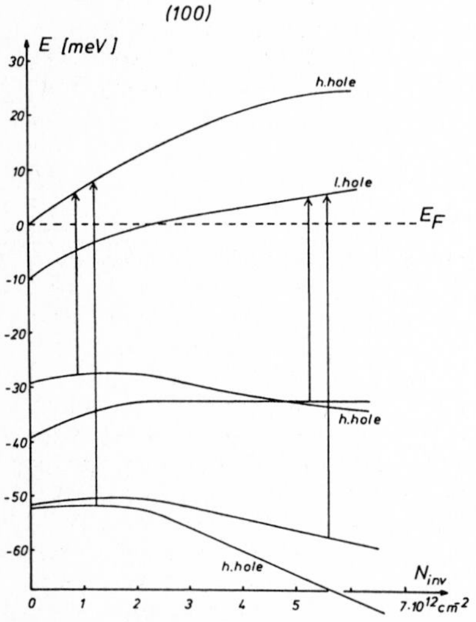

Fig. 5: Subband energy for (100) Si as a function of the surface hole concentration, E_F = Fermi energy. The arrows indicate allowed optical transitions. (polarization perp. surface)

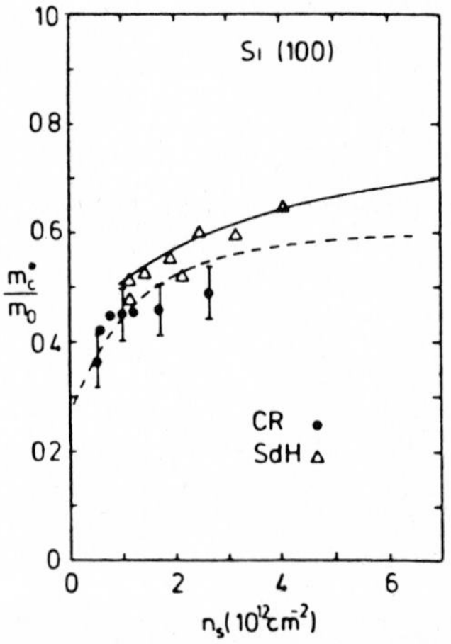

Fig. 6: Calculated cyclotron masses (solid line ref. /3/, dashed line ref. /4/) and experimental data (SdH, ref./10/; cyclotron res., ref. /16/) as a function of the surface hole concentration

calculations because the experimentally determined hole masses agree rather well with the theoretical predictions. This can be seen in Fig. 6, where calculated as well as experimentally determined cyclotron masses for the (100) orientation have been plotted. The upper solid curve was calculated by OHKAWA and UEMURA /4/ and the dashed one by BANGERT et al. /3/. The differences seem to arise mainly from somewhat different band structure parameters employed. The triangles show mass values derived from Shubnikov-de Haas experiments /10/. Cyclotron resonance measurements by KOTTHAUS and RANVAUD /15/ yielded masses shown by solid points. For the orientations (110) and (111) the agreement between theory and experiment is also reasonably good /10,15/.

In the original subband calculations /3/ the image potential was neglected. Due to the substantial difference in the dielectric constants for Si and SiO_2 this contribution to the potential is not negligible. Inclusion of the image potential in subband calculations for both n- and p-type inversion layers /16,17/ have shown that the subband energies are raised if the polarization field at the interface is included. The effect is most pronounced for the ground state subbands. On the other hand, the effect of many-body interactions like exchange and correlation has been neglected in our calculations. Work on n-type inversion layers has clearly shown that these effects are relevant /18/. Their importance depends on the surface carrier concentration, they are especially significant for electron concentrations below $10^{12}/cm^2$. At higher carrier densities, the screening effect reduces the importance of many-body interactions. It turned out, however, that calculations of the effective mass are not very sensitive to many-body effects and to the inclusion of the image potential, because the shape of the constant energy contours is not substantially modified. Due to the interacting valence bands, the inclusion of many-body effects in subband calculations is considerably more complicated than in the n-type case. Nevertheless, OHKAWA /17/ has studied many-body effects for (100) silicon p channels. It was necessary to neglect the nonparabolicity of the subbands. It turned out that the many-body corrections were of comparable importance as in n-type inversion layers. Calculations for the light hole subbands have not been performed so far. It can be expected, however, that many-body effects should be important.

3. Subband Calculations for GaAs

A two-dimensional electron gas can be generated not only at the interface between Si and SiO_2 but also in a hetero-junction made of GaAlAs and GaAs. The method of modulation doping /19/ has allowed to manufacture structures with electron mobilities exceeding 10^6 cm^2/Vs at helium temperature. The high mobility is achieved because the donors from which the electrons originate are located in the GaAlAs and therefore are spatially separated from the conducting GaAs layer. Recently, it has been possible to produce high quality p-type layers by making use of the modulation doping principle /20/, employing Be as a dopant. The value of the mobility indicated that the hole scattering times are comparable with those obtained in the best n-type structures. The samples investigated in high magnetic fields at temperatures as low as 0.55 K showed pronounced Shubnikov-de Haas oscillations and a well-developed quantum Hall effect /5/. At not too high magnetic fields two different periods were observed. Cyclotron resonance experiments revealed the presence of two holes with effective masses of 0.38 m_0 and 0.60 m_0. There was a linear relation between magnetic field and resonance frequency for fields between 3 and 8 Tesla. Further analysis indicated that the two kinds of carriers can be attributed to the spin-split branches of the heavy hole subband.

Because up to now no self-consistent calculations of subbands in p-type GaAs 2D-layers have been published we have modified the procedure described in the previous section for Si and applied it to the situation given in a GaAlAs-GaAs hetero-structure. The following band parameters were taken from a paper of SKOLNIK et al. /21/: $L = -6.08.10^{-15}$ $eVcm^2$, $M = -0.94.10^{-15}$ $eVcm^2$, $N = -6.56.10^{-15}$ $eVcm^2$. The potential well at the interface was considered to be infinitely steep. Consequently, it was assumed that the envelope function characterizing the holes in the subbands vanishes at the interface as well as in the bulk of the GaAs. Although the difference in the dielectric constants of GaAs (ε = 12.56) and AlAs (ε = 10.06) is not especially large,we incorporated the image potential in our calculation. Because of the composition of the $Ga_{1-x}Al_xAs$ (x = 0.5) in the hetero-structures investigated by STÖRMER et al. /5/ we chose a mean value of ε = 11.3 on the alloy side. The free hole concentration was chosen as $5.10^{11}/cm^2$ in order to allow comparison with the data of ref. /20/. Because the GaAs in the structure studied by Störmer et al. was of the highest purity, we neglected any space charge on the GaAs side. We also neglected the k-linear terms (caused by the lack of inversion symmetry of the ZnS lattice) because they are not significant.

We calculated the E(k) relation for various wave vector directions in the (100) plane and found, as expected, a lifting of the spin degeneracy for finite wave vectors. The results are shown in the upper part of Fig. 7. On the right side, the energy has been presented as a function of the k vector in the direction [011] and on the left side k[010] has been plotted. It is obvious from comparison with Fig. 3 that the dispersion is much larger than in Si. For a carrier concentration of $5.10^{11}/cm^2$ only the split heavy hole band is occupied. The constant energy contours are shown in the

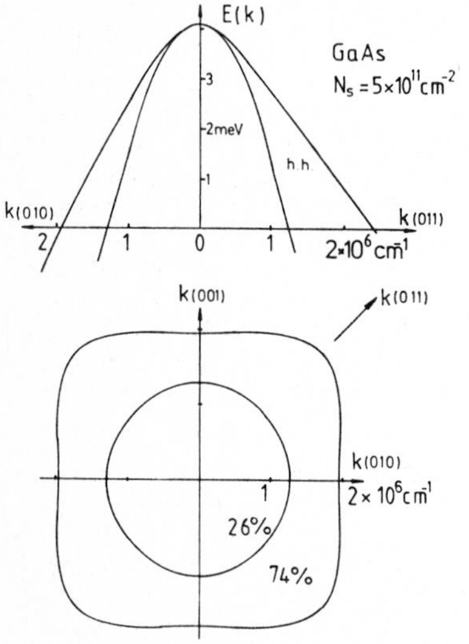

Fig. 7: Top: Energy vs. wave vector relation for the heavy hole subband in (100) GaAs for the [011] and [010] directions. Bottom: Constant energy contours with occupation ratio for the split heavy hole subband

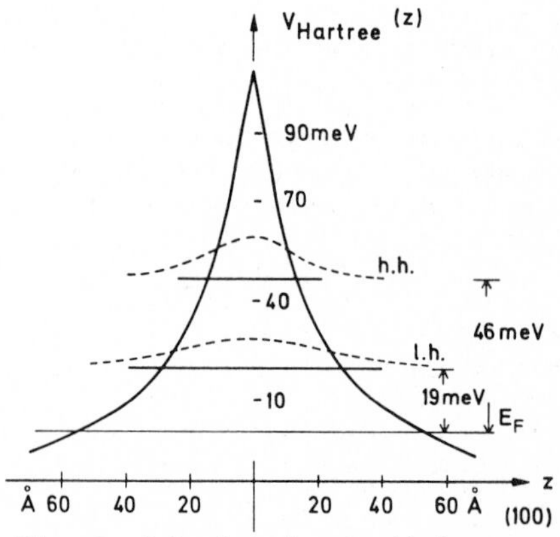

Fig. 8: Calculated potential energy in the valence band of a p-type Ge inversion layer as a function of the distance from the grain boundary with (100) orientation for a carrier concentration of $10^{13}/cm^2$ (Hartree approximation). Dashed lines: Hole density in the heavy hole (hh) and light hole (lh) subbands

lower part of Fig. 7. It is evident that the warping of the upper heavy hole branch is considerable whereas the lower branch is almost isotropic. For the semiclassical cyclotron masses we obtained values of 0.61 m_0 and 0.16 m_0 for the upper and lower heavy hole band, respectively. The occupancy of the lower subband turned out to be 26 % and 74 % for the upper one. The corresponding experimental data from ref. /5/, which were determined by the period of the Shubnikov-de Haas effect and Hall measurements, were 40 % and 60 %, respectively. Whereas the larger theoretical mass value of 0.61 m_0 agrees very well with the corresponding experimental one, there is a rather large discrepancy for the lower mass.

At present we have no plausible explanation for the small value of the calculated lower mass. Although many-body effects like exchange and correlation should be relevant at a carrier concentration of $5 \cdot 10^{11}/cm^2$, previous experience indicates that the effective masses should not be strongly affected. It is somewhat puzzling that the larger observed hole mass agrees very well with the calculated, whereas the experimental and theoretical values of the smaller mass differ by more than a factor of 2.

4. Electric Subbands in p-Type Germanium Inversion Layers

About 25 years ago, the transport properties of p-type Ge inversion layers adjacent to medium angle tilt grain boundaries were studied by LANDWEHR and HANDLER /6/. At low temperatures both conductivity and Hall effect were almost constant, indicating statistically degenerate behaviour. Calculations based on the Thomas-Fermi method /22/ showed that the thickness of the conducting layer for hole concentrations of a few times $10^{12}/cm^2$ should be less than 100 Å and that quantum effects should show up. The search for electric subbands was not successful at that time, because the magnetic fields applied were not high enough. Recently, the studies of the low temperature magneto-transport properties of space charge layers associated with medium angle grain boundaries were resumed with the purpose to investigate many-body effects which have been predicted for disordered two-dimensional systems /23/. It turned out that the resistivity increases logarithmically with decreasing temperature and that the magneto-resistance shows a logarithmic B dependence at high magnetic fields. The interpretation of the data in terms of recently developed theories /24/ requires the knowledge of the electronic structure of the inversion layers studied. Moreover, Shubnikov-de Haas oscillations were observed in a particular bicrystal with 15^0 tilt angle, indicating that quantum effects have to be taken into account /8/. Therefore, we have performed subband calculations for the Ge bicrystal system.

The procedure is essentially the same as outlined before, a 6x6 matrix effective mass Schrödinger equation was solved self-consistently together with Poisson's equation for zero temperature. Because the potential is symmetric with respect to the grain boundary, a different set of orthogonal envelope functions has to be employed. We chose the functions $f_n(z) = ch(z)^{-1} P_n (tgh(z))$, where P_n are Legendre polynomials. For reasons of simplicity the tilt angle was neglected, the grain boundary was assumed to be a (100) plane. The charged dislocations, which stabilize the grain boundary mechanically, were considered as scattering centres. It should be noted that the grain boundary problem has inversion symmetry so that the Kramers degeneracy is not lifted. Consequently, the electric subbands are twofold degenerate.

The results of the subband calculations are shown in Fig. 8. The Hartree potential energy has been plotted against the distance from the grain boundary plane. The free hole concentration in the calculation was $10^{13}/cm^2$ and the depletion layer charge arising from a bulk-donor concentration of $6.10^{15}/cm^3$ was taken into account. These numbers apply to the 15^0 bicrystal for which Shubnikov-de Haas oscillations were observed. One can recognize in Fig. 8 that two electric subbands are occupied, the upper one located 46 meV above the Fermi energy and the lower one 19 meV. The calculated wave functions of the subbands which are indicated by the dotted lines allow the identification as heavy (hh) and light hole (lh) band. It is evident that the band bending is substantial, the Fermi level is depressed about 100 meV below the potential maximum. A calculation of the constant energy contours revealed that the light hole subband is almost isotropic, whereas the heavy hole subband is considerably warped, the shape of the E = constant lines resembles those calculated for (100) Si (see Fig. 4). Also the cyclotron masses were calculated as a function of the total free hole concentration. The results have been plotted in Fig. 9. Whereas the heavy hole mass is not very concentration dependent, substantial variation of the light hole mass is predicted at not too high N values. The results indicate that the electric field at the interface has modified the band structure substantially. This becomes obvious by comparing the light hole mass in the bulk, which amounts to 0.042 m_0, with the calculated masses shown in Fig. 9.

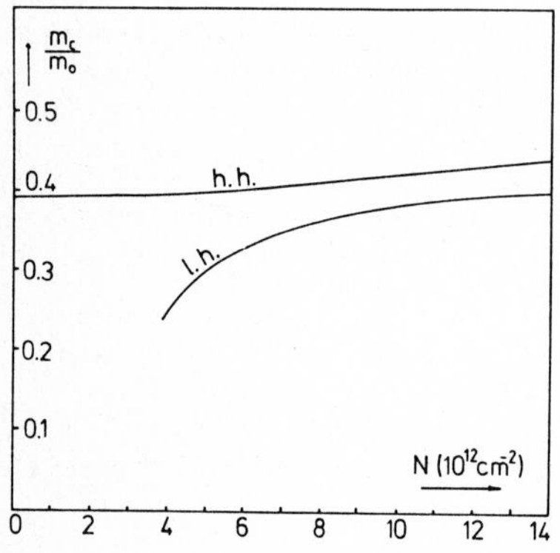

Fig. 9: Effective mass of the heavy hole (hh) and light hole (hl) subbands in a p-type Ge inversion layer adjacent to the grain boundary of a (100) tilted bicrystal as a function of total hole concentration

The Shubnikov-de Haas data /8/ allow a direct check of the subband calculations. From the period of the oscillations the concentration of the light holes can be determined as $3.10^{12}/cm^2$. This compares rather well with the predicted light hole concentration of $2.6.10^{12}/cm^2$. The mobility in the heavy hole subband was only about 400 cm^2/Vs so that no quantum oscillations could be observed. From the temperature dependence of the amplitude of the light hole oscillations an effective mass of 0.34 m_0 was deduced. This is somewhat lower than the calculated mass of 0.39 m_0. It should be kept in mind, however, that no many-body effects were incorporated in the calculations. Moreover, it was supposed that the interface is crystallographically undisturbed. It might be that this assumption is too idealized because it is known that in the immediate vicinity of dislocations the lattice is stressed.

The subband calculations have clearly indicated that quantum effects are significant and that they have to be taken into account for the interpretation of transport data.

References

1 A.D. Wieck, E. Batke, D. Heitmann and J.P. Kotthaus: Surf. Science (1984), in print
2 G. Abstreiter, U. Claessen and G. Tränkle: Sol. State Comm. 44, 673 (1982)
3 E. Bangert, K. v.Klitzing and G. Landwehr: Proc. 12th Int. Conf. Phys. Semicond. (M.H.Pilkuhn, Ed.), p.714, Teubner Verlag, Stuttgart (1974)
E. Bangert and G. Landwehr: Surf. Science 58, 138 (1976)
4 F.J. Ohkawa and Y. Uemura: Suppl. Progr. Theor. Physics 57, 164 (1975)
5 H.L. Störmer, A.M. Chang, Z. Schlesinger, D.C. Tsui, A.C. Gossard and W. Wiegmann: Phys. Rev. Lett. 51, 126 (1983)
6 G. Landwehr and P. Handler: J. Phys. Chem. Solids 23, 891 (1962)
7 S. Uchida and G. Landwehr: Application of High Magnetic Fields in Semiconductor Physics, Grenoble 1982 (G.Landwehr, Ed.), Springer Verlag, Lecture Notes in Physics 177, p. 65
8 S. Uchida, G. Landwehr and E. Bangert: Sol. State Comm. 45, 869 (1983)
9 F. Stern and W.E. Howard: Phys. Rev. 163, 816 (1967)
10 K. v.Klitzing, G. Landwehr and G. Dorda: Sol. State Comm. 15, 489 (1974)
11 G. Landwehr: in "Festkörperprobleme" (Advances in Solid State Physics) Vol. XV, p. 49, (H.J.Queisser, Ed.), Pergamon/Vieweg, Braunschweig (1975)
12 T. Ando, A. Fowler and F. Stern: Rev. of Modern Physics 54, 437 (1982)
13 J.M. Luttinger and W. Kohn: Phys. Rev. 97, 869 (1955)
14 K. v.Klitzing, G. Landwehr and G. Dorda: Sol. State Comm. 14, 387 (1974)
15 J.P. Kotthaus and R. Ranvaud: Phys. Rev. B15, 5758 (1977)
16 F. Stern, unpublished, see ref. 12
17 F.J. Ohkawa: J. Phys. Soc. Japan 41, 122 (1976)
18 See, e.g. ref. 12
19 R. Dingle, H.L. Störmer, A.C. Gossard and W. Wiegmann: Appl. Phys. Lett. 33, 665 (1978)
20 H.L. Störmer: Surf. Science (1984), in press
21 M.S. Skolnik, A.K. Jain, R.A. Stradling, J. Leotin, J.C. Ousset and S. Askenazy: J. Phys. C: Solid State Phys. 9, 2809 (1976)
22 G. Landwehr: Phys. Stat. Sol. 3, 440 (1963)
23 see, e.g. D.J. Thouless: Physica 109 & 110 B, 1523 (1982)
24 H. Fukuyama: in "Anderson Localization" (Y. Nagooka and H. Fukuyama Eds.), Springer Verlag (1982), p. 89

Interaction of a Two-Dimensional Electron Gas with Ionized Impurities and Polar Optical Phonons in High Magnetic Fields

R. Lassnig

Institut für Experimentalphysik, Universität Innsbruck, Schöpfstraße 41
A-6020 Innsbruck, Austria

The influence of various scattering center distributions on the Landau level width is discussed for GaAs-AlGaAs heterostructures in high magnetic fields. The filling factor dependent dielectric screening of long-range potentials leads to an oscillating behavior of the lavel width. The properties of interface longitudinal optical phonons and their interaction with electrons confined in double heterostructures are derived. A continuous transition from two- to three-dimensional interaction is demonstrated.

1. Introduction

The interaction of 2D electrons with ionized impurities and polar optical phonons is of fundamental interest for the understanding of the physical properties of the 2D electron gas. Although these two scattering mechanisms lead to quite different effects, the discussion of both gives a general insight into the main features of 2D effective interaction, with respect to the three-dimensional case. This is the main intention of the present paper.

In a quasi 2D system like a heterostructure or a superlattice the electron motion is confined in a plane parallel to the interfaces. The k_z wavevector (in the following all interfaces are taken in the (x,y)--plane) is replaced by the electric subband index as the characteristic quantum number. Through the application of a high magnetic field H transverse to the interface the electron system is completely quantized into nearly discrete Landau levels. In this context a "high magnetic field" means that the product of the cyclotron resonance frequency with the typical scattering time of the system is much larger than one ($\omega_c\tau >> 1$).

The discretization of the density of states leads to a number of pronounced resonance phenomena, such as cyclotron resonance (CR) and magnetophonon effect (typical transition energies are smaller than 50meV, which is the far-infrared regime). The investigation of these effects reveals the basic physical processes as well as the specific material properties. Therefore the theoretical results are discussed in relation to recent experimental observations at high magnetic fields.

In the first part of the paper the Landau level broadening due to ionized impurities is derived for GaAs-GaAlAs heterostructures. Special attention is paid to the influence of the scattering center distribution relative to the inversion layer. The basic theory of quantum transport and CR in 2D systems has been developed by ANDO and UEMURA |1,2|. ENGLERT et al. |3| observed an oscillation of the CR linewidth with the magnetic field. LASSNIG and GORNIK |4| explained the effect by the filling factor dependent screening of ionized impurities. Self-consistent calculations of the screening of ionized impurities in a 2DEG at high magnetic fields have also been performed by K. HEIFT and J. HAJDU |5|.

In the second part the change of the polar optical phonon structure due to the presence of interfaces is discussed. A semiclassical ansatz to determine the interface phonon properties has been developed by TZOAR |6|. Here a generalized approach is presented to calculate the interface phonon dispersion relations and the effective electron-phonon interaction |7|, and the results are compared with recent magnetophonon experiments |8,9|.

2. Landau Level Broadening due to Ionized Impurities

The calculation of the level broadening starts from the unperturbed Landau levels with harmonic oscillator wave functions $|n\rangle$ and energy $E_n = \hbar\omega_c \cdot (n + 0.5)$. Applying the theory of ANDO |2| the perturbations are treated in self-consistent Born approximation. For the z-dependent part of the wave functions the Stern-Howard functions

$$\zeta(z) = \sqrt{b^3/2}\; z e^{-bz/2} \tag{1}$$

are used. The parameter b is determined variationally and the mean distance of the electrons from the interface is $\bar{z} = 3/b$.

Neglecting image charge effects, the potential of an ionized impurity at z_0 is:

$$\phi(q_\perp, z-z_0) = u_0(q_\perp) \cdot e^{-q_\perp |z-z_0|} \tag{2}$$

$$u_0(q_\perp) = e/2q_\perp \varepsilon_0 \tag{3}$$

where ε_0 is the static dielectric constant of the medium. u_0 is the 2D Coulomb interaction and the exponential factor describes the influence of the third space component (normal to the interface).

The Landau level broadening Γ_n due to ionized impurity scattering is obtained as a sum over the 2D momentum transfer $q_\perp$ |2|:

$$\Gamma_n^2 = 4\sum_\mu N_\mu \sum_{q_\perp} |u_\mu(q_\perp)\cdot\langle n|e^{iq_\perp x}|n\rangle|^2 \tag{4}$$

where N_μ is the number of scattering centers of type μ. In the effective interaction $u_\mu(q_\perp)$

$$u_\mu(q_\perp) = \int_0^\infty \rho(z)\phi_\mu(q_\perp,z)dz/\varepsilon(q_\perp) = u_0(q_\perp)\cdot F_\mu(q_\perp,b,z_0)/\varepsilon(q_\perp) \tag{5}$$

the potential $\phi(q_\perp,z-z_0)$ is mapped with the electronic charge density $\rho(z)$. In order to separate the pure 2D interaction $u_0(q_\perp)$, the form factors $F_\mu(q_\perp)$ have been introduced in (5). The dielectric screening

$$\varepsilon(q_\perp) = 1 + u_0(q_\perp)\cdot\pi(q_\perp) \tag{6}$$

depending on the static polarizability $\pi(q_\perp)$ of the electron gas is also incorporated in the effective potential.

As can be seen in (5), two principal effects reduce the scattering strength $u_\mu(q_\perp)$ of ionized impurities relative to the bare Coulomb potential $u_0(q_\perp)$:

First, the long-range parts of the potential are screened by the inversion layer electrons. In the system under consideration the polarizability $\pi(q_\perp)$ oscillates strongly with the filling factor of the Landau levels |4|. This is qualitatively indicated in Fig. 1.

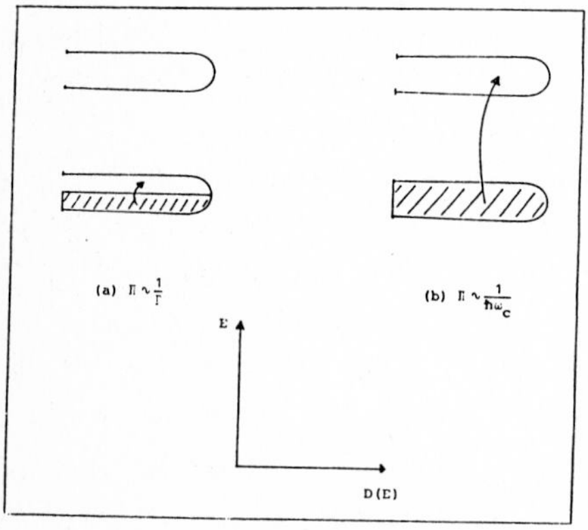

Fig.1: Schematic representation of the main contributions of virtual transitions to the polarization π: (a) half-filled, (b) completely filled Landau levels

Screening by the polarization of the 2D electron gas means that the electrons are pushed virtually from their unperturbed "original" state to unoccupied states. This process is enhanced in case (a), if the Landau level is half filled and not much virtual energy is required to push the electrons ($\pi \sim 1/\Gamma$). On the other hand, screening is considerably weaker if an integer number of Landau levels is fully occupied (case b) and the Fermi energy lies between adjacent Landau levels. In that configuration it is difficult to polarize the electron system ($\pi \sim 1/\hbar\omega_c$). Therefore maxima of the Landau level width and the CR linewidth are expected at integer values of the filling factor.

Second, the distance of the localized impurity from the electron gas plays an important role. With increasing separation, the short-range parts of the potential become less effective. Therefore, impurities in the GaAlAs contribute relatively weaker to the level width than scattering centers in the GaAs, which are directly in the range of the electron wave function.

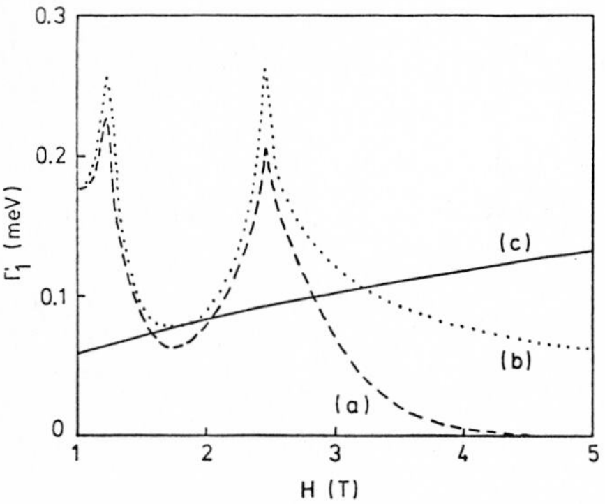

Fig. 2: Calculation of the level width Γ_1 for ionized impurities in the AlGaAs (a), in the GaAs (b) and for point-like scatterers with $\tau = 2\cdot10^{-10}$s (c)

In Fig. 2 the width Γ_1 of the first Landau level is calculated |4| for three different configurations of scattering centers and an inversion layer density of $N_{inv} = 1.2 \times 10^{11}/cm^2$, as a function of the magnetic field. The effective mass is taken to be $m^* = 0.07$ times the free electron mass and $\varepsilon_0 = 12.8$. In case (a) the AlGaAs doping is taken to be $N_2 = 5 \times 10^{14} cm^3$, and after a spacer layer of 100Å it is assumed to be strongly doped, so that there are $N_3 = N_{inv}$ charges located at this distance. In case (b), only a scattering center distribution in the GaAs with $N_1 = 5 \times 10^{14}/cm^3$ is assumed.

In both cases, the level width oscillates with the filling factor due to screening. Maxima are obtained for 2.48 T and 1.24 T, corresponding to a filling factor $\nu = 1$ and $\nu = 2$. The main differences for the two configurations arises at high magnetic fields, where only the lowest Landau level is partially occupied: In case (a), the level width decreases strongly with the magnetic field, since the screening becomes more effective. On the other hand, if the impurities are directly in the range of the wave function (case b), the short-range scattering contributions are more important. They are not damped by screening and the level width remains nearly constant at high magnetic fields.

Case (c) shows for a comparison the well-known result for an extremely short-range potential |2|:

$$\Gamma^2 = 2\hbar^2 \omega_c / \pi\tau \tag{7}$$

where τ denotes the zero field relaxation time. The level width shows no influence of screening and increases slowly with H.

From the above comparison of the different scattering types we find that potentials of long and intermediate range lead to an oscillation of the Landau level width. Increasing short-range contributions lead to larger level widths for high magnetic fields.

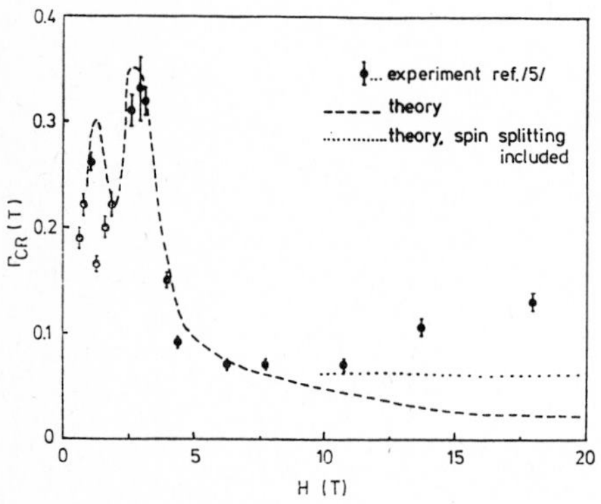

Fig. 3: Comparison of the experimentally observed cyclotron resonance linewidth oscillation with the theoretical calculations

In Fig. 3 the calculated CR linewidth Γ_{CR} is compared with the experimental results of ENGLERT et al. |3|. The agreement is very good and the oscillation of Γ_{CR} is described excellently. For $H > 5$ T, only scattering from impurities in the GaAs dominates, which leads to a constant CR linewidth. This allows one to determine an upper limit for the GaAs impurity concentration.

3. Polar-Optical Electron-Phonon Interaction

In layered systems the dispersion relation of polar optical phonons and their effective interaction with 2D-confined electrons can be strongly modified compared to the 3D case. Additionally to the restricted number of bulk-like modes (with nodes at the interfaces) there appears a series of 2D modes, so-called interface phonons.

For a semiconductor-insulator structure the interface optic phonon properties have been derived by TZOAR |6|. Here a generalized approach is presented for a Double-Heterostructure (DHS) made of arbitrary polar semiconductors |7|. The DHS has been chosen since on the one hand it is quite similar to a superlattice. On the other hand, a single heterostructure represents only a limiting case of a DHS. Therefore the essential physical features of both systems can be derived in a quite simple approach.

In order to determine the phonon structure in a DHS, the classical rate of energy of a test charge is derived in a first step. The interfaces are taken at $z = \pm d/2$ and the bulk dielectric functions of the layer and the surrounding material are denoted by ε_m and ε_s, respectively. The symmetry of the problem allows working in the $(q_\perp, z)$ space, which makes the calculations simple.

From Poisson's equation, the relation between potential, charge and dielectric function is

$$\Phi(q,\omega) = 4\pi\rho(q,\omega) / q^2\varepsilon(q,\omega) \tag{8}$$

where for a point charge moving in the $z = 0$ plane with velocity $v_\perp$ one obtains

$$\rho(q_\perp,\omega) = e\delta(\omega - q_\perp v_\perp) / 8\pi^3 . \tag{9}$$

In order to account for the dielectric discontinuities at $z = \pm d/2$, a modified image charge ansatz for the potential in a DHS is introduced:

$$\Phi(q_\perp,\omega,z) = \frac{e\delta(\omega-q_\perp v_\perp)}{8\pi^2 q_\perp} \cdot \begin{cases} [e^{-q_\perp z} + \sigma_m(q_\perp)(e^{-q_\perp(d-z)} + e^{-q_\perp(d+z)})]/\varepsilon_m & \ldots |z|<d/2 \\ [e^{-q_\perp z}(1+\sigma_s(q_\perp)) + \sigma_m(q_\perp)e^{-q_\perp(z-d)}]/\varepsilon_s & \ldots z>d/2 . \end{cases} \tag{10}$$

The value and position of the image charges (σ_m, σ_s) depend on the position of the observer at z (with respect to the observer, they are always placed on the opposite side of the interface). However, it should be mentioned that they are not conventional image charges, since they depend also on the Fourier wave vector $q_\perp$ parallel to the interface. With this remaining degree of freedom in (10), the Maxwell continuity conditions for both the potential and the dielectric displacement at the interfaces can be fulfilled:

$$\sigma_m(q_\perp) = -\sigma_s(q_\perp) = \frac{(\varepsilon_m - \varepsilon_s)}{\varepsilon_s(1+\gamma) + \varepsilon_m(1-\gamma)} \quad , \qquad \gamma = e^{-q_\perp d} . \tag{11}$$

The classical energy loss of the test charge is given by |6|:

$$\frac{dW}{dt} = \lim_{z \to 0} e\vec{v}_\perp \vec{\nabla}\Phi(x,z,t) = 2e\int d^2 q_\perp \int_0^\infty \omega d\omega \, \mathrm{Im}\Phi(q_\perp,\omega) \tag{12}$$

$$\mathrm{Im}\Phi = \frac{e\delta(\omega-q_\perp v_\perp)}{8\pi q_\perp}\cdot\left\{\frac{(1-\gamma)}{(1+\gamma)}\delta(\varepsilon_m) - \frac{4\gamma}{(1+\gamma)}\delta(\varepsilon_s(1+\gamma)+\varepsilon_m(1-\gamma))\right\} . \tag{13}$$

In the following we restrict ourselves to the situation when the layer as well as the surrounding material are of single-mode type:

$$\varepsilon_{m,s}(\omega) = \varepsilon^{\infty}_{m,s}(\hbar^2\omega^2-L^2_{m,s})/(\hbar^2\omega^2-T^2_{m,s}) . \tag{14}$$

ε^{∞}, L and T denote the high frequency dielectric constant, the longitudinal and the transverse optic phonon energy, respectively. In this case the ω integration in (12) can be performed analytically:

$$\frac{dW}{dt} = \frac{e^2}{8\pi\hbar}\int\frac{d^2q_\perp}{q_\perp}\Big\{\delta(L_m-\hbar q_\perp v_\perp)(L^2_m-T^2_m)(1-\gamma)/(1+\gamma) + \tag{15}$$

$$\pm\sum_{\pm}\delta(\omega_\pm-q_\perp v_\perp)(\hbar^2\omega^2_\pm-T^2_m)(\hbar^2\omega^2_\pm-T^2_s)\cdot 4\gamma/(1+\gamma)(\varepsilon'_s+\varepsilon'_m)\hbar^2(\omega^2_+-\omega^2_-)\Big\}$$

$$\omega^2_\pm = \left(p\pm\sqrt{p^2-(\varepsilon'_m+\varepsilon'_s)(\varepsilon'_m L^2_m T^2_s+\varepsilon'_s L^2_s T^2_m)}\right)/(\varepsilon'_m+\varepsilon'_s)\hbar^2 \tag{16}$$

$$p = (\varepsilon'_m(T^2_s+L^2_m)+\varepsilon'_s(T^2_m+L^2_s))/2 , \qquad \varepsilon'_m=\varepsilon^{\infty}_m(1-\gamma), \quad \varepsilon'_s=\varepsilon^{\infty}_s(1+\gamma). \tag{17}$$

Equation (15) is now compared with the quantum mechanical rate of energy loss. In terms of phonon destruction operators a_i and a_b the electron-phonon interaction is written in the following form (which already includes the specific form of the phonon wave function):

$$V^{(i)}_{ph} = \sum_{q_\perp} e^{iq_\perp x}\,(e^{-q_\perp(d/2-z)}+e^{-q_\perp(d/2+z)})v_i a_i + \text{h.c.} \tag{18}$$

$$\ldots|z|<d/2.$$

$$V^{(b)}_{ph} = \sum_{q_\perp} e^{iq_\perp x}\sum_{n=1}^{\infty}\sin((z/d+1/2)n\pi)v_b a_b + \text{h.c.} \tag{19}$$

Abbreviating the matrix element for one-phonon emission with $\langle H_{int}\rangle$, the Fermi golden rule gives:

$$\frac{dW}{dt} = \frac{2\pi}{\hbar}\sum_{q_\perp}\omega|\langle H_{int}\rangle|^2\cdot\delta(\omega-q_\perp v_\perp) . \tag{20}$$

From a comparison of the phonon energies it is found that the first term in (15) corresponds to the interaction with bulk layer phonons, with:

$$v^2_b = \frac{e^2}{A}\frac{L_m}{\varepsilon^{\infty}_m d}(1-T^2_m/L^2_m)/(q^2+(n\pi/d)^2) . \tag{21}$$

A is the normalization area. On the other hand, in the second term of (15) the phonon energies $\hbar\omega^{\pm}$ depend on the dielectric properties of both dielectric functions ε_s and ε_m. This term corresponds to the interface phonon modes and $v_i(q_\perp)$ is given by

$$v^2_{i,\pm} = e^2\hbar\omega_\pm f_\pm/Aq_\perp \tag{22}$$

$$f_\pm = \pm(\hbar^2\omega_\pm^2 - T_m^2)(\hbar^2\omega_\pm^2 - T_s^2)/4\hbar^4\omega_\pm^2(\omega_+^2-\omega_-^2)(\varepsilon_m' + \varepsilon_s')(1+\gamma)\,. \tag{23}$$

In the 2D system the motion of the electrons normal to the interface is practically replaced by discrete transitions between the electric subbands. In most cases only scattering within the zeroth subband or between the zeroth and the first subband is important. Therefore it is useful to define an effective 2D interaction, which means that V_{Ph}^μ is written as a sum over 2D momentum transfers, with appropriate form factors:

$$V_{Ph}^\mu = \sum_{q_\perp} e^{iq_\perp x}\, v_{2D} F_\mu(q_\perp d) \tag{24}$$

$$v_{2D}^2 = e^2 L(1-T^2/L^2)/4Aq_\perp\varepsilon^\infty\,. \tag{25}$$

The index μ includes the specific transition (e.g.,$0\rightarrow 0$) and the type of phonon involved. v_{2D} is the idealized Fröhlich interaction of a point charge moving in a plane and interacting with bulk phonons. The form factors F_μ carry all the information on the z dependence of the system.

We restrict ourselves to the $(0\rightarrow 0)$ transition, using simple sinusoidal electronic wave functions:

$$\Psi(z) = \sqrt{2/d}\,\sin((z/d+1/2)\pi). \tag{26}$$

Mapping the local electron-phonon interaction (18,19) with the wave functions (26), one obtains the effective interaction with the following form factors:

$$F_b = 16\sqrt{\alpha/(\alpha^2+\pi^2)}\,/3\pi\,, \qquad \alpha = q_\perp d \tag{27}$$

$$F_i^\pm = 2(1-e^{-\alpha})\sqrt{4\hbar\omega_\pm f_\pm \varepsilon^\infty L/(L^2+T^2)}\,/\alpha(1+\alpha^2/4\pi^2)\,. \tag{28}$$

Generally it can be said that the interface phonon energies as well as the form factors depend strongly on the dimensionless product of the layer thickness with the phonon wavevector ($\alpha = q_\perp d$). The numerical calculations are performed for a $Ga_{0.5}In_{0.5}As$ layer embedded in InP. The material constants are $L_m = 34$, $T_m = 31$, $L_s = 43$, $T_s = 37.6$ (all in meV), $\varepsilon_m^\infty = 11.35$ and $\varepsilon_s^\infty = 9.56$.

Figure 4a shows the phonon energy spectrum as a function of α, revealing the characteristic features of DHS phonons. The full line corresponds to the bulk phonon energy, the dashed and the dotted lines to the higher (InP) and lower 2D phonons. In contrast to the constant dispersion relation of the bulk, the dependence of the LO phonon energies on the wavevector $q_\perp$ is a new feature of DHS phonons. For $\alpha \ll 1$, i.e.,for relatively long phonon wavelengths, the DHS phonon corresponding to the layer material (GaInAs) is damped down to the TO phonon energy T_m. At the same time the

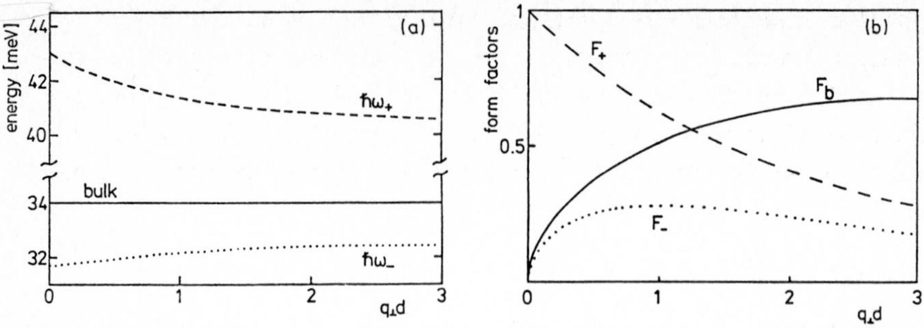

Fig. 4: Phonon energies (a) and form factors (b) for a $Ga_{0.5}In_{0.5}As$-InP double heterostructure as a function of $\alpha = q_{\perp}d$

InP-like mode approaches the bulk energy value. On the other hand, for relatively small wavelengths ($\alpha \to \infty$), the DHS phonon energies converge to the single interface phonon energies, which are the single heterostructure limit.

Figure 4b shows the 2D form factors $F_{\pm}$ and F_b. They represent a reasonable measure for the effective coupling. Just as for the energies, the interaction of the InP-like phonon approaches the three-dimensional value for $\alpha \to 0$, describing the transition from two- to three-dimensional electron-phonon interaction. It can be seen that, at least for relatively long wavelengths, the phonons of the surrounding material couple into the layer regime via the interface phonons.

For the GaInAs modes over the whole regime the bulk interaction is considerably stronger than the interface interaction. This is due to the fact that the DHS phonon energy $\hbar\omega_-$ is closer to the TO phonon energy, which corresponds to weaker polarity.

The resonant scattering of electrons between different Landau levels via LO phonons leads to an oscillatory magnetoresistance and is called magnetophonon effect. In contrast to optical transitions there is no selection rule for electron-phonon scattering and the resonance condition is given by $\hbar\omega_L = n\hbar\omega_c$.

The magnetophonon effect in 2D systems has first been observed in GaAs-GaAlAs heterostructures |10|, and the theory has been developed by LASSNIG and ZAWADZKI |11|. PORTAL et al. |9| investigated magnetophonon effects in $Ga_{0.47}In_{0.53}As$-InP superlattices (for d = 80, 100 and 150Å and InP layers of 400Å) and detected two series of oscillations. For the lower (GaInAs) series they found $\hbar\omega$ = 33.6 (32.9) meV for a 150 (80) Å sample. The decrease towards lower d values can be well explained by the increasing influence of interface phonons for $d \to 0$.

For the higher mode the energy was found around 43meV, which is the bulk InP value. Although the existence of an InP-like DHS phonon coupling through the interface to the electrons has been well explained in the above discussion, a somewhat lower energy is expected from the theoretical viewpoint. The discrepancy is possibly due to shortcomings in the description of nonparabolicity.

Acknowledgements

The author wishes to thank Prof. Erich Gornik, Prof. Wlodek Zawadzki and Prof. Jean-Claude Portal for good cooperation and helpful discussions.

References

1 T. Ando: J. Phys. Soc. Jpn. 36, 1521 (1974); 37, 662 (1974); 37, 1233 (1974); 38, 989 (1975)
2 T. Ando, Y. Uemura: J. Phys. Soc. Jpn. 37, 1044 (1974)
3 T. Englert, J.C. Maan, C. Uihlein, D.C. Tsui, A.C. Gossard: 16th Int. Conf. on the Physics of Semiconductors, Montpellier 1982
4 R. Lassnig, E. Gornik: Solid State Comm. 47, 959 (1983)
5 K. Heift, J. Hajdu: Lecture Notes in Physics 177, 139 (1983)
6 N. Tzoar: Surf. Sci. 84, 440 (1979)
7 R. Lassnig, W. Zawadzki: Proc. 5th Int. Conf. on Electronic Properties of Two-Dimensional Systems, Oxford 1983
8 R.J. Nicholas, M.A. Brummel, J.C. Portal, M. Razeghi, M.A. Poisson: Int. Conf. on the Appl. of High Magnetic Fields in Semiconductors, Grenoble 1982 (Springer 1983)
9 J.C. Portal, J. Cisowski, R.J. Nicholas, M.A. Brummel, M. Razeghi, M.A. Poisson: J. Phys. C16, L573, (1983)
10 D.C. Tsui, Th. Englert, A.Y. Cho, A.C. Gossard: Phys. Rev. Lett. 44, 1980 (1980)
11 R. Lassnig, W. Zawadzki: J. Phys. C16, 5435 (1983)

FIR Investigations of GaAs/GaAlAs Heterostructures

E. Gornik, W. Seidenbusch, R. Lassnig

Institut für Experimentalphysik, Universität Innsbruck, Schöpfstraße 41
A-6020 Innsbruck, Austria

H.L. Störmer, A.C. Gossard, and W. Wiegmann

AT & T Bell Laboratories, Murray Hill, NJ 07974, USA

Experimental and theoretical investigations of the CR linewidth and line position are presented. Oscillations of the linewidth due to screening effects are observed. A theoretical analysis of the linewidth gives important information on residual doping levels and interface qualities in the GaAs/GaAlAs heterostructures. Cyclotron emission experiments reveal impurity associated transitions at energies above the cyclotron energy. Significantly reduced polaron effects have been found in a two-dimensional electron gas in GaAs as compared to the bulk case.

1. Introduction

Far infrared optical techniques are applied to investigate electric properties of a two-dimensional electron gas (2DEG) as present at the GaAs/GaAlAs interface. Cyclotron resonance (CR) studies reveal the dominant scattering mechanism and the effective carrier mass. CR linewidth studies have been performed by several groups with partly contradictory results |1-6|. Samples with low densities ($n_s < 2 \times 10^{11} cm^{-2}$) and high mobilities ($\mu$ (at 4.2K) $> 2 \times 10^5 cm^2/Vs$) show an oscillatory behaviour of the linewidth as a function of the Landau level (LL) filling factor ν ($\nu = \frac{n_s}{2(eB/\hbar)}$ neglecting spin-splitting) |1,2|. Samples with lower mobilities or higher densities do not reveal oscillations |3| but partly lineshifts and linesplittings |4-6|. From the CR line position the LL splitting as a function of energy is determined. This splitting reveals the band nonparabolicity and the electron-polar-optical phonon interaction (polaron interaction) as was demonstrated by LINDEMANN et al. |7| in bulk GaAs.

Recently, DAS SARMA |8| has predicted that polaron effects are enhanced for a 2DEG as compared to the 3D case. HORST et al. |9| have performed CR studies in InSb inversion layers for the $0 \rightarrow 1$ Landau transition through the reststrahlen-region and confirmed the predicted increased polaron interaction. However, this is in contradiction to subband resonance studies performed by SCHOLZ et al. |10| in CdHgTe accumulation layers, who found reduced polaron effects. SEIDENBUSCH et al. |2| and LINDEMANN et al. |11| investigated polaron effects in a 2DEG in GaAs and also found a strong reduction of the polaron effect. DAS SARMA |12| and LASSNIG |13| have suggested that these different behaviours are due to an interplay of screening, occupation and levelwidth effects.

In this paper we report CR transmission and emission experiments investigating screening effects on the CR linewidth and the polaron interaction. As an additional variable we introduce the heating of the electron gas with an electric field to study the electron temperature influence on these effects.

2. Cyclotron Resonance Linewidth

Cyclotron resonance experiments performed on Si-MOS inversion layers |14| have revealed several specific effects, especially in the quantum limit. The most prominent is a line narrowing and a line shift at low densities ($\nu < 1$), attributed to electron-electron interaction |15|, pinned charge density waves |16| or trapped electrons |17,18|. However, scattering due to defects near the interface limits the mobility in Si to modest values and the lines are correspondingly broad.

In the 2DEG in GaAs a combination of molecular beam growth techniques |19| and selective doping |20| are used to reduce impurity scattering to extremely low levels. In these relatively clean systems a variety of phenomena associated with correlation or coherence may become observable.

Several studies of the CR linewidth in GaAs/GaAlAs heterostructures have been published. VOISIN et al. |3| investigated the linewidth in rather low mobility heterostructures (μ (at 4.2K) $\sim$ 71 000 cm^2/Vs) and attributed the observed linewidth to short-range scatterers (mainly ionized impurities). A linewidth increase with $\sqrt{B}$ for fields up to 2T (B is the magnetic field) is in agreement with the theory of ANDO and UEMURA |21|, assuming a range of 50Å for a Gaussian scattering potential.

A somewhat different behaviour of the CR linewidth in similar samples was observed by NARITA et al. |4| and MURO et al. |5|, who have investigated the linewidth as a function of density and temperature. A linesplitting at low density and low temperature (< 15K),which disappears with increasing density or temperature, was found and explained by a pinned charge density wave |22|. The measured linewidth is explained in terms of long-range Gaussian scatterers in the higher magnetic field range (> 4T) due to ionized impurities in the GaAlAs material.

Great impact was given to the field by the paper of ENGLERT et al. |1|, who observed an oscillation of the CR linewidth as a function of filling factor and ascribed it to screening effects. The linewidth shows maxima at integer filling factors (completely filled levels) and minima in between. A screening dependent LL width was first predicted by DAS SARMA |23|.

A detailed calculation of the LL width and CR linewidth was performed by LASSNIG and GORNIK |24| including impurity scatterers at different sites in the GaAlAs and GaAs. The experimental results of ENGLERT et al. |1| are well explained in terms of a filling factor dependent CR linewidth due to screening of the ionized impurity scatterers by the 2DEG.

3. Cyclotron Resonance Transmission Experiments

We have performed CR studies on several GaAs/GaAlAs heterostructures (summarized in Table 1) as a function of density and wavelength at 4.2K.

Table 1.

Sample	Electron density (cm^{-2}) dark	Mobility at 4.2K (cm^2/Vs)
B1	4.6×10^{11}	1.3×10^5
B2	2.6×10^{11}	2.1×10^5
B3	7.1×10^{11}	1.5×10^4
B4	2.2×10^{11}	4.5×10^5

Figure 1 shows transmission spectra for a given density for 3 different wavelengths. It is apparent that the lines are Lorentzians. Sample B1 has a larger linewidth in the whole magnetic field range. The linewidth is

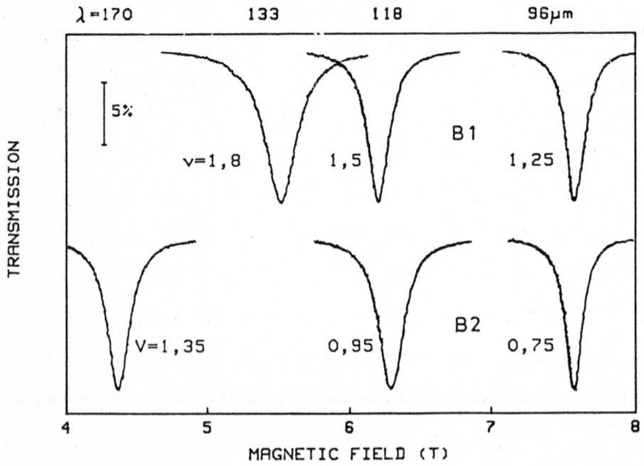

Fig. 1: Cyclotron resonance transmission spectra for several wavelengths and Samples B1 and B2. The filling factor ν is given for every curve

filling factor dependent but much weaker than observed by ENGLERT et al. |1|. A detailed investigation of the linewidth was performed for fixed wavelengths as a function of filling factor for Sample B2. The concentration and thus ν was changed by light illumination with a LED. In Fig. 2 the observed linewidth is plotted as a function ν for a wavelength of 96μm (open circles). An increase in linewidth between $\nu = 0.6$ and $\nu = 1$ is found. In this range transitions between the lowest and the first LL ($n = 0 \rightarrow n = 1$) are dominant. A further increase of ν induces additional LL transitions ($n = 1 \rightarrow n = 2$), which are shifted in energy due to nonparabolicity. Therefore the linewidth does not show a pronounced minimum for $\nu = 1.5$ but rather a flat dependence due to the superposition of two lines. The splitting of the lines is in the order of the linewidth.

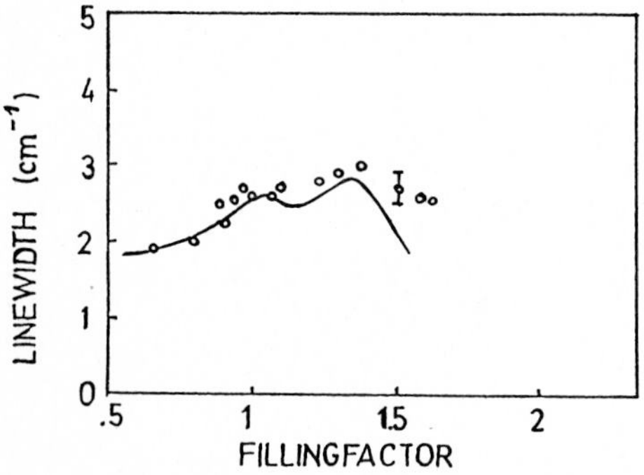

Fig. 2: Measured CR linewidth dependence on filling factor for Sample B2 and a wavelength of 96μm (open circles). The full curve shows a calculation according to |24| with the parameters given in the text

An application of the theory from |24| is given by the full curve which includes the superposition of two different transitions. A total impurity content of $5 \times 10^{14} cm^{-3}$ is assumed for the GaAs material, resulting in a depletion charge of $10^{11} cm^{-2}$, a doping of $5 \times 10^{14} cm^{-3}$ for the spacer and $2 \times 10^{18} cm^{-3}$ for the high doped GaAlAs layer. However, to account for the rather small changes in linewidth between $\nu = 0.6$ and $\nu = 1.0$ we have also to include neutral impurity scattering with $\tau = 7 \times 10^{-10}$s at B = 0. These scatterers are presumably defects at the interface or in the vicinity of the interface. The agreement between theory and experiment is reasonable within experimental error.

What are the conclusions we can draw from this analysis? LASSNIG and GORNIK |24| have shown that the long-range impurity scatterers as present in the GaAlAs can be completely screened for filling factors $\nu < 1$. The residual linewidth for $\nu \sim 0.5$ is only due to impurities in the range of the wavefunction of the 2DEG if no neutral scatterers are present. This means that we can get information on the residual doping level in the GaAs material. The analysis of the data of |1| shows that in the investigated material only ionized impurity scatterers are dominant and no neutral impurity (point defect) scattering was necessary to get a good fit. However, in Sample B2 a significant contribution of neutral impurity scattering is present, which means that this sample has a relatively large number of point defects which might give some evidence for a lower interface quality.

However, for samples with higher concentrations ($n_s > 4 \times 10^{11} cm^{-2}$) the analysis becomes difficult since $\nu < 1$ is reached at rather large magnetic fields where the spin-splitting comes into play and the condition of a single half-filled level is reached only at very high fields. In addition, for samples with higher doping levels the lines become broader and the effect of linewidth oscillations diminishes. An analysis of material properties should therefore be performed only for low concentrations, which can be realized in most of the samples with a back gate |25|.

4. Cyclotron Resonance Emission Experiments

A complementary method of CR transmission is CR emission or Landau emission. In this experiment the 2DEG is heated by electric field pulses. The resulting recombination radiation is measured and analyzed with a narrowband GaAs detector |26|. This method has first been applied to a low mobility GaAs accumulation layer (Sample B3) |27|. Two emission lines were observed: one due to the 2DEG, one due to bulk electrons. A linewidth analysis has revealed the total doping level for the bulk and a somewhat narrower width for the 2DEG. The electron heating showed a saturation behaviour for electric fields higher than 50 V/cm in the bulk case but considerably weaker heating in the 2D case with no saturation tendency up to 100 V/cm.

The emission technique was recently applied to high mobility GaAs/GaAlAs heterostructures. Figures 3a and 3b show emission spectra as obtained with a GaAs photoconductive detector (at B = 0)|26|. The detector has a peak sensitivity at 4.4 meV (35 cm^{-1}) with a linewidth of 2 cm^{-1}. In the experiment the emission signal is monitored as a function of the 2DEG magnetic field for various electric fields and densities. The most prominent feature is that the emission spectra differ considerably from the transmission spectra, obtained from identical samples (see Fig. 1). This effect is extremely pronounced for Sample B1 (Fig. 3a): for very low electric fields the emission spectrum consists of a broad line, which is well below the bulk cyclotron position (corresponding to a higher energy). With increasing electric field this peak becomes smaller and a line at the position of the transmission resonance grows until it dominates the spectrum at high fields. The

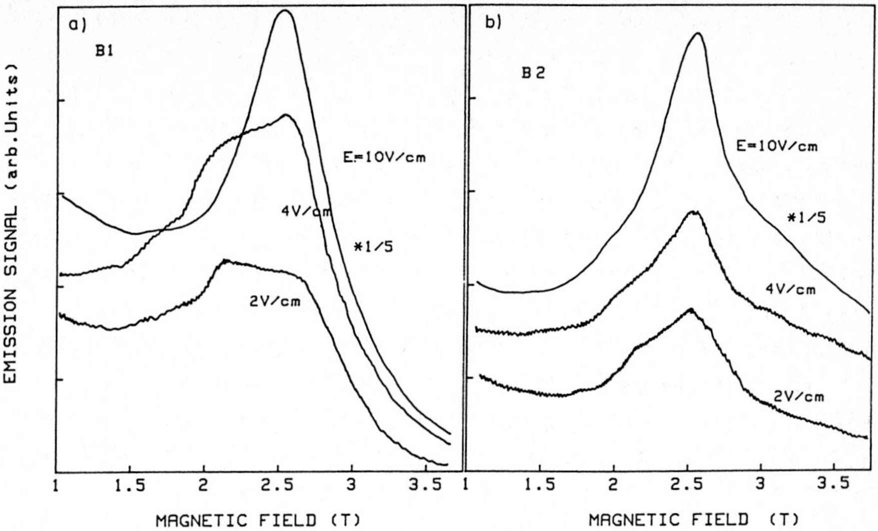

Fig. 3: Cyclotron emission signal detected with a GaAs detector (at B = 0) as a function of the magnetic field of the 2DEG for several electric fields: Sample B1 (3a), Sample B2 (3b)

2D character of the structure was tested by tilting the magnetic field. For Sample B2 (Fig. 3b) the peak at lower magnetic field is not that strong. Sample B2 has a higher mobility and a narrower linewidth. A shoulder on the low magnetic field side in the emission spectrum has been found in all investigated samples. Some correlation with the d.c. mobility and the CR transmission linewidth was found. The higher the mobility the less pronounced the additional peak becomes. For the sample with the highest mobility (B4), a rather narrow continuous shoulder was observed. This feature is not restricted to the 3T magnetic field range. We have found this behaviour in the whole investigated magnetic field range up to 8T.

The broad peak, which we want to assign as "impurity peak" can also be influenced by changing n_s. It becomes smaller with increasing n_s but does not vanish at the highest densities we were able to achieve ($n_s \sim 5.5 \times 10^{11}$ cm^{-2}) for sample B2 and for the lowest electric fields applied (1 V/cm).

As the additional peak is observed only in emission it must be correlated to the special excitation technique used: the electric field excites carriers to the next higher LL. The carriers first thermalize within the upper LL and occupy only states at the lower edge. If localized states exist below the edge, the electrons will get trapped in these states. As the number of excited electrons is very small at low electric fields, the recombination radiation will mainly be due to transitions out of these localized states. The emission frequency will be shifted to higher energy due to the binding potential similar to the bulk case, where a so-called impurity shifted CR line is observed. With increasing electric field the binding potentials are washed out and the pure CR line from extended states is observed. These binding potentials are in our opinion mainly due to impurities in the GaAlAs spacer layer or at the interface, since the impurities in the GaAs are negatively charged.

At present we are trying to correlate the impurity emission peak with the doping level in the GaAlAs. From the present understanding we can speculate that the linewidth limiting factor for $\nu < 1$ in high mobility samples is the residual doping of the bulk GaAs and the quality of the interface. For $\nu > 1$ and samples with lower mobility, the doping distribution of the GaAlAs material also contributes to the linewidth.

5. Polaron Effects

From the position of the LL transitions the energy level structure in a magnetic field is determined and usually expressed in an energy-dependent effective mass. LINDEMANN et al. |7| have performed a detailed analysis of LL transition energies in bulk GaAs. As a result a decrease in LL energy splitting with increasing energy was found and well assigned to non-parabolicity and polaron effects.

Recently this method was extended to the 2DEG case in GaAs |2,11|. Figure 4 shows a comparison of a bulk CR spectrum (obtained for high purity GaAs) with a spectrum obtained for Sample B2. Two CR transitions ($n = 0 \rightarrow n = 1$ and $n = 1 \rightarrow n = 2$) are seen. In the bulk case the population of the $n = 1$

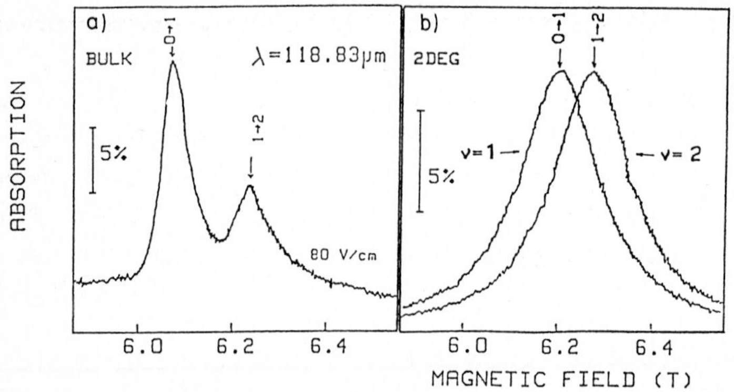

Fig. 4: Cyclotron resonance transmission spectra for a bulk GaAs sample (4a) and Sample B2 (4b). Two lines are observed in the bulk case due to electric field heating. In the 2D case single lines for different ν are shown

LL is achieved by heating the electron gas with electric field pulses. In the 2D case the n = 1 LL is populated by changing the filling factor from $\nu = 1$ to $\nu = 2$ with light illumination. It is obvious that the splitting between the lines is considerably smaller for the 2D case. Experiments at several wavelengths have been performed. Figure 5 shows a plot of the line splittings obtained in the bulk and 2D cases as a function of the cyclotron energy. In the whole range the splitting is considerably smaller in the 2D case. The curves represent theoretical calculations of the splitting in the 3D case |7| including polaron effects and nonparabolicity (upper curve) and only nonparabolicity (lower curve) for the 2D case |2|.

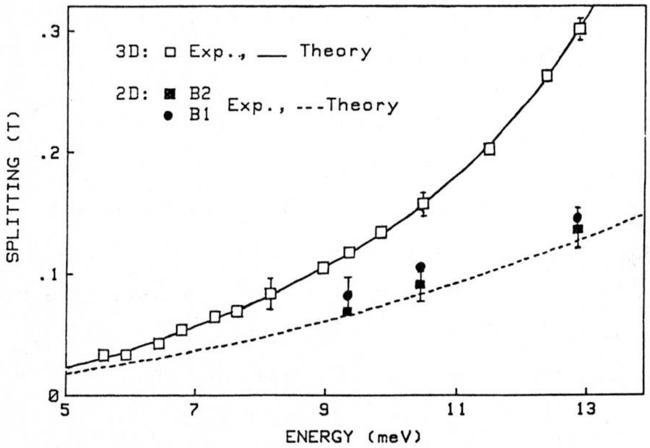

Fig. 5: Observed Landau transition energy difference (corresponding to the splitting of the lines in Fig. 4) plotted as a function of laser energy. The full curve is a theoretical fit including nonparabolicity and polaron effects. The dashed curve includes only nonparabolicity

It is evident that the 2D results are well described by nonparabolicity alone and that we cannot identify a significant contribution of the polaron effect. This strong reduction of the polaron effect can be qualitatively ascribed to screening and occupation effects.

Recent CR emission experiments |28| on high mobility GaAs/GaAlAs heterostructures ($\mu \gtrsim 5 \times 10^5 cm^2/Vs$) show evidence for an increased polaron effect with increasing electron heating. Electric fields up to 300 V/cm were applied resulting in electron temperatures up to 100K. Above an electron temperature of 50K a significant polaron effect becomes observable which reaches bulk values at the highest fields.

Acknowledgements

The work was partly supported by the Fonds zur Förderung der wissenschaftlichen Forschung, the Jubiläumsfonds der Österreichischen Nationalbank and the European Research Office London.

References

1 Th. Englert, J.C. Maan, Ch. Uihlein, D.C. Tsui, A.C. Gossard: Physica 117B & 118B, 631 (1983)
2 W. Seidenbusch, G. Lindemann, R. Lassnig, J. Edlinger, E. Gornik: Proc. 5th Int. Conf. on "Electronic Properties of Two-Dimensional Systems", Oxford, Sept. 1983, to appear in Surf. Sci. (1984)
3 P. Voisin, Y. Guldner, J.P. Vieren, M. Voos: Appl. Phys. Lett. 39, 982 (1981)
4 S. Narita, K. Muro, S. Mori, S. Hiyamizu, K. Nanbu: Proc. Int. Conf. on "The Application of High Magnetic Fields to Semiconductor Physics", Lecture Notes in Physics 177, 194 (Springer Verlag 1983)
5 K. Muro, S. Mori, S. Narita: Proc. 5th Int. Conf. on "Electronic Properties of Two-Dimensional Systems", Oxford, Sept. 1983, to appear in Surf. Sci. (1984)
6 Z. Schlesinger, S.J. Allen, J.C.M. Hwang, P.M. Platzmann: Proc. 5th Int. Conf. on "Electronic Properties of Two-Dimensional Systems", Oxford, Sept. 1983, to appear in Surf. Sci. (1984)
7 G. Lindemann, R. Lassnig, W. Seidenbusch, E. Gornik: Phys. Rev. B28, 4693 (1983)
8 S. Das Sarma: Phys. Rev. B27, 2590 (1983)
9 M. Horst, U. Merkt, J.P. Kotthaus: Phys. Rev. Lett. 50, 754 (1983)
10 J. Scholz, F. Koch, J. Ziegler, H. Maier: Sol. State Comm. 46, 665 (1983)
11 G. Lindemann, W. Seidenbusch, R. Lassnig, J. Edlinger. E. Gornik: Physica 117B & 118B, 649 (1983)
12 S. Das Sarma: Proc. 5th Int. Conf. on "Electronic Properties of Two-Dimensional Systems", Oxford, Sept. 1983, to appear in Surf.Sci.(1984)
13 R. Lassnig, W. Zawadzki: Proc. 5th Int. Conf. on "Electronic Properties of Two-Dimensional Systems", Oxford, Sept. 1983, to appear in Surf. Sci. (1984)
14 T. Ando, A.B. Fowler, F. Stern: Rev. Mod. Phys. 54, 437 (1982)
15 R.J. Wagner, T.A. Kennedy, B.D. Mc Combe, D.C. Tsui: Phys. Rev. B22, 945 (1980)
16 B.A. Wilson, S.J. Allen, Jr., D.C. Tsui: Phys. Rev. B24, 5887 (1981)
17 H.J. Mikeska, H. Schmidt: Z. Physik B20, 43 (1975)
18 T. Ando: Surf. Sci. 113, 182 (1982)
19 A.Y. Cho, J.R. Arthur: Prog. Solid State Chem. 10, 157 (1975)
20 H.L. Störmer, R. Dingle, A.C. Gossard, W. Wiegmann, M. Sturge: Solid State Comm. 29, 705 (1979)
21 T. Ando, Y. Uemura: J. Phys. Soc. Jpn. 36, 959 (1974)
22 H. Fukuyama, P.A. Lee: Phys. Rev. B18, 6245 (1978)
23 S. Das Sarma: Solid State Comm. 36, 357 (1980)
24 R. Lassnig, E. Gornik: Solid State Comm. 47, 959 (1983)
25 H.L. Störmer, A.C. Gossard, W. Wiegmann: Appl.Phys.Lett. 39, 493 (1981)
26 E. Gornik: Proc.Int.Conf. on the "Appl. of High Magnetic Fields in Semiconductor Physics", Lecture Notes in Physics 177, 248 (Springer Verlag 1983)
27 E. Gornik, R. Schawarz, D.C. Tsui, A.C. Gossard, W. Wiegmann: J.Phys.Soc. Jpn. 49, Suppl.A, 1029 (1980) and Solid State Comm. 38, 541 (1981)
28 E. Gornik, H.L. Störmer, R. Lassnig, A.C. Gossard, W. Wiegmann: to be published

The Magnetophonon Effect in Two-Dimensional Systems

R.J. Nicholas and M.A. Brummell

Clarendon Laboratory, Parks Rd., Oxford, OX1 3PU, United Kingdom

J.C. Portal

Laboratoire de Physique des Solides, I.N.S.A., Ave. de Rangueil
F-31077 Toulouse, France, and
S.N.C.I.-C.N.R.S., 166X, F-38042 Grenoble Cedex, France

Magnetophonon oscillations due to L.O. phonon absorption have been observed in GaAs-GaAlAs, GaInAs-InP and GaInAs-AlInAs heterojunctions and superlattices. This has been used to show that the two-dimensional behaviour persists up to 300 K. GaAs-GaAlAs and GaInAs-InP heterojunctions are dominated by scattering from 'GaAs-like' L.O. phonons. GaInAs-AlInAs heterojunctions show a predominant interaction with 'InAs-like' phonons. For thin quantum wells of GaInAs between InP layers, scattering from both GaAs and InP phonons is observed, with the InP phonons becoming the stronger for well thicknesses below 100 Å.

1. INTRODUCTION

The magnetophonon effect is the result of a resonant inelastic scattering process, under conditions where the electronic energy levels are quantized by a large magnetic field. This usually involves longitudinal optic (L.O.) phonons. The effect was first predicted in 1961 by GUREVICH and FIRSOV (1) and by KLINGER (2), and was first observed experimentally in the magnetoresistance of n-type InSb at 90 K (3,4). It is characterized experimentally by an oscillatory component of the magnetoresistance, periodic in 1/B, whose amplitude reaches a few per cent of the monotonic component, at maximum. Subsequently the effect has been observed in various transport coefficients of a large number of different semiconductors, as has been reviewed by HARPER et al.(5) and PETERSON (6).

The normal magnetophonon effect arises from the absorption of optic phonons, which cause transitions between the Landau levels produced by a magnetic field. This results in a resonant relaxation of the electron momentum. The conservation of crystal momentum limits the wavevector of the phonons involved to a region close to the centre of the Brillouin zone, and hence the optic phonon energy is essentially monoenergetic. The dominant electron-phonon coupling is with L.O. phonons, due to the large electric polarization associated with these modes. In high magnetic fields the conductivity σ_{xx} is proportional to the scattering rate, and hence to the joint density of states of the levels separated in energy by $\hbar\omega_{L.O.}$. The conductivity thus exhibits a maximum each time the L.O. phonon energy is equal to the separation of one or more Landau levels, giving the resonance condition:

$$N\hbar\omega_c = \hbar\omega_{L.O.} \tag{1}$$

with N = 1, 2, 3, ... The resistivity also shows a similar series of maxima since:

$$\rho_{xx} = \frac{\sigma_{xx}}{\sigma_{xy}^2 + \sigma_{xx}^2} ; \tag{2}$$

and usually $\mu B > 1$, $\sigma_{xy} > \sigma_{xx}$, so that $\rho_{xx} \propto \sigma_{xx}$. The resulting oscillations are periodic in 1/B, with the positions of the individual extrema given by:

$$N B_N = \frac{m^* \omega_{L.O.}}{e} , \qquad (3)$$

where the effective mass m^*_{MPR} must be adjusted from the band edge value, to take account of the effects of non-parabolicity, and the resonant polaron coupling. This latter effect increases the mass by a factor of $(1 + \alpha/4)$ for bulk material, where α is the Frölich polaron coupling constant. In two dimensions (2-D) DAS SARMA has shown (7) that the polaron mass enhancement is dependent upon the Thomas-Fermi screening, but for typical experimental conditions, it is about half of the bulk value. This correction is therefore of order 1% for a 2DEG in GaAs.

The increase in mass due to non-parabolicity can be quite large for a 2-D system. Using three band k.p theory (8) the increase above the band edge mass (m^*_0) is given by:

$$\frac{1}{m^*_{MPR}} = \frac{1}{m^*_0} \{ 1 + \frac{K_2}{E_g} [(2L + N + 1) \hbar\omega_{L.O.}/N + 2E_{||}] \} \qquad (4)$$

where E_g is the band gap, L is the quantum number of the initial Landau level, and $E_{||}$ is the kinetic energy due to motion in the potential well giving rise to the two-dimensional behaviour. This is not the same as the bound state energy in the well, and for a triangular potential $E_{||} = E_i/3$ (8). The coefficient K_2 depends upon the band positions of the material under consideration, and has values of 0.85 for $Ga_{.47}In_{.53}As$ and of 0.95 for GaAs, the two materials under consideration in this work. For samples with a low electron concentration we may usually put $L = 0$, and for more heavily doped material, under the conditions $\hbar\omega_c$, $\hbar\omega_{L.O.} < E_F$, the Fermi energy, equation [4] may be reduced to:

$$\frac{1}{m^*_{MPR}} = \frac{1}{m^*_0} \{ 1 + \frac{K_2}{E_g} [2E_F + 2E_{||}] \} . \qquad (5)$$

The other important modification to the resonance condition [3], is that for 2-D systems the Landau energies are determined only by the component of magnetic field parallel to the surface normal of the 2-D plane, $B\cos\theta$, where θ is the angle between the field direction and the surface normal.

In order to detect the usually rather weak oscillatory components of the magnetoresistance, it is necessary to suppress the monotonic magnetoresistance. In the high field regime, this may be done by subtraction of a voltage proportional to the magnetic field, however this is effective only over a rather limited range of field, and for strong oscillations. A more sensitive technique is to take the second derivative of the magnetoresistance, which strongly enhances the rapidly varying oscillatory components. This is mainly done by the use of high pass electronic filters, but has also worked successfully with both field modulation and digital processing. The action of taking the second derivative changes the sign of the oscillations, so that $-d^2R/dB^2$ is displayed to give maxima at the positions of the maxima in the original resistance.

To date 2-D magnetophonon resonance has been ovserved only in heterostructures with electrons confined in GaAs or $Ga_{.47}In_{.53}As$. Several workers have made careful studies of silicon inversion layers, however the many valleyed conduction band minima, lower mobilities and non-polar coupling have meant that any structure is much weaker, and it has not been observed experimentally. Searches for magnetophonon structure in narrow gap semiconductors have been hampered by the persistence of Shubnikov-deHaas peaks at high temperatures.

2. GaAs-GaAlAs

The first observations of two-dimensional magnetophonon resonance were made by TSUI et al.(9), and were extended to higher magnetic fields by ENGLERT et

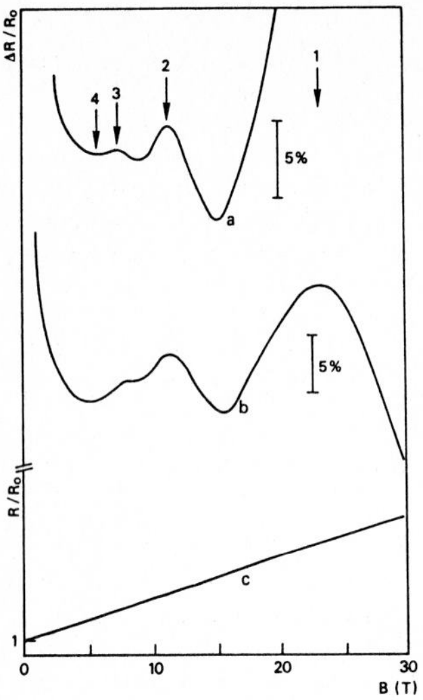

Fig. 1: Magnetophonon resonance in a GaAs-GaAlAs heterojunction. Curve a was taken using a Bitter solenoid, and b and c using pulsed magnetic fields. After (10)

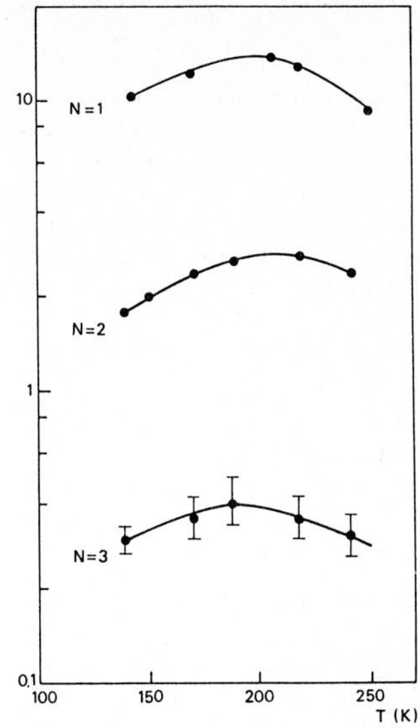

Fig. 2: The temperature dependence of the amplitude (arbitrary units) of the oscillations in a GaAs-GaAlAs heterojunction. After (10)

al.(10) and KIDO et al.(11), in heterojunctions and superlattices of GaAs-$Ga_{1-x}Al_xAs$, with x of order 0.3. Some typical recordings are shown in Fig. 1 for a heterojunction, with an electron concentration of 3.4 x 10^{11} cm^{-2} and a mobility of 300,000 $cm^2V^{-1}s^{-1}$. The position of the fundamental resonance, N = 1, is at 22.5 $\pm$ 0.5 T, and weaker resonances for N = 2, 3, 4 are observed at lower fields. Using a frequency of 293 cm^{-1}, the L.O. phonon frequency of bulk GaAs, equation [3] gives an effective mass m^*_{MPR} = (0.071 $\pm$ 0.0015) m_o. Equation [4] shows this to be in excellent agreement with the known band edge mass of 0.0665 m_o for bulk material, thus indicating that the effect of polaron corrections on the effective mass is rather small, as suggested by the calculations of DAS SARMA (7). The presence of the interface has an almost negligible effect upon the energies of the phonons for this system (12), so that it is to be expected that the bulk phonon energy should be applicable.

The amplitude of the oscillations is determined by two factors. There must be a sufficient population of L.O. phonons to cause scattering, and so the amplitude decreases at low temperatures, while at higher temperatures the increased broadening of the Landau levels smears out the resonances in the joint density of states. There is therefore a maximum in amplitude, as shown in Fig. 2, which occurs around 200 K. A final important feature of the oscillations was that they were shown to follow the perpendicular magnetic field component, B cos θ, at these high temperatures, and could therefore only originate from the 2DEG. There has been one recent report (13) of the observation of magnetophonon oscillations in GaAs-GaAlAs at 4.2 K, due to the emission of L.O. phonons by hot electrons. This was detected as a weaker additional periodicity in a Fourier analysis of the magnetoresistance oscillations, in the presence of strong Shubnikov-deHaas structure. The identification remains unconfirmed to date.

3. GaInAs-InP

Several reports have appeared recently of the formation of a 2DEG in GaInAs-InP heterojunctions (14,15) and superlattices (16,17), in which it has been shown that very high quality samples may be produced by the LP-MOCVD technique (18). The magnetophonon spectra of these materials show several very interesting properties (19-21), due to the two phonon modes of the GaInAs, and the rather polar nature of the InP. The optic phonon modes of $Ga_{.47}In_{.53}As$ have been studied by a number of workers (22-25), who have found two L.O. phonons at 233 cm^{-1} and 271 cm^{-1}. The dielectric constant for a material with two sets of optic phonon modes is:

$$\varepsilon = \varepsilon_\infty + \{\varepsilon' - \varepsilon_\infty\}\frac{\omega_{TO}^2(\omega_{TO}^2 - \omega^2)}{(\omega_{TO}^2 - \omega^2)^2 + \gamma^2\omega^2} + \{\varepsilon_0 - \varepsilon'\}\frac{\omega_{TO'}^2(\omega_{TO'}^2 - \omega^2)}{(\omega_{TO'}^2 - \omega^2)^2 + \gamma'^2\omega^2} \qquad (6)$$

where ω_{TO} and $\omega_{TO'}$ are the two T.O. phonon frequencies, ε_0 and ε_∞ are the low and high frequency dielectric constants, γ and γ' are damping parameters, and ε' is defined by the generalised Lyddane-Sachs-Teller relations:

$$\frac{\varepsilon'}{\varepsilon_\infty} = \frac{\omega_{LO}^2}{\omega_{TO}^2}, \qquad \frac{\varepsilon_0}{\varepsilon'} = \frac{\omega_{LO'}^2}{\omega_{TO'}^2}. \qquad (7)$$

The oscillator strengths associated with each mode are $(\varepsilon_0 - \varepsilon')$ and $(\varepsilon' - \varepsilon_\infty)$. For GaInAs this gives 0.82 for the 'InAs-like' mode at 233 cm^{-1} and 1.58 for the 'GaAs-like' mode at 271 cm^{-1}. It is thus not surprising that magnetophonon measurements on bulk GaInAs have shown that the dominant scattering is by 'GaAs-like' phonons (26), although recent measurements have detected some trace of oscillations due to 'InAs-like' modes (27).

Magnetophonon measurements were reported by PORTAL et al.(21) on several GaInAs-InP heterojunctions grown at the Thomson-C.S.C. Laboratories (18), with typical electron concentrations of 3 - 4 x 10^{11} cm^{-2} and mobilities of 8,000 $cm^2V^{-1}s^{-1}$ at 300 K and 30,000 $cm^2V^{-1}s^{-1}$ at 77 K. Oscillations were detected by standard second derivative techniques, at temperatures between 80 K and 300 K, and a typical series of curves is shown in Fig. 3. Rotation of the sample relative to the field at several temperatures showed that the resonances were two-dimensional up to 300 K.

The oscillations observed form a single series with a fundamental field, $N B_N$, of 14.4 T. Using the cyclotron mass of 0.048 m_0 measured by BRUMMELL et al.(28), with a small correction for non-parabolicity to give m^*_{MPR} (20), the phonon energy was deduced from [3] to be 278 cm^{-1}. The resonant polaron correction will reduce this value by $\sim$1%, although there is some uncertainty over the exact magnitude of the correction to be used, as discussed above. The dominant scattering would thus appear to be by 'GaAs-like' phonons in the GaInAs. It can be seen from Fig. 3 that an additional peak at 10.2 T appears at low temperatures. The amplitudes of this peak and the N = 2 magnetophonon oscillation are plotted in Fig. 4. The magnetophonon peak is visible over the whole range of temperatures studied, and has a maximum at 150 K, a behaviour very similar to that seen in the GaAs-GaAlAs system, and in most bulk material. The extra peak appears only below 220 K, and its amplitude increases monotonically with decreasing temperature. This is thought to be the remains of the N = 1 Shubnikov-deHaas oscillation, which is observed very strongly at this field at 4.2 K; at such high temperatures the Shubnikov-deHaas amplitude will fall off less quickly than expected from the classic $X/\sinh X$ relation ($X = 2\pi^2 kT/\hbar\omega_c$), since kT is no longer much less than $\hbar\omega_c$.

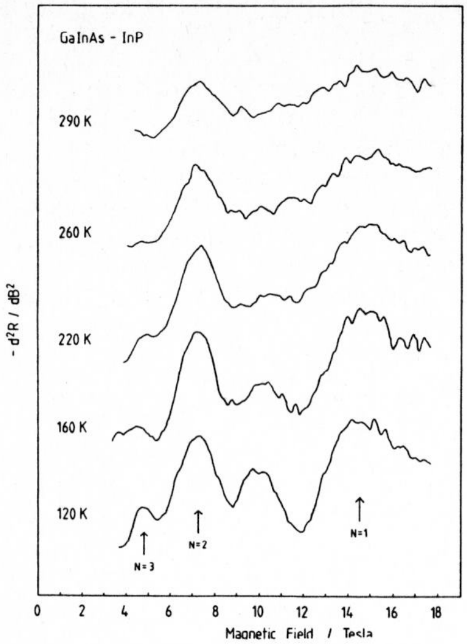

Fig. 3: Magnetophonon oscillations in a GaInAs-InP heterojunction as a function of temperature. The extra peak at 10.2 T is the fundamental Shubnikov-deHaas oscillation

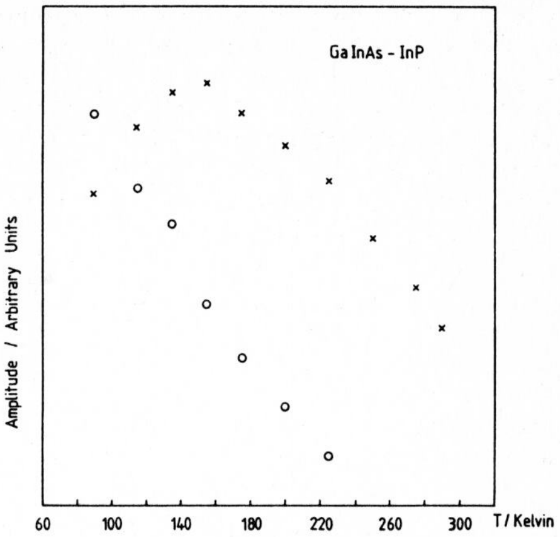

Fig. 4: The amplitudes of the N = 2 magnetophonon resonance (x), and the Shubnikov-deHaas peak (o), as a function of temperature

Two reports (19,21) have appeared recently of the observation of magnetophonon oscillations in GaInAs-InP superlattices. The samples studied consisted of ten nominally undoped layers of GaInAs with n $\sim 10^{15}$ cm^{-3}, with thicknesses of either 80 Å, 100 Å or 150 Å, separated by 400 Å thick layers of InP with n $\sim 3 \times 10^{16}$ cm^{-3}, grown on semi-insulating substrates. The two-dimensional electron concentrations as determined by Shubnikov-deHaas measurements are shown in Table 1. The effective masses of the electrons bound in the GaInAs have been determined by cyclotron resonance (19), and are found to increase from 0.041 m_o to 0.054 m_o as the layer thickness is decreased due to non-parabolicity and the increased zero point energy.

The magnetophonon resonances in these structures are shown in Fig. 5, and illustrate once again that conduction is two-dimensional up to high temperatures. It is also clear that there are two series of resonances present in all three samples studied. The series with the higher fundamental field, which is shown by primed indices in Fig. 5, increases in relative intensity as the layer thickness of GaInAs is decreased; while for the 150 Å layers this second series is rather weak compared with the main resonances, by 80 Å both series are of comparable intensity. The phonon energies were deduced from the fundamental fields by the use of the experimentally determined cyclotron masses, after suitable correction for non-parabolicity, and equation [3]. As shown in Table 1, there are two optic phonon modes involved in scattering, with frequencies of approximately 265 cm^{-1} and 350 cm^{-1} respectively. The lower energy phonon is the 'GaAs-like' mode of GaInAs, as seen in the GaInAs-InP heterojunctions. The L.O. phonon frequency of the intervening InP layers is 349 cm^{-1} (23). The conclusion would thus appear to be obvious: that the electrons bound in the GaInAs potential wells are being scattered both by the GaAs-like L.O. mode of the GaInAs and by the InP L.O. phonon. The exact mechanisms and nature of the phonon modes are not, however, completely clear.

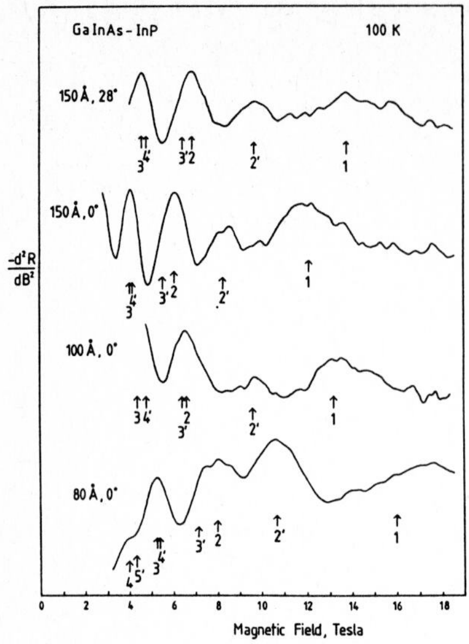

Fig. 5: Magnetophonon oscillations in the three GaInAs–InP superlattices at 120 K. The unprimed numbers are the resonance indices of the oscillations due to the 'GaAs-like' L.O. mode of the GaInAs, while the primed numbers show the resonance indices of the peaks due to the InP L.O. phonon. Rotation of the 150 Å superlattice shows that both series follow the $B\cos\theta$ dependence characteristic of a two-dimensional system, as is shown in the top two recordings

HESS and VOGL (29) proposed that a long-range polar phonon interaction could occur in silicon MOS devices, due to the interface optic phonon modes of the silicon-silicon dioxide interface. This type of mechanism would seem to be an excellent candidate to explain the scattering due to InP optic phonons.

Table 1. Magnetophonon results (uncorrected for polaron mass enhancement)

Sample thickness (Å)	n (10^{11}cm^{-2})	m_c^*/m_o (119 µm, 125 K)	m_{MPR}^*/m_o (N = 2)	N B_N (T)	$\hbar\omega_{LO}$ (cm^{-1})
80	1.7	0.0543	0.0563	16.0	265
			0.0575	21.4	347
100	2.2	0.0458	0.0475	13.2	260
			0.0485	19.2	370
150	1.1	0.0407	0.0422	12.3	271
			0.0431	16.3	353

The energies of the interface phonons are given by the condition that the real part of the dielectric constant at the interface $(\varepsilon_1 + \varepsilon_2)/2$ is equal to zero, where ε_1 and ε_2 are the real parts of the dielectric constants of the two materials. For GaInAs we may use [6] as described above, and for InP with only one optic mode we simply put $\varepsilon' = \varepsilon_\infty$. The dielectric constants ε_{GaInAs} and $-\varepsilon_{InP}$ are plotted in Fig. 6, with the damping parameter taken as $\gamma/\omega_T = 0.02$. The frequencies of the interface modes will be given by the crossings of the two curves, giving interface phonons at 260 cm^{-1} and 328 cm^{-1}. LASSNIG and ZAWADZKI (30) have reported calculations of the dispersion of the interface modes in this system, showing that the values above are the limiting cases for large well thicknesses, and that the InP mode is higher in energy for narrow wells and small q vectors, while the GaInAs mode is lower in energy. They also showed that the form factor, and hence coupling strength,was larger for the InP mode in narrow wells, and that of the GaInAs mode goes to zero for narrow wells at q = 0. Nevertheless it would still seem that the InP-like interface

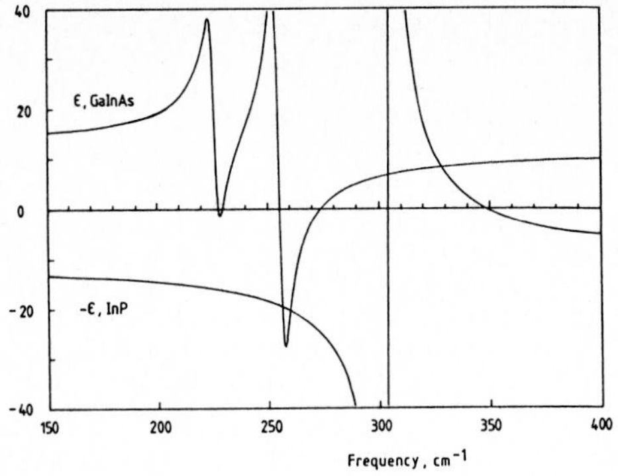

Fig. 6: The dielectric constants of GaInAs and InP plotted as ε_{GaInAs} and $-\varepsilon_{InP}$. The bulk modes occur at the points where the individual dielectric constants are zero, and the interface modes are found at the frequencies where $\varepsilon_{GaInAs} = -\varepsilon_{InP}$

mode is rather too low in energy to be the source of the series of oscillations of higher fundamental field. It is more probable that the bulk InP phonon would be involved, since the heterostructure interfaces grown by M.O.C.V.D. have been shown to be graded over approximately 40 Å (31), possibly smearing out the interface modes. For the thinner layers, the energies deduced for the 'GaAs-like' modes are lower than the bulk energy, and closer to the interface phonon energy. This may also be due to grading of the interface. Both Raman (24) and reflectivity (23) measurements have shown that as InP is admixed into GaInAs to form GaInAsP, the energy of the 'GaAs-like' mode is reduced. Due to these difficulties in a definitive assignment of the origin of the resonances, no attempt has been made to correct the phonon energies given in Table 1 for resonant polaron enhancement of the mass. This effect may be expected to reduce the values quoted by approximately 1%.

Several factors therefore contribute to the scattering of electrons in the GaInAs by optic phonons in the InP. Firstly there will be only a finite screening of the InP optic phonon electric field in the GaInAs and coupling to InP interface modes becomes very strong for thin wells; secondly there will be some penetration of the carrier wavefunction into the InP; thirdly, the presence of some phosphorus atoms in the graded interface will aid the propagation of vibrations around the InP phonon frequencies; and finally InP is a more polar material than GaInAs. These factors combine so that by the time the layer thickness is 80 Å, the strength of the scattering by the InP modes, judged by the relative amplitudes of the resonances, is comparable to the 'GaAs-like' mode.

4. GaInAs-AlInAs

CHENG et al.(32) have shown that it is possible to grow good heterojunctions of $Ga_{.47}In_{.53}As$-$Al_{.48}In_{.52}As$ by M.B.E., with mobilities in excess of 90,000 cm^2/Vs at 4.2 K, and electron concentrations of $\sim 10^{12}$ cm^{-2}. Magnetophonon oscillations in these samples, reported by BRUMMELL et al.(33), are shown in Fig. 7. A single series was observed, which accurately obeyed the $B \cos\theta$ behaviour due to 2-D conduction. This feature is important, since some parallel conduction also takes place in the bulk GaInAs layer, which is 1.5 μm thick. For one sample magnetophonon oscillations were observed even at 90°, due to this bulk conduction, and these oscillations had an almost identical periodicity to those of the 2DEG.

The fundamental field of the two-dimensional magnetophonon oscillations is 12.6 T. Cyclotron resonance on these samples (34) gives $m^* = 0.049\ m_0$ and so [3] and [5] then give an L.O. phonon frequency of 239 cm^{-1}. Polaron correction will reduce this to approximately 236 cm^{-1}, which is very close to the 'InAs-

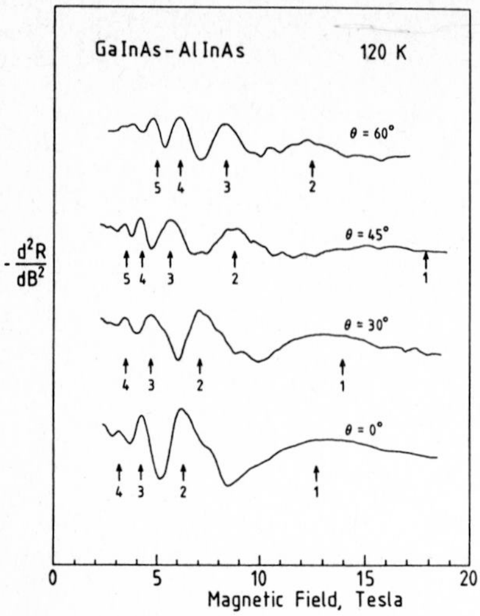

Fig. 7: Magnetophonon oscillations in a GaInAs-AlInAs heterojunction at 120 K. Rotation of the sample shows two-dimensional behaviour

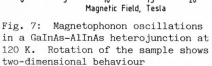

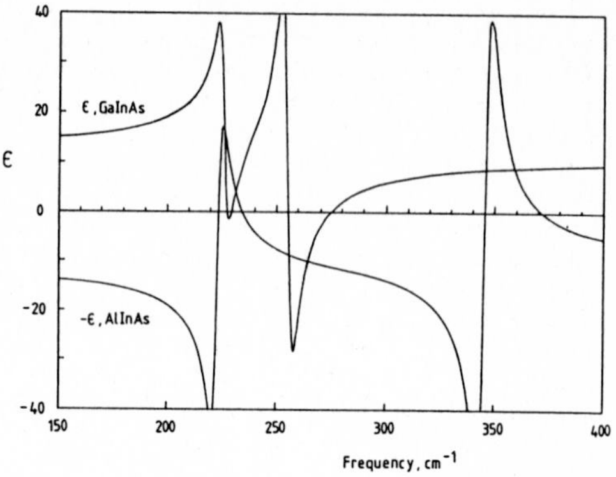

Fig. 8: Dielectric constants of GaInAs and AlInAs as a function of frequency. The interface modes are found at the frequencies where the two curves cross

like' L.O. mode of GaInAs at 233 cm^{-1}. The dominant electron-optic phonon interaction has thus been completely altered from the case of the GaInAs-InP heterostructures, where only GaAs-like and InP modes were seen. The phonon properties of AlInAs have not been extensively studied to date, however it is reasonable to assume that there will be two-mode behaviour across the whole range of alloy compositions, since the optic phonons of InAs are well separated from those of AlAs. Raman measurements on $Al_{.48}In_{.52}As$ have shown L.O. phonon peaks at 236 cm^{-1} and 369 cm^{-1}. Both components of the heterojunction thus possess L.O. modes in the region of 235 cm^{-1}, and there should be an interface mode around this energy. This may be seen by plotting the dielectric constants of the two materials, shown in Fig.8, in the same way as described above for GaInAs-InP, using the values given by BRUMMELL et al.(33). The values of the interface phonon energies and oscillator strengths are shown in Table 2.

It is thus to be expected that a heterojunction of GaInAs and AlInAs will show a strong interaction with a phonon of around 234 cm^{-1}, since the 'InAs-like' modes in the two materials are both degenerate with the interface phonon. In addition the oscillator strength of the 'InAs-like' mode in the AlInAs is almost double that of the GaInAs. This is because of the presumed full 'two-mode' behaviour of the AlInAs, leading to a greater T.O. - L.O. mode gap for

Table 2. The frequency in cm^{-1} followed by the oscillator strength for all bulk and interface L.O. phonon modes in the InP-GaInAs-AlInAs system

mode	'InAs'	'GaAs'	'InP'	'AlAs'
Bulk phonons				
GaInAs	233/0.8	271/1.6		
InP			349/3.0	
AlInAs	236/1.2			369/1.4
Interface phonons				
GaInAs-InP	228/0.4	261/0.8	328/1.5	
GaInAs-AlInAs	234/1.0	265/0.8		359/0.7

the middle of the alloy range. The net result of this is that the 'InAs-like' modes have become the dominant scattering mechanism, in contrast to both bulk GaInAs (26) and the GaInAs-InP structures described above (19-21).

5. CONCLUSION

The magnetophonon effect in two-dimensional systems shows several important differences from bulk material. The work on GaAs-GaAlAs has shown that the strength of the polaron coupling is significantly less in 2-D systems due to screening. The InP-GaInAs-AlInAs family of structures demonstrates that the strength of the coupling can be very strongly modified by the structure studied. The phonon properties of this system are summarised in Table 2 , calculated on the assumption of a single isolated heterojunction. For the GaInAs-InP system the interface phonon energies seem significantly too low to account for the resonances observed, however the remarkable strength of the InP mode scattering and the dominance of the 'InAs-like' mode in the GaInAs-AlInAs would seem to prove that interface modes play a significant role. More detailed calculations (30) including a finite well thickness and dispersion would seem to support this conclusion. One complicating factor which remains is the influence of interface grading, which is significantly greater in the GaInAs-InP structures.

Finally it may be said that the scattering of a 2DEG in GaInAs by L.O. phonons may be controlled so that the dominant interaction is with one of the three phonon modes 'InAs', 'GaAs' or 'InP'. These three cases correspond to:

(i) 'InAs' dominance in GaInAs-AlInAs heterojunctions and probably superlattices.

(ii) 'GaAs' dominance in GaInAs-InP heterojunctions or quantum well of thickness greater than ∿100 Å.

(iii) 'InP' dominance in GaInAs-InP quantum wells less than ∿80 Å thick.

ACKNOWLEDGEMENTS

The authors would like to thank the following for their contributions to this work: J. Beerens, J. Cisowski, K.Y. Cheng, A.Y. Cho, J.P. Duchemin, Th. Englert, M.A. de Forti-Poisson, G. Gregoris, A.C. Gossard, T.P. Pearsall, M. Razeghi and D.C. Tsui.

REFERENCES

1. V.L. Gurevich and Y.A. Firsov: Sov. Phys. J.E.T.P. 13 137 (1961)
2. M.I. Klinger: Sov. Phys. Solid State 3 974 (1961)
3. S.M. Puri and T.H. Geballe: Bull. Am. Phys. Soc. 8 309 (1963)
4. Yu. A. Firsov, V.L. Gurevich, R.V. Parfeniev and S.S. Shalyt: Phys. Rev. Lett. 12 660 (1964)
5. P.G. Harper, J.W. Hodby and R.A. Stradling: Rep. Prog. Phys. 36 1 (1973)
6. R.L. Peterson: in 'Semiconductors and Semi-metals' (Academic, New York) 10 221 (1975)
7. S. Das Sarma: Phys. Rev. B27 2590 (1983)
8. E.D. Palik, G.S. Picus, S. Teitler and R.F. Wallis: Phys. Rev. 122 475 (1961); T. Ando: J. Phys. Soc. Japan 51 3893 (1982)
9. D.C. Tsui, Th. Englert, A.Y. Cho and A.C. Gossard: Phys. Rev. Lett. 44 341 (1980)
10. Th. Englert, D.C. Tsui, J.C. Portal, J. Beerens and A.C. Gossard: Solid State Commun. 44 1301 (1982)
11. G. Kido, N. Miura, H. Ohno and H. Sakaki: J. Phys. Soc. Japan 51 2168 (1982)
12. R. Lassnig and W. Zawadzki: J. Phys. C: Solid State Phys. 16 5435 (1983)
13. M. Inoue, H. Hida, M. Inayama, Y. Inuishi, K. Nanbu and S. Hiyamizu: Physica 117B & 118B 720 (1983)

14. R.J. Nicholas, M.A. Brummell, J.C. Portal, M. Razeghi and M.A. Poisson: Solid State Commun. 43 825 (1982)
15. Y. Guldner, J.P. Vieren, P. Voisin, M. Voos, M. Razeghi and M.A. Poisson: Appl. Phys. Lett. 40 877 (1982)
16. M. Razeghi, M.A. Poisson, J.P Larivain, B. deCremoux, J.P. Duchemin and M. Voos: Electron. Lett. 18 339 (1982)
17. J.C. Portal, R.J. Nicholas, M.A. Brummell, M. Razeghi and M.A. Poisson: Appl. Phys. Lett. 43 293 (1983)
18. J.P. Duchemin, M. Razeghi, J.P. Hirtz and M. Bonnet: in GaAs and Related Compounds, Oiso, Japan (Bristol, England, I.O.P.) Conf. Ser. 62 89 (1982)
19. M.A. Brummell, R.J. Nicholas, J.C. Portal, M. Razeghi and M.A. Poisson: Physica 117B & 118B 753 (1983)
20. J.C. Portal, J. Cisowski, R.J. Nicholas, M.A. Brummell, M. Razeghi and M.A. Poisson: J. Phys. C: Solid State Phys. 16 L573 (1983)
21. J.C. Portal, G. Gregoris, M.A. Brummell, R.J. Nicholas, M. Razeghi, M.A. di Forte-Poisson, K.Y. Cheng and A.Y. Cho: Surf. Sci. in press (1984)
22. M. Brodsky and G. Lucovsky: Phys. Rev. Lett. 21 990 (1968)
23. C. Pickering: J. Electron. Mater. 10 901 (1981)
24. A. Pinczuk, J.M. Worlock, R.E. Nahory and M.A. Pollack: Appl. Phys. Lett. 33 461 (1978)
25. K. Kakimoto and K. Katoda: Appl. Phys. Lett. 40 826
26. R.J. Nicholas, S.J. Sessions and J.C. Portal: Appl. Phys. Lett. 37 178 (1980)
27. C.K. Sarkar: D.Phil. thesis, University of Oxford (1983)
28. M.A. Brummell, R.J. Nicholas, L.C. Brunel, S. Huant, J.C. Portal, M. Razeghi, M.A. diForte-Poisson, K.Y. Cheng and A.Y. Cho: Surf. Sci. in press (1984)
29. K. Hess and P. Vogl: Solid State Commun. 30 807 (1979)
30. R. Lassnig and W. Zawadzki: Surf. Sci. in press (1984)
31. R. Bisaro, G. Laurencin, A. Friederich and M. Razeghi: Appl. Phys. Lett. 40 978
32. K.Y. Cheng, A.Y. Cho, T.J. Drummond and H. Morkoc: Appl. Phys. Lett. 40 147 (1982)
33. M.A. Brummell, R.J. Nicholas, J.C. Portal, K.Y. Cheng and A.Y. Cho: J.Phys. C: Solid State Phys. 16 L579 (1983)

Thermodynamics of Two-Dimensional Electron Gas in a Magnetic Field

Wlodek Zawadzki

Institute of Physics, Polish Academy of Sciences, 02668 Warsaw, Poland

Magnetization, specific heat and thermo-electric power of a two-dimensional electron gas in the presence of a quantising magnetic field are calculated and compared with existing experiments.

1. Electron Density, Fermi Energy, Magnetization

We consider thermodynamic properties of a two-dimensional gas of noninteracting electrons in a parabolic, spherical energy band at a finite temperature T in the presence of a quantising magnetic field H. We include the spin degeneracy but assume the spin-splitting factor $g^*=0$. Inversion layers and superlattices based on GaAs satisfy quite well these assumptions, if the g^* - value enhancement is neglected. An incorporation of the spin splitting into the theory is straightforward. We assume further that only one electric subband is populated. The energetic density of states is taken in the form of a sum of Gaussian peaks

$$\rho(\varepsilon) = \frac{2}{2\pi L^2} \sum_n \sqrt{\frac{2}{\pi}} \frac{1}{\Gamma} \exp\left[-2\left(\frac{\varepsilon-\lambda_n}{\Gamma}\right)^2\right] \tag{1}$$

where $L = (\hbar c/eH)^{1/2}$; $\lambda_n = \hbar\omega_c(n + \frac{1}{2})$ and Γ is the broadening parameter (the level width $\Delta\varepsilon \simeq 2\Gamma$). A dependence of broadening on magnetic field is neglected. The electron density in 1 cm^2 is

$$N = A \sum_n \sqrt{\frac{2}{\pi}} \frac{1}{\gamma} \int_0^\infty \frac{1}{1+e^{z-\eta}} e^{-2y_n^2} dz \tag{2}$$

where $A = (1/\pi)(eH/\hbar c)$, $y_n = (z-\theta_n)/\gamma$ and $z = \varepsilon/kT$, $\eta = \zeta/kT$, $\theta_n = \lambda_n/kT$, $\gamma = \Gamma/kT$ are the reduced quantities. The filling factor of the system is defined as $\nu = N/A$, denoting the number of occupied Landau levels. The condition of a constant electron density in the sample leads to an integral equation for the Fermi energy $\zeta(H)$. Fig. 1 shows this dependence calculated for $m^* = 0.0665\ m_o$, $N_o = 8\times10^{11}\ cm^{-2}$, $\Gamma = 0.5$ meV and T = 6K.

The free energy of the system is

$$F = N\zeta - kT \int \rho(\varepsilon) \ln\left[1 + \exp\left(\frac{\zeta-\varepsilon}{kT}\right)\right] d\varepsilon . \tag{3}$$

It is convenient to write $\rho(\varepsilon)$ in the form

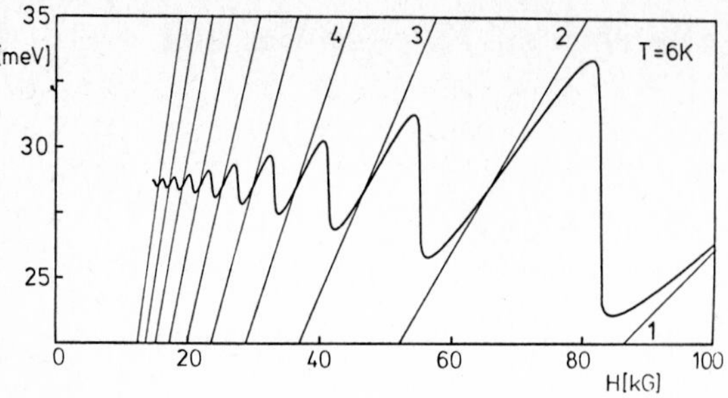

Fig.1 The Fermi energy versus magnetic field, calculated for 2D electron gas in GaAs at a constant electron density and T=6K. The Landau levels are also indicated. After ref. |1|

$$\rho(\varepsilon,H) = a \cdot H \cdot R(\varepsilon,H) \tag{4}$$

where $a = (1/\pi)(e/\hbar c)$. A magnetization of the system is $M = -dF/dH$. After some manipulation one obtains

$$M = kTa \sum_n \sqrt{\frac{2}{\pi}} \frac{1}{\gamma} \int_0^\infty \ln(1 + e^{\eta - z}) e^{-2y_n^2} (1 + 4 \frac{\theta_n}{\gamma} y_n) dz. \tag{5}$$

In order to calculate a contribution to magnetization from one completely filled Landau level one should put $\eta - z \gg 0$ and set the integration limits $\pm\infty$. The integral can then be calculated to give $M_r = kTa(\eta - 2\theta_r)$. Figure 2 shows the magnetization calculated according to Eq.(5) for the above m^*, N_0, Γ and T=4.2K. It can be seen that the orbital magnetism of the 2D electron gas oscillates symmetrically around the zero value, vanishing in the limit of $H \to 0$. The inclusion of the spin splitting does not change this picture - it simply doubles the number of levels. As follows from Figs 1 and 2, the magnetization oscillations follow quite closely those of the Fermi level. On the other hand

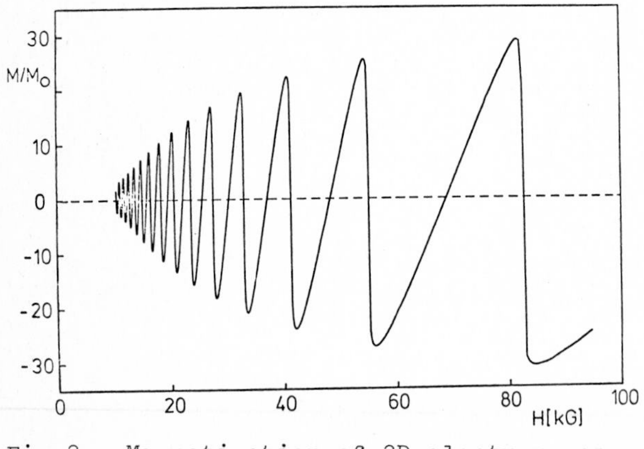

Fig.2 Magnetization of 2D electron gas versus magnetic field, calculated for the same conditions as in Fig.1 and T = 4.2 K. $M_0 = kT(2e/\hbar c)$. After ref. |1|

the Shubnikov-deHaas oscillations of magnetoresistance have maxima when the Landau levels cross the Fermi energy. It follows then that there should be a phase shift between the magnetization and the magneto-resistance oscillations. In fact, this has been observed in the magnetization measurements on GaAs - GaAlAs superlattice, see Fig. 3.

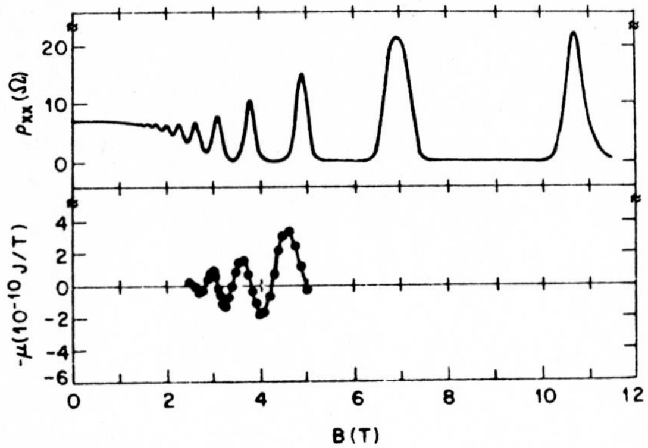

Fig.3 Magnetic moment μ of a GaAs - GaAlAs superlattice (172 layers) versus magnetic field as measured at T = 1.5K. The ShdH data ζ_{xx} taken at T = 0.2K on a piece of the same sample are also shown. Note the phase shift between the magneto-resistance and the magnetization oscillations. After ref. |3|

The first theory of magnetization for a 2D electron gas has been carried out using the delta-like density of states, with the results quite similar to those presented in Fig.2 |2|. It is also of interest to note that in the first explanation of the dHvA effect in 3D metals, Peierls used a two-dimensional model of a metal at T=0. Already this early consideration exhibited certain features presented above (cf. |4|).

Orbital and spin magnetic susceptibilities of a 2D electron gas have been calculated recently for the limit of vanishing magnetic fields |5|. It turns out that the orbital susceptibility is diamagnetic (negative) and the spin susceptibility is paramagnetic (positive). The ratio of the two susceptibilities is independent of the degree of degeneracy: $|\chi_{sp}/\chi_{orb}| = (3/4)(g^*m^*/m_0)^2$, which is similar to the three-dimensional case.

2. Specific Heat

The specific heat is given in general as

$$C_v = \int_0^\infty \frac{df}{dT}(\varepsilon-\zeta)\rho(\varepsilon)\,d\varepsilon \tag{6}$$

in which

$$\frac{df}{dT} = -\frac{\partial f}{\partial \varepsilon}\left(\frac{\varepsilon-\zeta}{T}+\frac{\partial \zeta}{\partial T}\right). \tag{7}$$

The dependence $\partial\zeta/\partial T$ at a constant concentration is determined by differentiating Eq. (2) with respect to T and using Eq. (7). This leads to

$$\frac{\partial\zeta}{\partial T} = -\frac{L_1}{L_o} \quad \text{and} \tag{8}$$

$$C_v = kA(L_2 - \frac{L_1^2}{L_o}) \quad \text{where} \tag{9}$$

$$L_r = \sum_n \sqrt{\frac{2}{\pi}}\frac{1}{\gamma} \int_0^\infty \frac{e^x}{(1+e^x)^2} x^r e^{-2y_n^2} dz \tag{10}$$

in which $x = z-\eta$ and the other quantities are defined above. The Fermi energy is first calculated from the condition N=const. and then all the integrals computed as functions of the magnetic field for the corresponding ζ values. Fig. 4 shows the specific heat calculated for the above parameters at T=6K.

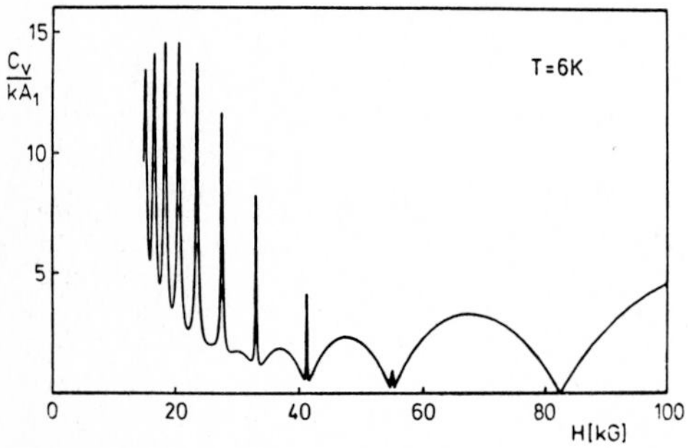

Fig.4 Specific heat of 2D electron gas versus magnetic field, calculated for the same conditions as in Fig. 1. Intralevel and interlevel regimes of thermal excitations are clearly seen. $A_1 = (1/\pi)(eH/\hbar c)$ for H = 1 kG. After ref. |1|

The specific heat of 2D electron gas is seen to consist of two contributions. At high magnetic fields, where $\hbar\omega_c >> kT$, only the intralevel thermal excitations contribute to C_V. When the Fermi energy is between two Landau levels, the lower levels are completely filled, the upper ones completely empty, and C_V vanishes. At such a magnetic field the system can not absorb low energy excitations. At weaker magnetic fields the interlevel excitations begin to come into play if the temperature is not too low. They are of importance when the Fermi energy lies between two Landau levels. The interlevel contribution to C_V is seen in Fig. 3 in the form of sharp spikes since, as follows from Fig. 1, the Fermi energy "jumps" between two Landau levels within a narrow range of magnetic field strength. Thus, Fig. 4 illustrates a continuous transition of the specific heat from the intralevel to the interlevel behavior, which is characterised, among other, by a change of phase of magneto-oscillations.

At very high magnetic fields the Fermi energy is forced below the lowest Landau level and the electron statistics becomes non--degenerate. In this limit only the lowest level is occupied and the specific heat as well as the electron concentration can be calculated analytically, to give at a finite temperature

$$C_V = \frac{1}{4} \gamma^2 kN . \tag{11}$$

It can be seen that the intralevel part of C_V depends crucially on the level broadening, going to zero at vanishing Γ.

If the above effects are to be observable, the specific heat of the electron gas should be comparable to that of the lattice. A typical period of a superlattice is 200 Å. For this value we calculate the electronic specific heat $C_V/k = 2.42\times10^{15}$ |...| (cm^{-3}), where the values in brackets are plotted in Fig. 4. At low temperatures the specific heat of the lattice, due to three acoustic phonon branches, is $C_V^l = 234k(N_a/2)(T/\theta_D)^3$, where N_a is the number of atoms and θ_D is the Debye temperature |4|. The factor 1/2 accounts for two atoms per unit call in GaAs-type materials. Taking for GaAs $N_a = 2.21\times10^{22}$ cm^{-3} and θ_D = 426K we obtain $C_V^l/k = 3.34\times10^{16}T^3$ (cm^{-3}), so that at T = 1K the electronic and the lattice specific heats are, in fact, well comparable.

3. Thermoelectric Power

A quantum theory of thermo-magnetic transport phenomena offers some serious difficulties, since in the presence of a temperature gradient the system is not homogeneous. As a consequence, the automatic application of the Kubo method led in the past to results which did not satisfy the Onsager symmetry relations, violated the third law of thermodynamics, etc. These paradoxes and puzzles were resolved by Obraztsov |6,7|, who showed that in order to obtain a correct description of the off-diagonal components of thermo-magnetic tensors, one should explicitly include in the theory a contribution of magnetization. This is related to the fact that the microscopic surface currents, which determine the Landau magnetization of conduction electrons, make a significant contribution to the macroscopic current density when a temperature gradient is present. At high magnetic fields, i.e.,for $\omega_c\tau \gg 1$, the diagonal components of the transport tensor may be neglected with respect to the off-diagonal ones. The latter do not depend on electron scattering in the high-field limit. Taking into account the contribution of magnetization M one obtains for the off-diagonal component of the macroscopic thermoelectric tensor

$$\beta_{xy} = \beta^o_{xy} + c\,\frac{dM}{dT} = \frac{c}{H}\,S \tag{12}$$

where β^o_{xy} determines the microscopic current density. When β^o_{xy} is calculated using the standard methods of the density matrix, the equality (12) is obtained, in which S is the entropy of the electron gas. The thermoelectric power becomes

$$\alpha(H) \simeq \frac{\beta_{xy}}{\sigma_{xy}} = -\frac{S}{eN} . \tag{13}$$

The entropy: $S = -(\partial F/\partial T)_V$, can be calculated from Eq. (3) to give

$$S = k \cdot A \sum_n \sqrt{\frac{2}{\pi}} \frac{1}{\gamma} \int_0^\infty \left[\ln(1+e^{-x}) + \frac{x}{1+e^x} \right] e^{-2y_n^2} dz \qquad (14)$$

where, as before, $x = z - \zeta$. It can be easily seen that completely filled levels (for which $x \ll 0$) give vanishing contribution to the entropy. N is given in Eq. (2), so that $\alpha(H)$ can be readily computed in the no-scattering limit. Measuring α at a constant electron concentration one determines directly the entropy of the electron gas.

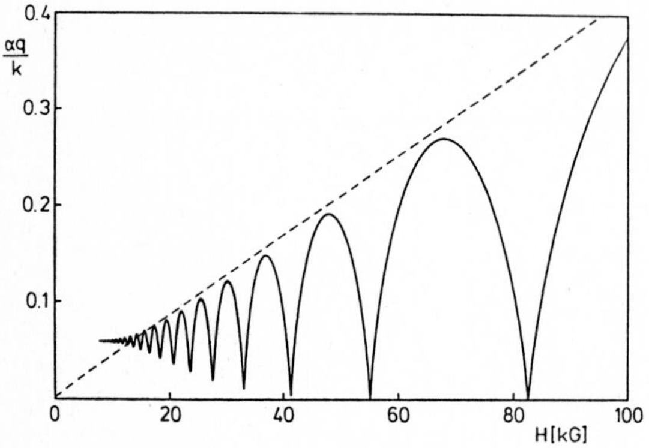

Fig.5 Thermoelectric power of 2D electron gas versus magnetic field, calculated for GaAs parameters and T = 6K. The dashed line indicates maxima values of $(-e/k)(\ln 2)/\nu$. After ref. |1|

Fig. 5 shows the thermo-power of the 2D gas in a strong transverse magnetic field, calculated for the above parameters and T = 6K. At high fields, as long as the interlevel contribution to C_V is negligible (cf. Fig. 4), the entropy (and consequently α) vanishes when the Fermi energy is between two Landau levels. At lower fields the interlevel contribution to C_V becomes of importance and the entropy as well as α do not reach the zero values. These general predictions agree quite well with the first experimental observations of the thermo-power oscillations in GaAs-GaAlAs heterostructures, cf. Fig. 6.

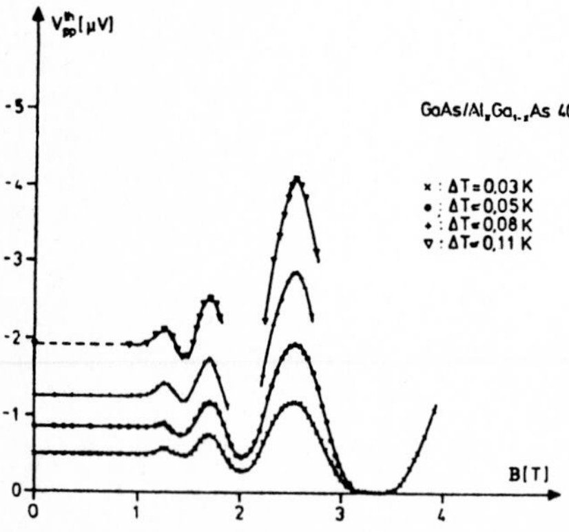

Fig.6 Thermal voltage versus magnetic field at different temperature gradients ΔT, as measured on a GaAs-GaAlAs heterostructure. The bath temperature T_B = 4.2K. After ref. |8|

Acknowledgments

The theoretical work presented here has been performed at the University of Innsbruck. I am grateful to Dr Rudolf Lassnig for the fruitful collaboration on the subject and to Prof.Erich Gornik for generous hospitality during my stay in Austria.

References

1. W. Zawadzki and R. Lassnig: Proceed. 5th Intern. Conf. Electronic Properties of Two-dimensional Systems, Oxford 1983, Surface Science (in print)
2. W. Zawadzki: Solid State Commun. 47, 317 (1983)
3. T. Haavasoja, H.L. Störmer, D.J. Bishop, V. Narayanamurti, A.C. Gossard and W. Wiegmann: in ref. |1|
4. C. Kittel: Introduction to Solid State Physics, 3d Edition (J.Wiley)
5. W. Zawadzki: J. Phys. C: Solid State Physics, Lett. to Editor (in print)
6. Yu.N. Obraztsov: Sov. Phys. Sol. State 6, 331 (1964)
7. Yu.N. Obraztsov: Sov. Phys. Sol. State 7, 455 (1965)
8. H. Obloh, K.von Klitzing and K. Ploog: in ref. |1|

Part II

Growth and Devices

Molecular Beam Epitaxy of GaAs and AlGaAs for Optoelectronic Devices and Modulation Doped Heterostructures

G. Weimann and W. Schlapp

Forschungsinstitut der Deutschen Bundespost, Am Kavalleriesand 3
D-6100 Darmstadt, Fed. Rep. of Germany

Abstract. Molecular beam epitaxy has been used to grow double heterostructure (DH) lasers, multiquantum well (MQW) lasers, and graded index separate confinement heterostructure (GRINSCH) lasers. Threshold current densities of 1 kA/cm^2 and 400 A/cm^2, respectively, have been measured. Optimization of the epitaxial growth process resulted in high quality interfaces, yielding mobilities of 725000 cm^2/Vs at 4.2 K in modulation doped (MD) heterostructures.

1. Introduction

Molecular beam epitaxy (MBE), owing to its slow, layer by atomic layer growth, allows close control of thickness, doping, and composition, making possible the growth of numerous devices involving $Al_xGa_{1-x}As$/GaAs heterostructures. Applications include DH lasers |1|, MQW lasers |2|, and GRINSCH lasers |3|, as well as MD heterostructures, which make use of the high mobility of the two-dimensional electron gas (2DEG) at the interface |4, 5|. Advantages of the MBE process include wafer size (typical diameter = 50 mm) and the spatial homogeneity of composition and doping in growth direction and across the wafer. Residual doping of MBE layers is generally higher than in liquid phase epitaxial films, although growth at high substrate temperatures (≃640°C for GaAs and ≃700°C for AlGaAs) yields material of device quality. Optimization of growth parameters and the properties of epitaxial layers will be given in the next chapter, followed by the performance of different types of injection lasers and modulation doped heterostructures.

2. Growth and Properties of GaAs and AlGaAs Epitaxial Layers

The epitaxial layers were grown in a commercial VARIAN GEN II growth system, which is equipped with a load lock, substrate rotation and cryoshrouds around the substrate. Elemental source materials of 6N purity (with the exception of Be, where single crystals of 4N purity were used) were evaporated from pyrolytic boron nitride crucibles. Furnaces driven at temperatures above 1100°C (Al at 1150°C and Si at 1250°C) have all ceramic parts of boron nitride.

The GaAs substrates were prepared by chemomechanical polishing in NaOCl on pellon pads. Approximately 100 µm of material were taken off, to remove sawing damage. Another 2 µm were then etched off in H_2SO_4:H_2O_2:H_2O (3:1:1); this etch was stopped with deionized water, the wafers were then spun dry. Liquid In was used to mount the wafers on molybdenum substrate carriers, which were introduced into the growth chamber via the load lock. The wafers were degassed by heating in ultrahigh vacuum at 200°C for one hour, then at 400°C for another hour. Finally, the native oxide was driven off from the growth surface by heating the sample for 10 min at 620-630°C in an As_4 beam.

Epitaxial growth is always carried out under As-rich conditions, this corresponds to beam equivalent pressure ratios (measured at the position of the substrate) of $P(As_4)/P(Ga) = 16$ for GaAs grown at 640-650°C and $P(As_4)/P(Al+Ga) = 20$ for AlGaAs grown at 690-710°C. Increasing the growth temperature resulted in a marked improvement of layer quality, reducing the density of deep levels, decreasing the compensation and improving mobility, and, above all, increasing the photoluminescence yield. We found the PL yield of our GaAs layers to be equivalent to that of similarly doped LPE material. Nominally undoped GaAs shows a residual carrier concentration of $p \approx 10^{14}$ cm^{-3}, with mobilities of 420 cm^2/Vs and 8000 cm^2/Vs at 300 K and 77 K, respectively.

We used the MBE specific dopants, namely Sn and Si as donors, and Be as acceptor. Sn gives free electron densities in the range of 10^{15}-10^{19} cm^{-3} with low compensation ratios $k = (N_D^+ + N_A^-)/n < 1.2$ |6|. It has, however, two disadvantages. Firstly, its incorporation is surface-rate limited |7|, leading to a Sn accumulation on the growth surface, which makes sharp high-low doping transitions problematic. Secondly, due to its high vapour pressure, Sn reevaporates from the growth surface at high growth temperatures, thus reducing the concentration of incorporated donors. Si is free of these disadvantages and can be used in a carrier concentration range from a few 10^{14} cm^{-3} to $2 \cdot 10^{18}$ cm^{-3}. Autocompensation is somewhat higher than with Sn; we measured $k < 1.5$. The highest mobilities found in our Si-doped GaAs layers were 100000 cm^2/Vs for $n = 5 \cdot 10^{14}$ cm^{-3} at 77 K. Be can be incorporated to give hole concentrations up to several 10^{19} cm^{-3}. Deep levels in GaAs layers have been measured with DLTS, optimized epitaxial layers have trap concentrations $N_T < 10^{12}$ cm^{-3}.

The MBE growth of AlGaAs is more problematic, due to the strong Al-O bond, which favours the reaction of Al with residual contaminants in the growth chamber (e.g. H_2O, CO) leading to the incorporation of oxygen, or, by reduction of CO, of free carbon |8| into the growing layer. Improved UHV technology, higher purity of starting materials and elevated growth temperatures have helped to overcome these difficulties. We found a PL yield of AlGaAs to increase by three orders of magnitude by going from 600°C growth temperature to 700°C, reaching the PL yield of comparably doped LPE AlGaAs layers. Figure 1 shows the photoluminescence of nominally undoped AlGaAs with an Al content $x = 0.35$, exhibiting the bound exciton and the excitation independent free-to-bound transition involving the carbon acceptor, separated by 23 meV from the exciton peak. This separation does not depend on x, as the comparison with the results of JUNG et al. shows (|9| for $x = 0.25$).

Undoped AlGaAs layers show high resistivities in v.d.Pauw measurements; they are compensated. Sn doping levels are hard to control, due to the reevaporation at the elevated growth temperatures necessary for AlGaAs. Si and Be are readily incorporated, the carrier concentrations follow the vapour pressure curves of the dopants, as in the case of GaAs. The lower ends of the doping ranges are difficult to reach, as the compensation and the deep level concentration are considerably higher than in GaAs. Table 1 lists the properties of several AlGaAs layers.

The DLTS results indicate that trap density and Si-doping level are related. The measured mobilities, on the other hand, compare favourably with mobilities of LPE material; especially the 77 K value in a MD heterostructure consisting of N^+-$Al_{0.35}Ga_{0.65}As$ and undoped $Al_{0.1}Ga_{0.9}As$ is equivalent to the mobility of very pure LPE AlGaAs |10|.

Crystal growth at temperatures around 700°C, i.e. considerably above the congruent sublimation temperature of GaAs, is influenced by reevaporation of

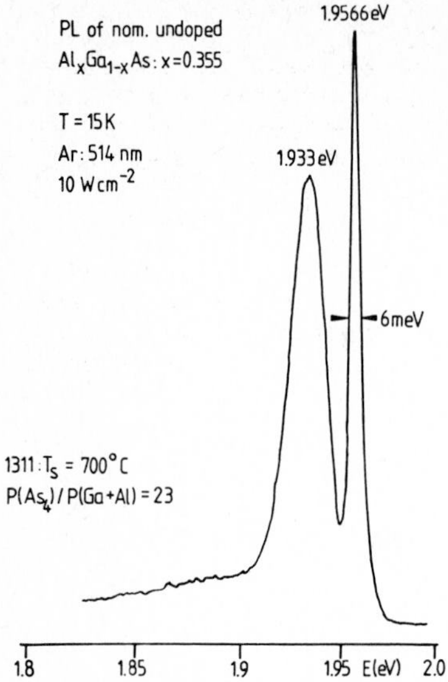

Fig.1. Photoluminescence of undoped AlGaAs, showing bound exciton and (e, C^0) band-acceptor transition. No (D, A) transitions are seen, indicating low residual donor concentration

Table 1. Mobilities and Trap Densities of Si-doped $Al_xGa_{1-x}As$

	x = 0.17	x = 0.26	MD-HS: x = 0.1
300 K : μ n	2170 cm^2/Vs $1.60\cdot10^{17}$cm^{-3}	2800 cm^2/Vs $1.50\cdot10^{15}$cm^{-3}	4930 cm^2/Vs $2.3\cdot10^{11}$cm^{-2}
77 K : μ n	2190 cm^2/Vs $1.47\cdot10^{17}$cm^{-3}		29000 cm^2/Vs $3.2\cdot10^{11}$cm^{-2}

Trap densities (DLTS):

x = 0.25: n = $1\cdot10^{16}$cm^{-3}, N_T = $4.0\cdot10^{15}$cm^{-3} (dom. trap: E_A = 0.52 eV)

x = 0.35: n = $8\cdot10^{16}$cm^{-3}, N_T = $2.5\cdot10^{16}$cm^{-3} (dom. trap: E_A = 0.32 eV)

Ga from the growth surface. Figure 2 gives the relative growth rates of GaAs and of GaAs in AlGaAs as functions of the growth temperature. These results were obtained from a series of MQW structures grown at different substrate temperatures with identical Ga and Al fluxes and the As_4 fluxes adjusted to always give As-rich growth. An increase in growth temperature from 600°C to 700°C reduces the effective sticking coefficient of Ga by 50% for a low temperature growth rate of 1.5 μm/h. The ternary AlGaAs is more stable, less Ga reevaporates from the ternary growth surface than from GaAs. The Ga loss is, however, still significant, leading to a change in composition, i.e. an increase of the Al content. The comparison with the results of FISCHER et al. |11|,who used a low temperature growth rate of 0.9 μm/h, shows that the absolute Ga reevaporation rate depends essentially only on the substrate temperature. The relative loss is more pronounced for lower growth rates. Increasing the As_4 flux partially suppresses the Ga reevaporation.

These optimized growth conditions for GaAs and AlGaAs - essentially high growth temperatures and V/III flux ratios as low as possible, while still

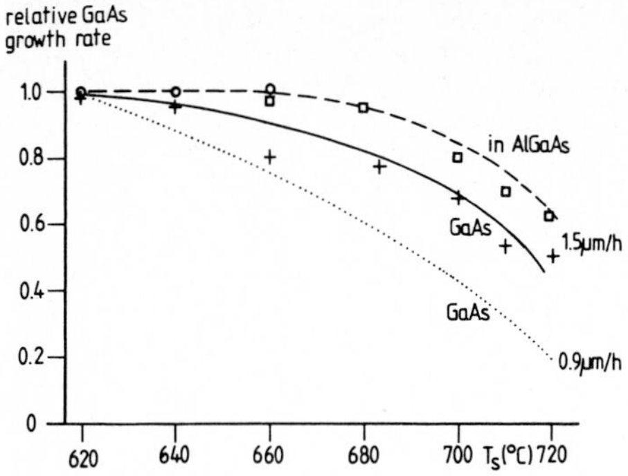

Fig.2. Reevaporation of Ga from growth surfaces of GaAs and AlGaAs, as function of substrate temperature

maintaining As-rich growth conditions - have been used to grow different laser structures, MQW structures and MD heterostructures.

3. GaAs-AlGaAs Double Heterostructure Lasers

DH lasers rely on high quality AlGaAs for good device properties. Liquid phase epitaxy especially, organometallic vapour phase epitaxy, and MBE more recently have produced laser devices with low threshold currents and long lifetimes. Only two laboratories have demonstrated MBE grown DH lasers with $Al_xGa_{1-x}As$ active layers ($x < 0.1$) so far |12, 13|. We have grown DH lasers with active layers of GaAs and AlGaAs following the growth procedures outlined in the first part of this paper.

The epitaxial structures consist of five layers grown on Si-doped substrates. After an n-doped GaAs buffer layer (1 μm thick, $n = 10^{18}$ cm^{-3}, $T_s = 620°C$), the N-AlGaAs confinement layer is grown (2 μm, $N = 2\text{-}4 \cdot 10^{17}$ cm^{-3}, Si- or Sn-doped, $x = 0.3\text{-}0.45$), followed by the active layer (0.1-0.15 μm thick) and the P-AlGaAs layer (1.0-1.5 μm, $P = 2\text{-}4 \cdot 10^{17}$ cm^{-3}, Be-doped). These layers are grown at elevated substrate temperatures around 700°C. The GaAs contact layer follows, being grown at 640°C with increasing Be doping reaching 10^{19} cm^{-3} at the surface to facilitate ohmic contact formation. Care has to be taken if Sn is used as donor, as its accumulation on the growth surface can lead to a misplacement of the p-n junction. This can be accounted for by reducing the Sn flux as the active layer is approached. Figure 3 shows the reduction of threshold current densities with increasing growth temperature; lowest values of 1 kA/cm² are reached for $T_s = 690°C$.

We investigated different types of stripe geometry DH lasers, working usually with shallow mesa lasers structured by chemical etching into the P-AlGaAs |14|. Ohmic contacts were evaporated (40 nm Ti, 40 nm Pt, 300 nm Au) and alloyed at 400°C giving contact resistances of typically 10^{-5} Ωcm² to the highly doped GaAs cap layer, thus defining the active stripe. The resistance to the lower doped AlGaAs outside the stripe is higher by 3-4 orders of magnitude. This structure (Fig.4) has the advantage of simple manufacture, and, additionally, gain guiding can be achieved by etching the mesa to within 0.5 μm of the active layer. The device performance of this laser type is similar to other stripe geometry lasers. Proton implanted and oxide defined devices made from the same wafers had unchanged characteristics.

Threshold current densities of 0.9-1.0 kA/cm² have been obtained with active layers of GaAs. T_0 ranged between 200 K and 230 K. External differential

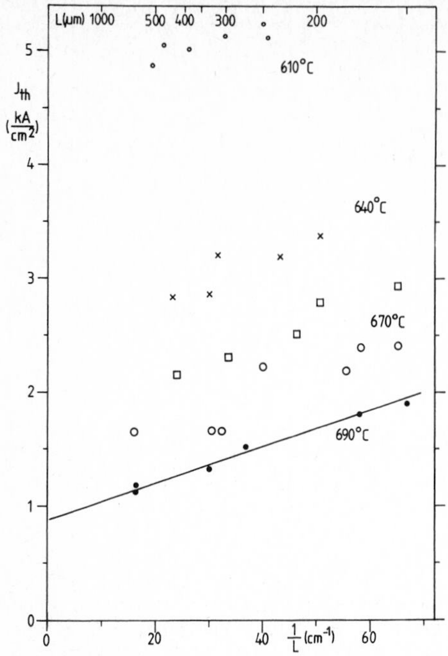

Fig.3. Threshold current densities of 25 μm wide stripe lasers as function of inverse diode length; laser diodes are made from DH wafers grown at temperatures between 610°C and 690°C

quantum efficiencies η_D were measured to be 20-25% per mirror. DH lasers with Al-containing active layers showed only insignificantly increased threshold current densities: 1.05-1.2 kA/cm² were measured for Al contents of 0.09 (lasing wavelength λ = 810 nm) and 2.5 kA/cm² for 0.14 (λ = 775 nm), respectively. These values demonstrate the high quality of the AlGaAs in the active region.

4. GRINSCH Lasers and MQW Lasers

The capability of MBE to grow ultra-thin layers and atomically smooth interfaces, together with the easy variation of composition of the ternary AlGaAs,

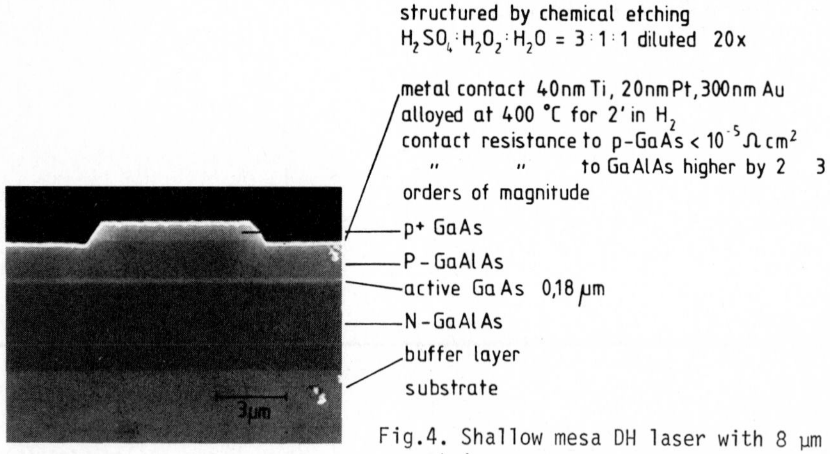

SEM 7000x

Fig.4. Shallow mesa DH laser with 8 μm wide stripe

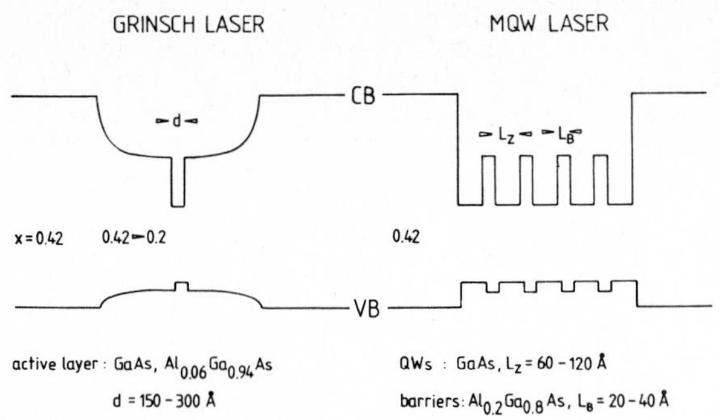

Fig.5. Schematic band structures of GRINSCH and modified MQW lasers

makes this growth technique particularly suited for novel types of lasers, such as GRINSCH and MQW lasers. Figure 5 gives the schematic band structures of GRINSCH and modified MQW lasers and the parameters of the devices grown in our laboratory.

GRINSCH lasers had active layers of 150 Å and 300 Å width, the optical confinement regions were 0.4 µm thick with parabolic refractive index profiles. This was achieved by varying the Al flux so that the Al content was altered from 0.42 to approximately 0.2 and vice versa. Lasers with 150 Å well width yielded threshold current densities of 400 A/cm^2 for GaAs and 450 A/cm^2 for $Al_{0.06}Ga_{0.94}As$ wells. This corresponds to threshold currents of 30 mA for laser diodes 5 µm wide x 300 µm long. Measured η_D values were 30% per mirror.

Modified MQW lasers with up to 10 GaAs wells were made. The lowest threshold current densities were obtained with optimized structures |2| consisting of 5 potential wells of 75 Å width, separated by 20 Å wide $Al_{0.2}Ga_{0.8}As$ barriers. The confinement layers consisted of $Al_{0.42}Ga_{0.58}As$, the active MQW regions were undoped. Measured values of J_{th} = 400 A/cm^2 again resulted in cw-threshold currents of 30 mA for 300 µm long diodes. The light vs. current characteristics were linear up to output powers of 30 mW/mirror; η_D typically exceeded 35%. Temperature dependence of threshold currents was weak, T_0 values

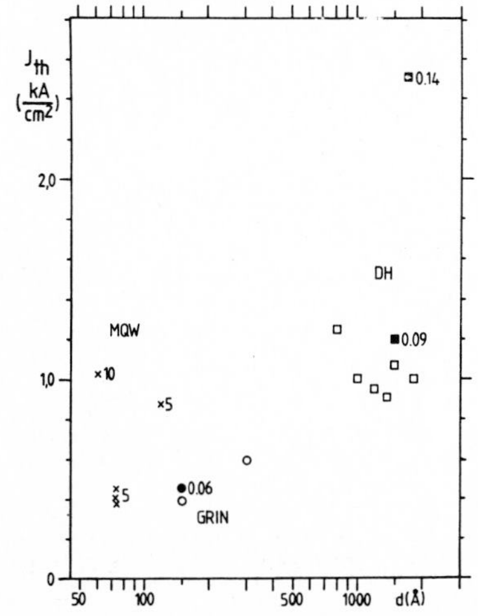

Fig.6. Threshold current densities measured on 25 µm broad stripe lasers as function of active layer thickness or potential well width. Squares represent DH lasers, dots GRINSCH lasers and crosses MQW lasers with number of coupled GaAs wells. Filled symbols give Al content of active layers

higher than 300 K were found. Low threshold currents and high values of T_0 allowed upside-up mounting of the laser diodes on heat sinks in cw operation, making mounting technology easier. Figure 6 summarizes the threshold current densities obtained with GaAs-AlGaAs lasers grown in our laboratory. These values were all measured on 25-30 μm wide stripes, which still show some current spreading.

5. MQW Structures and Superlattices

MQW structures with coupled and uncoupled potential wells have been grown and investigated by photoluminescence in the optimization of the growth parameters for MQW lasers. The GaAs potential wells were 20-200 Å wide, the AlGaAs barriers had thicknesses ranging from 100 Å to 200 Å for uncoupled wells, and they were as thin as 4 atomic layers (≈11.2 Å) for superlattices. High growth temperatures and V/III flux ratios as low as possible, while still retaining As-rich growth, improved the quality of MQW structures as in the case of bulk AlGaAs and double heterostructures. The photoluminescence yield increased steadily with rising growth temperatures, without reaching a maximum even at the highest used growth temperatures of 730-740°C. The linewidths of the transitions between confined particle states, however, had optimal values for growth temperatures of 675-710°C. Higher substrate temperatures resulted in very strong but very broad PL peaks, with additional peaks appearing indicative of inhomogeneous samples. We believe that growth temperatures substantially higher than 700°C lead to deterioration of the interfaces, due to interdiffusion or three-dimensional growth. Additionally, due to reevaporation of Ga, it becomes more and more difficult to control layer thicknesses. Our MQW structures were therefore grown at substrate temperatures between 680°C and 700°C, the beam equivalent pressures, measured at the growth position of the substrate, being $1.0\text{-}1.2\cdot10^{-5}$ Torr for As_4, $5\cdot10^{-7}$ Torr for Ga and $1.7\cdot10^{-7}$ Torr for Al. These pressures nominally correspond to fluxes of $9\cdot10^{14} As_4/cm^2s$, $5\cdot10^{14} Ga/cm^2s$ and $4\cdot10^{14} Al/cm^2s$ or nominal growth rates of 1 μm/h for GaAs and 1.7 μm/h for $Al_{0.45}Ga_{0.55}As$. The apparent As_4 fluxes measured by the ion gauge are higher than the actual fluxes effective in growth, as As_4 molecules which have passed the measuring gauge are reflected back into it and are so measured twice.

Figure 7 shows that we found no evidence of alloy clustering in the ternary AlGaAs. The given PL spectrum of a superlattice of 40 GaAs wells (L_z = 35 Å)

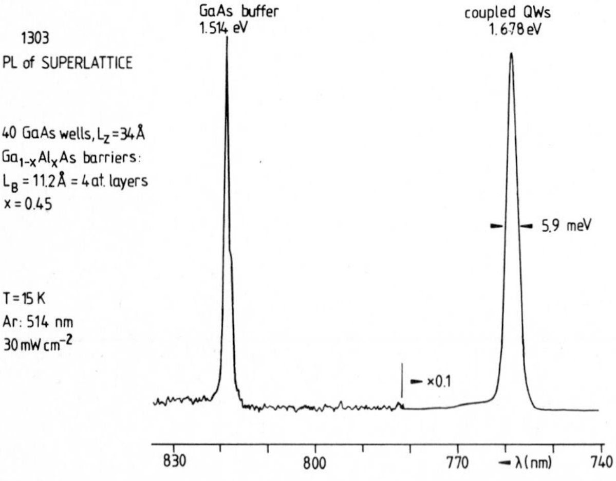

Fig.7. 15 K photoluminescence of superlattice consisting of coupled wells 35 Å wide

separated by very thin $Al_{0.45}Ga_{0.55}As$ barriers (L_B = 11.2 Å) exhibits only peaks from the GaAs buffer layer at 1.514 eV and from the coupled potential wells at 1.678 eV. In particular, there is no luminescence corresponding to well widths of $2L_z + L_B$, indicative of alloy clustering or barrier disruption. As both wells and barriers are very thin in the investigated sample, even monoatomic thickness variations would mean large relative variations of well widths or barrier widths, thus leading to either variations of the binding energy in the wells or, even more strongly, influencing the coupling between wells. The narrow PL line at 1.678 eV does not allow for such variations, indicating that well and barrier thicknesses vary neither within one well nor from well to well.

MQW structures usually show intrinsic PL lines typical of pure samples and smooth interfaces. The GaAs potential well grown first on top of a thick AlGaAs layer, however, shows an altogether different PL spectrum, characteristic of weak extrinsic photoluminescence due to impurities. The spectrum of 3 uncoupled potential wells grown on top of 1.5 µm AlGaAs, which is shown on the left side of Fig.8, clearly demonstrates the difference between the well grown first and the following two wells. These show only weak extrinsic transitions (marked by arrows) on the low energy side of the e-hh peak. The contrasting PL spectrum of a MQW structure again containing potential wells of three different widths, but with the two thinner wells grown on top of 10 identical QWs, is shown in the right part of Fig.8. The extrinsic luminescence of the first well is masked, as its contribution to the PL signal is small. The impurities seen in this extrinsic luminescence are either carbon, which

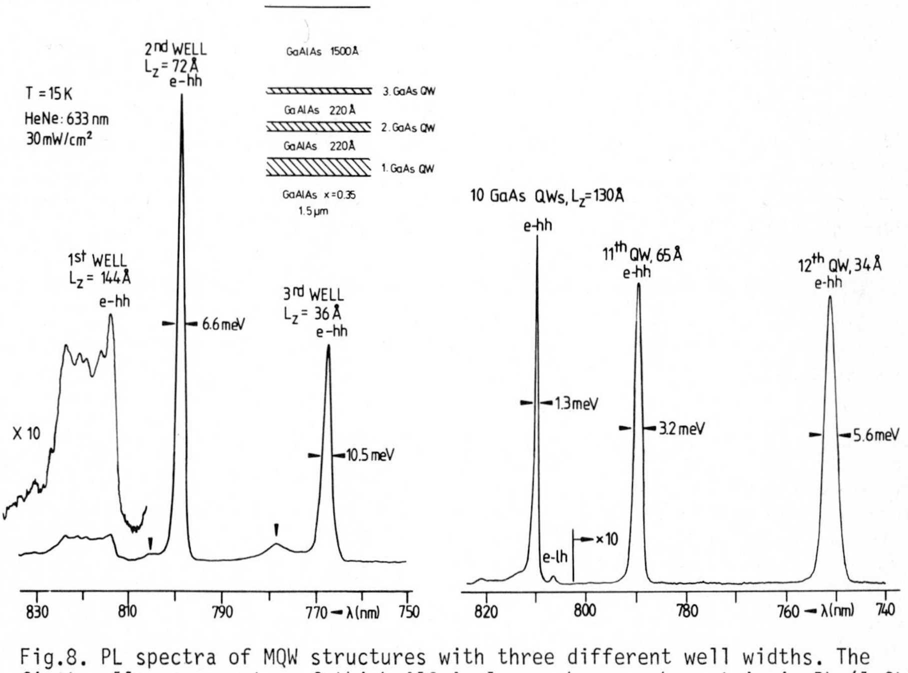

Fig.8. PL spectra of MQW structures with three different well widths. The first well grown on top of thick AlGaAs layer shows weak, extrinsic PL (left). A similar structure on top of 10 QWs shows higher PL yield and narrower lines (right, detection sensitivity left = 1, right = 0.25)

accumulates on the AlGaAs growth surface to be incorporated into the crystal at the first interface to GaAs |15|, or species outdiffusing from the substrate |16|.

The spectra of MQW structures grown under optimized conditions usually show well-split e-hh and e-lh transitions at room temperature with linewidths of 10-15 meV and very narrow PL lines with 1-2 meV HMFWs at 15 K. The best samples show additional peaks shifted to lower energies by approximately 1 meV, which are attributed to biexcitons |17|. Figure 9 shows such narrow lines.

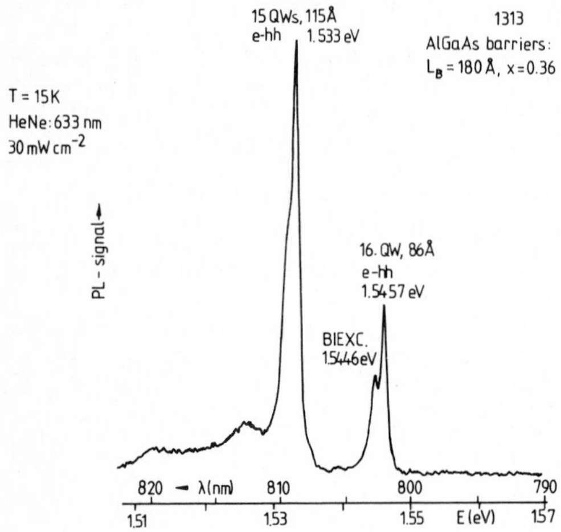

Fig.9. 15 K-PL of MQW structure consisting of 15 QWs (L_Z = 115 Å) with thinner QW (L_Z = 86 Å) on top. Linewidths are ≃1 meV, biexciton structure is resolved for thin QW

6. Modulation Doped Heterostructures

Selectively doped heterostructures have received considerable attention in the past years. Peak electron mobilities reported in the GaAs-AlGaAs system have exceeded 10^6 cm^2/Vs recently |4, 5|. The smaller gap GaAs has a higher electron affinity than AlGaAs, and, as only the wide gap AlGaAs is doped, free electrons drop into the GaAs, forming a two-dimensional electron gas (2DEG) at the interface. The free electrons are spatially separated from the ionized parent donors, which, in turn, increases the mobility, as ionized impurity scattering is no longer effective. Consequently, the advantage of selective doping is most marked at low temperatures (<80 K), where ionized impurity scattering is the dominant mechanism limiting mobility. At higher temperatures the mobility is limited by phonon scattering, thus reducing the mobility enhancement of modulation doping. However, at 300 K one still obtains mobilities of 8000 cm^2/Vs for sheet carrier densities of 10^{12}cm^{-2}, i.e. carrier densities typical of material doped to 10^{17}-10^{18}cm^{-3} are combined with mobilities specific to very pure material of 10^{14}cm^{-3} impurity concentration. Another advantage of MD heterostructures for the device physicist is that, in a quasiplanar technique, both normally-on and normally-off FETs can be fabricated on the same wafer by simply varying the thickness of the doped AlGaAs layer with etching.

An undoped AlGaAs spacer layer between the Si-doped AlGaAs and the GaAs reduces the coulombic interaction with the parent donors further, so increasing mobility at the price of decreasing the electron transfer. Thus the MD heterostructure usually consists of a nominally undoped GaAs layer (d > 1 µm), followed by an undoped AlGaAs spacer (d_S = 2-20 nm), a Si-doped AlGaAs layer ($N \simeq 10^{18}$ cm^{-3}, d_N = 40-80 nm, x = 0.25-0.4), and a thin GaAs cap layer to facilitate ohmic contact formation.

The characteristics of MD heterostructures depend on thickness, composition and doping level of the individual layers. The undoped GaAs is generally p-conducting with $p \simeq 10^{14}$ cm^{-3}. The thickness of the spacer layer d_S is a tradeoff between mobility and carrier density, whereas thickness and doping level of the Si-doped AlGaAs are chosen to give complete depletion of this layer and so prevent a bypass. A number of MD heterostructures have been grown in our laboratory with two objectives: epitaxial structures with high sheet carrier densities ($N_S = 10^{12}$ cm^{-2}) for FET applications on one hand, and samples with high mobilities, on the other. The essential difference between these two types lies in the spacer thickness. Table 2 lists a selection of these structures together with relevant dimensions.

Table 2. Parameters and Properties of MD Heterostructures

sample	1237	1242	1319	1338	1339	1353	1355	1342
growth temp. (°C)	700	700	620	640	640	640	640	655
contact, p^- (Å)	200, n^+	200	240	250	250	250	250	250
N^+-AlGaAs, d_N (Å)	800	800	600	380	380	420	390	400
Al content x	0.33	0.33	0.31	0.33	0.33	0.37	0.38	0.355
Si doping (cm^{-3})	$1.2\cdot10^{18}$	$2\cdot10^{17}$	$1.2\cdot10^{18}$	$6\cdot10^{17}$	$3\cdot10^{17}$	$1.2\cdot10^{18}$	$1.2\cdot10^{18}$	$7\cdot10^{17}$
spacer d_S (Å)	60	60	110	175	175	170	200	60
p^--GaAs (µm)	1.3	1.3	1.4	1.0	1.0	3.2	3.6	1.55 x=0.10
300K: N_S (cm^{-2})	$7\cdot10^{11}$		10^{12}	$3.4\cdot10^{11}$		$4.5\cdot10^{11}$	$4.4\cdot10^{11}$	$2.3\cdot10^{11}$
µ (cm^2/Vs)	6700		6600	8300		8460	8100	4900
77K: N_S (cm^{-2})	$5\cdot10^{11}$		$9\cdot10^{11}$	$4\cdot10^{11}$		$4\cdot10^{11}$	$4.7\cdot10^{11}$	$3.2\cdot10^{11}$
µ (cm^2/Vs)	110 000		111 000	162 000		162 500	192 000	29 000
4.2K: N_S (cm^{-2})			$5.1\cdot10^{11}$	$3.1\cdot10^{11}$		$3.6\cdot10^{11}$	$3.3\cdot10^{11}$	$\sim3\cdot10^{11}$
µ (cm^2/Vs)			433 000	295 000		517 000	725 000	~30000
" N_S (cm^{-2})			$5.7\cdot10^{11}$			$6.8\cdot10^{11}$	$5.9\cdot10^{11}$	
illumin. µ (cm^2/Vs)			473000			$1.05\cdot10^{6}$	$1.32\cdot10^{6}$	
remarks	BYPASS	DEPLET.			DEPLET.			AlGaAs

Figure 10 summarizes mobility and carrier density data measured at 4.2 K and shows the effect of illumination. The results given in Table 2 and Fig.10 allow, at least in part, determination of optimal MBE growth parameters for MD heterostructures.

Structures grown at 640-650°C were superior to those grown at 700°C, i.e. it seems to be necessary to optimize the GaAs more so than the AlGaAs, which, as thicker epitaxial layers have shown, turns out to be best when grown at 700°C. This holds independently of the Al content x, in fact, the sample with the highest mobilities of 192000 cm^2/Vs at 77 K and 725000 cm^2/Vs at 4.2 K also had the highest Al content (x = 0.38) investigated. We did not, however, find an unequivocal relation between mobility and Al content.

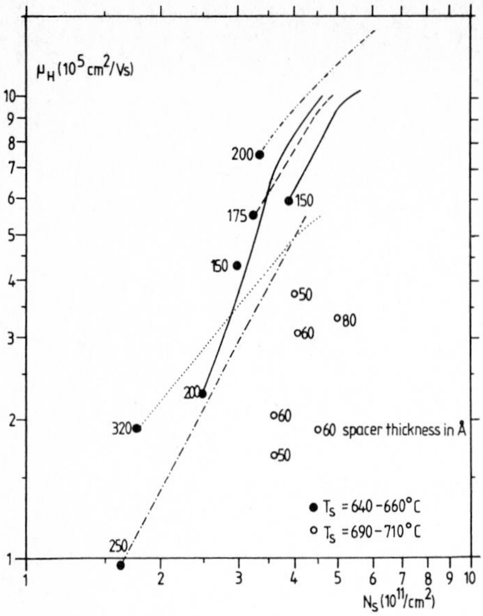

Fig.10. Hall mobilities μ_H and sheet carrier densities N_S of MD heterostructures at 4.2 K. Dots give measurements in dark, lines increase of N_S and μ_H with illumination. Spacer thicknesses are given in Å

The mobility values measured at 77 K (110000-192000 cm^2/Vs) indicate that the total ionized impurity concentration in GaAs is around 10^{14} cm^{-3}. The highest values have been reached with samples having undoped GaAs layers thicker than 3 µm. The equivalent residual doping levels are around $4 \cdot 10^{13}$ cm^{-3}, indicating that impurities outdiffusing from the substrate, which are less effective in thicker layers, possibly determine the ultimate background doping.

Spacer thicknesses between 2 nm and 30 nm have been investigated, mobilities increased monotonically with d_S upto 20 nm. Thicker spacers led to marked drops in mobility, probably due to the strongly decreased carrier density, which makes the electron scattering more effective |18|. Sheet carrier densities of 10^{12} cm^{-2} were reached only with spacers of 10 nm or thinner.

Total thicknesses of 60-70 nm for the AlGaAs layers (d_S + d_N) usually resulted in structures without bypass, i.e. for typical doping levels and compositions ($1.2 \cdot 10^{18}$ cm^{-3} and 0.33) full depletion of the doped AlGaAs was achieved. Our MD heterostructures served as starting material for depletion |19| and enhancement |20| field effect transistors. Extrinsic transconductances of 230 mS/mm and 150 mS/mm, respectively, have been measured.

The purity of undoped AlGaAs is shown by the high mobility values obtained on the $Al_{0.1}Ga_{0.9}As/N^+$-$Al_{0.35}Ga_{0.65}As$ MD heterostructure (sample 1342 in Table 2).

7. Summary

The optimization of the molecular beam epitaxial growth of GaAs and AlGaAs yielded epitaxial layers of device quality. DH lasers and laser structures with very thin active layers, such as GRINSCH and MQW lasers, had threshold current densities of 1 kA/cm^2 and 400 A/cm^2, respectively. Devices with active layers of AlGaAs performed equally well. MD heterostructures had mobilities close to the best values reported recently. Peak mobilities observed were 725000 cm^2/Vs without illumination and $1.32 \cdot 10^6$ cm^2/Vs with illumination, with corresponding carrier densities of $3.3 \cdot 10^{11}$ cm^{-2} and $5.5 \cdot 10^{11}$ cm^{-2}.

References

1. W.T. Tsang, R.L. Hartmann, H.E. Elder, W.R. Holbrook: Appl. Phys. Lett. 37, 141 (1980)
2. W.T. Tsang: Appl. Phys. Lett. 39, 786 (1981)
3. W.T. Tsang: Appl. Phys. Lett. 40, 217 (1982)
4. J.C.M. Hwang, A. Katalsky, H.L. Störmer, V.G. Keramidas: 2nd Intern. Symp. on MBE, Tokyo 1982
5. S. Hiyamizu, K. Nanbu, J. Saito, T. Ishikawa, T. Mimura, H. Hashimoto: ibid.
6. G. Weimann: Phys. Stat. Sol.(a) 53, K173 (1979)
7. C.E.C. Wood, B.A. Joyce: J. Appl. Phys. 49, 4854 (1978)
8. P.D. Kirchner, J.M. Woodall, J.L. Freeauf, D.J. Wolford, G.D. Pettit: J. Vac. Sci. Technol. 19, 604 (1981)
9. H. Jung, A. Fischer, K. Ploog: Appl. Phys. A33, 9 (1984)
10. A. Chandra, L.F. Eastman: J. Appl. Phys. 51, 2669 (1980)
11. R. Fischer, J. Klem, T.J. Drummond, R.E. Thorne, W. Kopp, H. Morkoç, A.Y. Cho: J. Appl. Phys. 54, 2508 (1983)
12. W.T. Tsang: J. Appl. Phys. 51, 917 (1980)
13. D.M. Collins, D.E. Mars, S.J. Eglash: J. Vac. Sci. Technol. B 1, 170 (1983)
14. M.C. Amann: El. Lett. 15, 441 (1979)
15. R.C. Miller, W.T. Tsang, O. Munteanu: Appl. Phys. Lett. 41, 374 (1982)
16. P.A. Maki, S.C. Palmateer, G.W. Wicks, L.F. Eastman, A.R. Calawa: Journ. El. Mat. 12, 1051 (1983)
17. D.A. Kleinman: Phys. Rev. B28, 871 (1983)
18. H.L. Störmer: Surface Science 132, 519 (1983)
19. H. Dämbkes, K. Heime: this meeting
20. D. Fritzsche: private communication

MOCVD Growth for Heterostructures and Two-Dimensional Electronic Systems

M. Razeghi and J.P. Duchemin

THOMSON-CSF-Laboratoire Central de Recherches, Domaine de Corbeville
BP No. 10, F-91400 Orsay Cedex, France

The LPMOCVD technique has been successfuly used to grow heterojunctions and superlattices of $Ga_xIn_{1-x}As_yP_{1-y}$ lattice-matched to InP for the complete compositional range between InP (λ = 0.91 um, Eg = 1.35 eV) and the ternary compound $Ga_{0.47}In_{0.53}As$ (λ = 1.67 um, Eg = 0.75 eV). We have observed Shubnikov-de Haas oscillations in heterojunctions and superlattices of $Ga_{0.47}In_{0.53}As$-InP and $Ga_{0.25}In_{0.75}As_{0.5}P_{0.5}$ - InP showing evidence of two-dimensional behaviour

1 INTRODUCTION

Following the work of Esaki and Tsu in 1970 (1), some fascinating studies have been made on heterostructures and multi-layer structures composed of thin layers of different semiconductors, with the electrons bound in quantum wells within the layers. Most of this work has been carried out on heterojunctions of GaAs and GaAlAs or of InAs and GaSb.

The evolution of MOCVD as a technique for the growth of multilayers of high quality semiconductors has allowed access to new device phenomena and to produce multilayer structures with extremely fine dimensional and compositional control.

In optoelectronic and microwave devices, the dimensions and composition of thin epitaxial layers strongly effect both their optical and electronic properties.

In electrical devices, charge carriers in finely layered structures can be separated from impurity regions by selective doping of the layers, thus producing enhanced mobilities. Using the MOCVD growth technique, a wide range of potentially useful new phenomena based on heterostructure epitaxy is thus becoming available and with imaginative implementation may be expected to lead to further novel device structures.

The MOCVD process, based on the pyrolysis of alkyls of group III elements in an atmosphere of the hydrides of group V elements, would seem to be a widely applicable growth technique as it is well adapted to the growth of submicron layers and heterostructures.

Open tube systems are used at atmospheric or reduced pressures in producing the III-V alloys. The process requires only one hot temperature zone for the in situ formation and growth of the semiconductor compound directly on a heated substrate.

Low pressure MOCVD growth offers an improved thickness uniformity and compositional homogeneity, reduction of autodoping, reduction of parasitic decomposition in the gas phase and allows the growth of good quality material over a large surface area. The technique is versatile, numerous starting compounds can be used and the growth is controlled by fully independent parameters.

Growth by MOCVD takes place far from a thermodynamic equilibrium and growth rates are determined generally by the arrival rate of material at the growing surface rather than by temperature-dependent reactions, between the gas and solid phases. The hot susceptor has a catalytic effect on the decomposition of the group III alkyls and the group V hydrides and the growth rate is proportional to the partial pressure of the group III species, but is independent of the partial pressure of group V species. The gas molecules diffuse across the boundary layer (which is adjacent to the substrate) to the substrate surface where the metal alkyls and hydrides decompose to produce the group III and group V elemental species. The elemental species move on the hot surface until they find an available lattice site where growth then occurs.

This talk will cover the feasibility of using various metalorganic sources of group III elements with hydride sources of group V species for the following.

- The growth across the entire compositional range of GaInAsP lattice matched to InP substrate exhibiting high photoluminescence efficiency comparable with LPE grown material, with a (2) high degree of uniformity over an area of 10 cm^2.

- The growth of non-intentionally doped InP with residual carrier concentration of $N_D-N_A = 2 \times 10^{14}\ cm^{-3}$. Electron Hall mobility of $u_H(300) = 5400\ cm^2\ V^{-1}\ S^{-1}$ and $u_H(77) = 70\ 000\ cm^2\ V^{-1}\ S^{-1}$ for a $N_D-N_A = 2 \times 10^{15}\ cm^{-3}$ have been obtained (3).

- The growth of non-intentionally doped $Ga_{0.47}In_{0.53}As$ lattice matched to InP substrate with $N_D-N_A = 2x10^{14}\ cm^{-3}$ and electron Hall mobility of $u_H(300) = 12000\ cm^2\ V^{-1}\ S^{-1}$ and $u_H(77) = 60000\ cm^2\ V^{-1}S^{-1}$ for $N_D - N_A = 2 \times 10^{15}\ cm^{-3}$ have been achieved (4).

- The growth of $Ga_{0.47}\ In_{0.53}$ As-InP superlattices (5).

- The growth of $Ga_{0.25}\ In_{0.75}\ As_{0.5}\ P_{0.5}$/InP superlattices with 21 alternate layers of $Ga_{0.25}In_{0.75}As_{0.5}P_{0.5}$ wells of 75 Å thickness and InP barrier with 75 Å thickness.

- The observation of a two-dimensional electron gas in $Ga_{0.47}\ In_{0.53}$ As/InP and $Ga_{0.25}\ In_{0.75}\ As_{0.5}P_{0.5}$/InP heterojunction and superlattices (6;7).

- The demonstration of quantum size effects (QSE) in multiple and single ultrathin quantum well structures of GaInAs/InP and GaInAsP/InP (8).

2 EXPERIMENTAL PROCEDURE

The reactor and associated gas distribution scheme used for this study are shown in fig.1 . This reactor is similar to that described by Duchemin (9). The Ga and In sources are (TEG), triethylindium (TEI), respectively, the group V hydrides being pure arsine (AsH_3) and pure phosphine (PH_3). Palladium diffused H_2 and pure N_2 are used as the carrier gases. The presence of N_2 is necessary

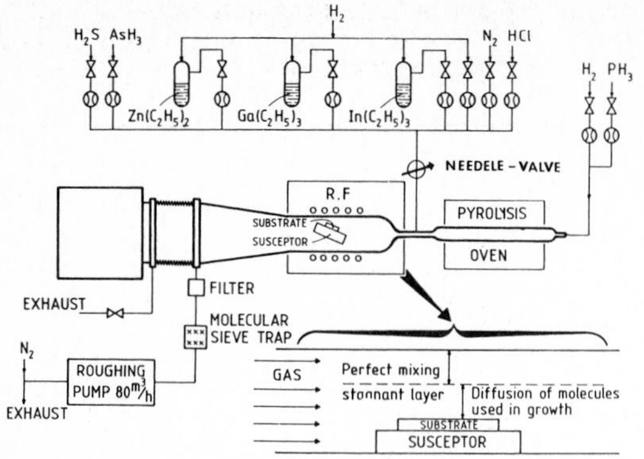

Fig.1. Schematic diagram of LP MOCVD reactor for the growth of InP and related compounds

to avoid the parasitic reaction between TEI and AsH_3 or PH_3. H_2 is necessary to avoid the deposition of carbon. The substrates are single crystal InP doped with Sn or Fe and are oriented 2° toward (110) from the (100) plane.

InP

InP layers can be grown at 76 torr at low temperature between 500 and 650°C using TEI and PH_3 in a H_2-N_2 carrier gas. The growth conditions and characterization of InP layers grown by LP MOCVD have been reported by RAZEGHI et al. (3). Fig.2 shows the electron carrier concentration of InP layers grown by LP MOCVD as a function of growth temperature and varies between 2×10^{14} cm^{-3} at 520°C to 2×10^{16} cm^{-3} at 650°C. SIMS analyses show that the dominant impurity in InP layers is Si (8).

A typical doping trace obtained using an electro-chemical profiler (using a 0.1 cm^2 area and 0.5 m HCl) of an InP Gunn diode structure ($n^+/n/n^+$) grown by LP-MOCVD is shown in fig.3.

Fig.4 shows the PL spectra measured at 6 K of an undoped InP layer grown by LP MOCVD using TEI from Ventron and PH_3 from Matheson. In a similar way to GaAs, several elementary recombination mechanisms occur and cause near band gap emission lines free exciton (x = 1.4181 eV), and excitons to shallow impurities (D°, X = 1.4169 eV), (A,X : 1.4147 eV).
Si is the dominant donor and Zn is the acceptor in undoped InP samples grown by LP MOCVD.

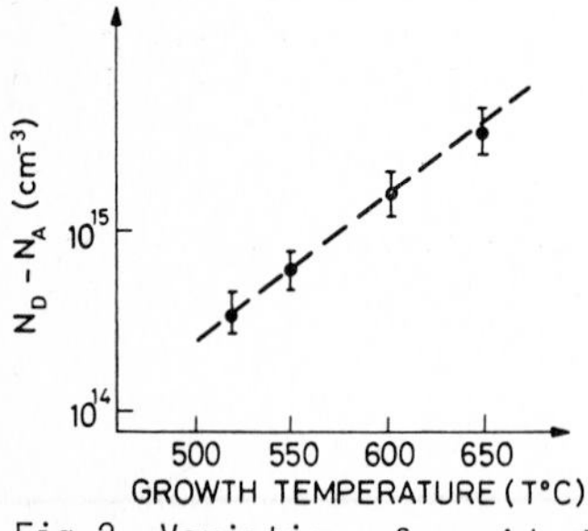

Fig.2. Variation of residual carrier concentration of InP layers grown by LP MOCVD as a function of growth temperature

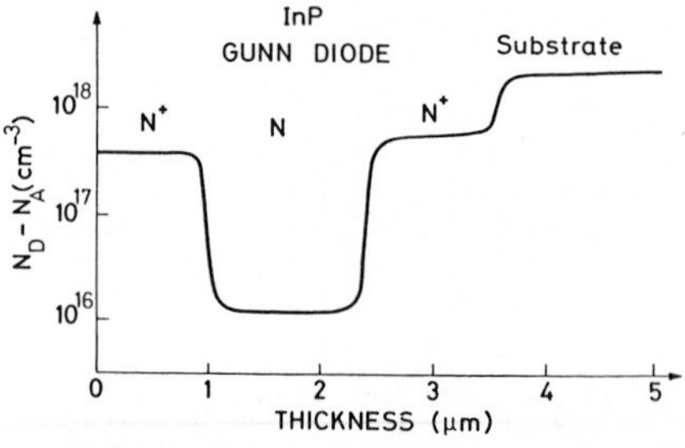

Fig.3. Electrochemical profile of an InP Gunn diode structure

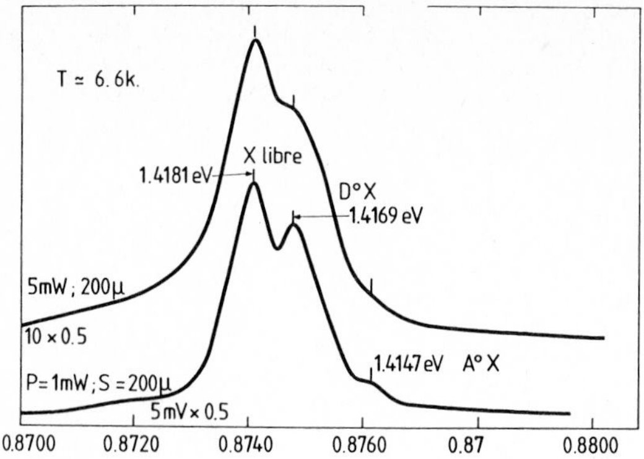

Fig.4. PL spectra at 6 K of InP undoped layer grown by LP MOCVD

GaInAs

In the same way $Ga_{0.47}In_{0.53}As$ layers lattice matched to InP can be grown by using TEI, TEG and AsH_3 in the same $H_2 + N_2$ carrier gas.

The growth conditions and characterization of GaInAs layers grown by LP-MOCVD have been reported in detail (10-11).The growth rate depends linearly upon the TEI + TEG flow rate , and is independent of that of AsH_3 within 60 and 90 CC/min, which suggests that, as in the case of InP, the epitaxial growth is controlled by the mass transport of the group III species. Uniform composition $Ga_{0.47}In_{0.53}As$ over an area of 10 cm^2 of InP substrate has been obtained.

3 OPTICAL PROPERTIES

The photoluminescence and absorption spectra of high purity GaInAs grown by different epitaxial techniques (LP MOCVD, LPE, VPE) have been compared (12). They showed resolved near band gap bound exciton lines,donor-acceptor transitions due to silicon and zinc-acceptor (identification made by SIMS) in the case of LPE and VPE material while carbon and zinc were found to be present in all MOCVD samples. All three elements are substitutional and do not form complexes. The different bands in our samples are somewhat narrower,indicating a smaller concentration of impurities and higher mobility. The spectral positions of all bands did not vary between one MOCVD sample and another, and for different points on the surface on the same sample, indicating the uniform chemical composition of the host.

The 2 K PL intensity of GaInAs layers on InP substrate grown by LP MOCVD depends directly on the purity of starting material. Fig. 5 shows the 2 K PL spectra of GaInAs-InP layers grown by :

a) using high purity starting materials,
b) using ordinary starting materials.

The PL intensity of (a) is 40 times higher than layer (b). In the case of layer (a), the near gap spectrum does not show any indication of recombination of donor-acceptor bounds.

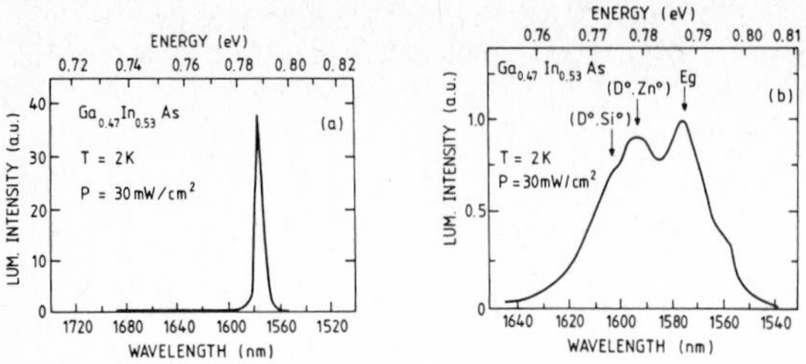

Fig.5. 2 K photoluminescence spectrum of: (a) a very high mobility LPMOCVD GaInAs layer using high purity starting material, no impurity band occur, (b) an undoped GaInAs layer using ordinary starting materials. Three pair bands due to Zn,c, and Si acceptors are resolved. The near band gap exciton line appears only weakly

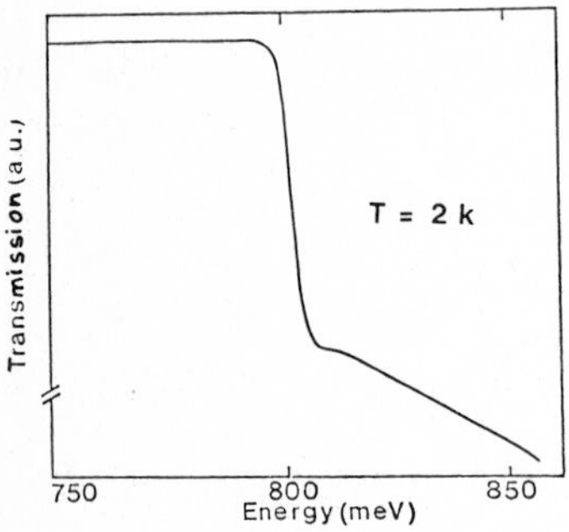

Fig.6. Transmission spectrum measured for a $Ga_{0,47}In_{0.53}As$-InP heterojunction at 1.5 K

Fig.6 presents a typical absorption spectrum (13) at 1.5 K, the light beam being perpendicualar to the interface. The absorption edge occurs around 807 meV, and is characteristic of free exciton absorption. The corresponding luminescence spectrum exhibits two lines around 805 and 788 meV, which are likely to be due, as in bulk GaInAs grown by LPE (14,15), to donor to valence band and to donor-acceptor recombination processes, respectively. These optical results show that the quality of the bulk material is good, and at least quite comparable to that obtained by liquid phase epitaxy.

4 ELECTRICAL PROPERTIES

The Hall mobilities were measured in a magnetic field of 4000 Gauss on clover leaf samples cut from the epitaxial wafers. Ohmic contacts were formed by evaporating approximately 2500 A of 12% Ge in Au, then annealing for 4 min at 460°C under nitrogen. This procedure yielded contacts which were ohmic at 300 and 77 K.

Typical values of mobility for undoped $Ga_{0.47}In_{0.53}As$ grown at temperatures between 550 and 650 C as a function of the thickness of the layer, composition, electron carrier concentration and temperature are summarized in table (1). The electron carrier concentration remained roughly constant between 300 and 77K. The mobilities increase and backgrounds (all n-type) decrease at lower growth temperature.

Table 1. Typical mobility for GaInAs/InP grown by LPMOCVD as a function of the thickness of the layer, composition, electron carrier concentration and growth temperature

GaInAs layer	T (c)	Thickness (μm)	(N_D-N_A) cm^{-3}	μ (300) cm^2 $V^{-1}s^{-1}$	μ (77) cm^2 $V^{-1}s^{-1}$	(Δa/a) strained
1	550	6 μm	1.8×10^{15}	11800	48000	$+5\ \times10^{-3}$
2	550	3.4 μm	2.7×10^{15}	9732	32150	-4.9×10^{-3}
3	550	1 μm	4.8×10^{15}	8700	41000	$+2.6\times10^{-3}$
4	650	0.78 μm	1.2×10^{16}	8100	17600	$+4\ \times10^{-4}$
5	650	1.21 μm	1.1×10^{16}	9020	21000	$+10^{-4}$
6	550	0.9 μm	4.2×10^{15}	9000	37300	-10^{-3}
7	550	0.53 μm	1.3×10^{16}	8700	32400	-1.4×10^{-3}
8	550	0.66 μm	2.5×10^{15}	11900	60000	$+10^{-4}$

The increased mobility in $Ga_{0.47}In_{0.53}As$ is comparable to the best reported mobility achieved by LPE, VPE and MBE for this composition. In table (1), the change of mobility at 77K with temperature shows that the material is apparently less compensated at 550 than at 650°C.

Compensation and mismatch effects would reduce the mobility for a particular electron concentration (5-8).

5 TWO-DIMENSIONAL ELECTRON GAS IN A $Ga_{0.47}In_{0.53}As$/InP HETEROJUNCTION AND SUPERLATTICES GROWN BY LP MOCVD

Heterojunction $Ga_{0.47}In_{0.53}As$-InP.

The samples, whose geometry is shown in fig.7, were grown on (100) 2° off semi-insulating Fe-doped InP substrates by LP MOCVD.

Fig. 8-a shows pronounced Shubmikov-de Haas oscillations at low magnetic fields observed at 4.2 K with the magnetic field B perpendicular to the heterojunction interface.

Fig.8-b gives the reciprocal magnetic field corresponding to the magneto-oscillation maxima as a function of the Landau index for different values of θ, which is the angle between B and the perpendicular to the interface. The oscillations are periodic in 1/B, and they follow the expected $(\cos\theta^{-1})$ dependence of a TDEG (16).

Fig. 9 presents typical cyclotron resonance data obtained at 2 K for $\theta=0$. The cyclotron frequency ($\omega_c = eB/m^*$) is found to vary as $(\cos\theta^{-1})$, as it should for a TDEG (17).

The quantum Hall effect (QHE) has been observed in modulation doped GaInAs-InP heterojunctions grown by LP MOCVD (18). Fig.10 shows data obtained at 1.85 K and 55 mK for the Hall resistance ρ_{xy} and the magnetoresistance ρ_{xx} as a function of B for a current equal to 10^{-8} A. These results are characteristic of the QHE, since they show clearly quantized plateaus in the Hall resistance ρ_{xy} and the vanishing of the magneto-resistance ρ_{xx}.

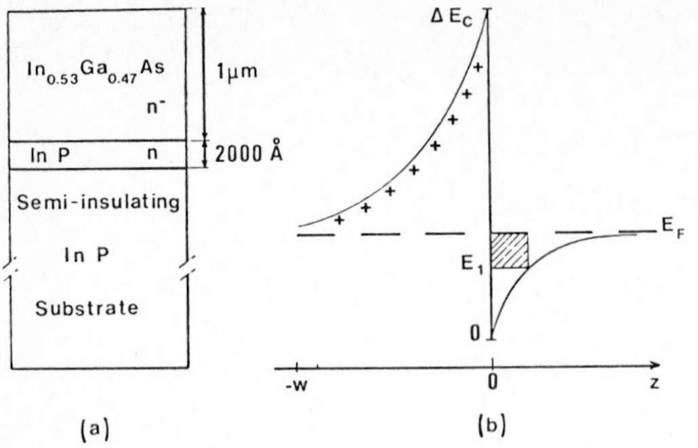

Fig.7. (a) Schematic geometry of the heterojunction under consideration here. (b) Energy band diagram for a modulation doped heterojunction - InP and GaInAs are on the left and right band side respectively of the interface which is at $Z = 0$. E_F and E_1 are the Fermi level and the first energy level in the potential well respectively, while ΔE_C is the conduction band discontinuity at the interface. The crosses represent ionized donors and the hatched area corresponds to the TDEG

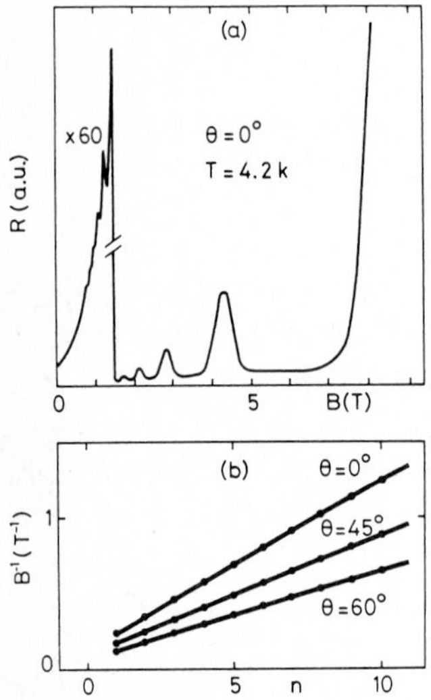

Fig. 8

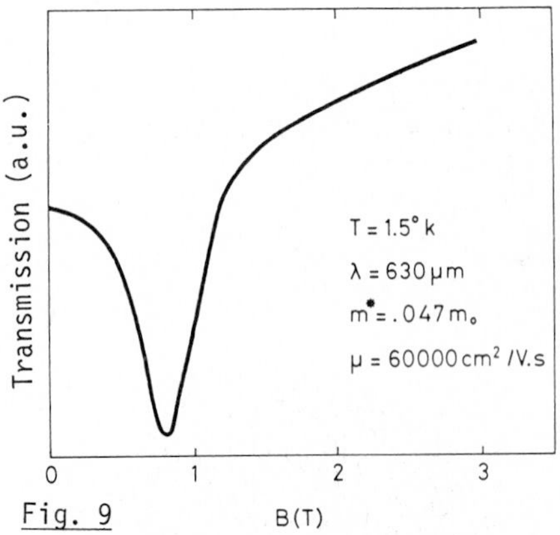

Fig. 9

Fig.8. (a) Magnetoresistance oscillations as a function of the magnetic field B. (b) Reciprocal field at maxima of those magnetooscillations VS Landau quantum number for different values of θ

Fig.9. Cyclotron resonance data at 2 K for an infrared wavelength equal to 630 μm

Magnetophonon measurements were made on several GaInAs-InP heterojunctions grown by LP MOCVD (19). The oscillations were detected by standard second-derivative techniques and a typical series of curves, taken at lattice temperatures between 80 K and 300 K, is shown in fig.11 (17). Rotation of the sample relative to the field at several temperatures showed that the resonances were two-dimensional up to 300K. A single series of oscillations due to scattering by the "GaAs-like" mode of GaInAs is seen.

Superlattice InP-GaInAs

21 alternate layers of InP and $Ga_{0.47}In_{0.53}As$ were grown on semi-insulating substrates by LP MOCVD. The InP layers were 300 Å thick while the GaInAs was 80 Å, 100 Å and 150 Å thick in the three superlattices. Fig. 12 shows a photo-

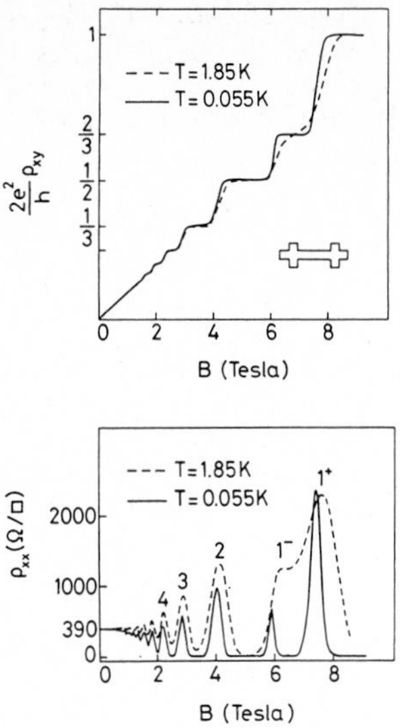

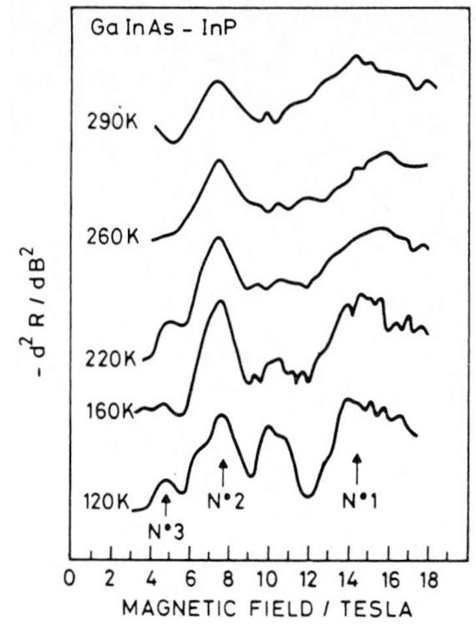

Fig.10. (a) Hall resistance ρ_{xy} as a function of B at different temperatures. (b) Magnetoresistance ρ_{xx} as a function of B at different temperatures. The corresponding Landau levels are denoted $1^+, 1^-, 2...$

Fig.11. Magentophonon oscillations in a GaInAs-InP heterojunction as a function of temperature. The extra peak at lower temperatures is thought to be N = 1 Shubnikov-de Haas peak

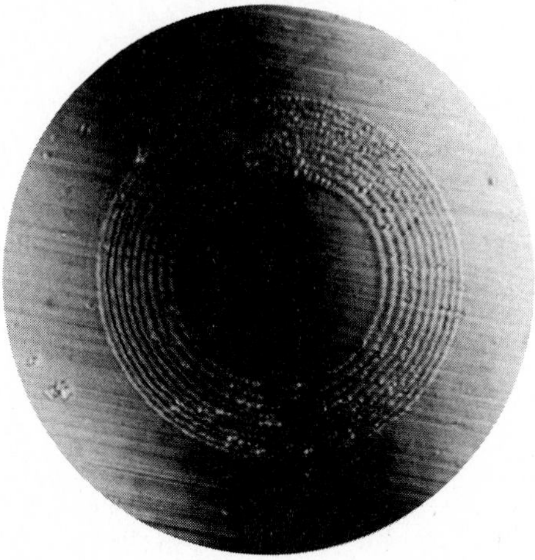

Fig.12. Photograph of 21 alternating layers of InP (300 A) and GaInAs (100 A) LP MOCVD superlattices

graph of a superlattice containing 21 alternating layers of GaInAs (100 A) and InP (300 A), which was revealed by milling a spherical hole in the sample by means of a rotating stainless steel ball, followed by a chemical etch.

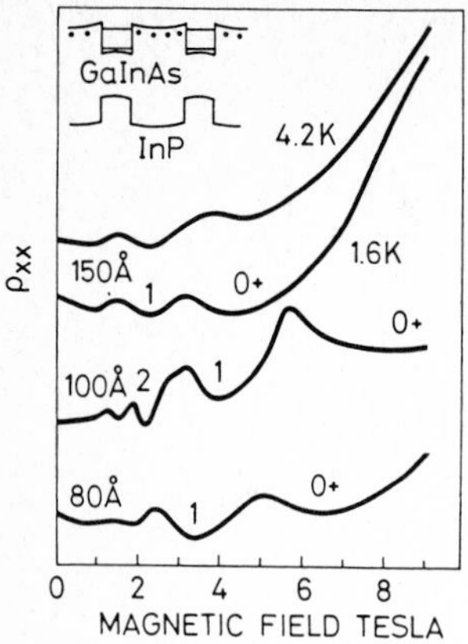

Fig.13. Experimental recordings of the magnetoresistance of the three superlattice samples at 1.6K a,d of the 150 A sample at 4.2 K. The magnetic field is applied parallel to the surface normal. The resistivity minima (and hence conductivity maxima) are labelled by the Landau index of the last unfilled state

Shubnikov-de Haas measurements for the three superlattices are shown in fig 13 . The periodicities give two-dimensional electron concentrations of 1.1, 2.2 and 1.7 x 10^{11} cm^{-2} for the 150, 100 and 80 Å superlattices, respectively (20).

Cyclotron resonance experiments were also performed over the temperature range 15 K - 150 K. Typical experimental recordings are shown in fig. 14. The effective masses of the confined electrons were measured by cyclotron resonance at lattice temperatures up to 150 K. These were a strong function of layer thickness increasing from 0.0407 m_0 for 150 Å, 0.0458 m_0 for 100 Å and 0.0543 for 80 Å. The effective mass is expected to rise for thin layers due to the increased energy of the lowest bound state in the potential well, and the resulting non-parabolicity. The first energy level for square wells of 150, 100 and 80 Å is at 40, 85 and 120 meV, respectively.

Magnetophonon resonance experiments were also performed for TDEG confined in the GaInAs layers of GaInAs-InP superlattices. Two series of oscillations are observed : one due to scattering by the "GaAs like" LO phonon mode of GaInAs, and the second due to interaction with InP LO phonons. The strength of the latter series (see fig.15) increases relative to the former as the GaInAs layer thickness is reduced. This is evidence for a long-range phonon interaction, with the InP phonon field extending into the GaInAs to couple significantly with the electrons bound in the quantum wells. No evidence of interface phonons is seen.

Quantum size effect in GaInAs-InP

Single and multiwell structures of GaInAs-InP have been grown by LP MOCVD (6). The multiwell structure consists of 25,50,100 and 200 Å quantum wells (GaInAs layers) separated by 500 A barrier (INP layers). Auger measurements indicate the presence of four distinct wells with abrupt boundaries (fig. 16) Photoluminescence measurements are consistent with the existence of four wells (fig. 17). However, deviations are noted between experimentally determined and theoretically predicted recombination energies (fig.18). An analogous situation exists for the single (50 and 100 Å) quantum well structures. Possible explanations, including variation of well composition, variation of well thickness, and participation of impurities in the recombination process, are suggested.

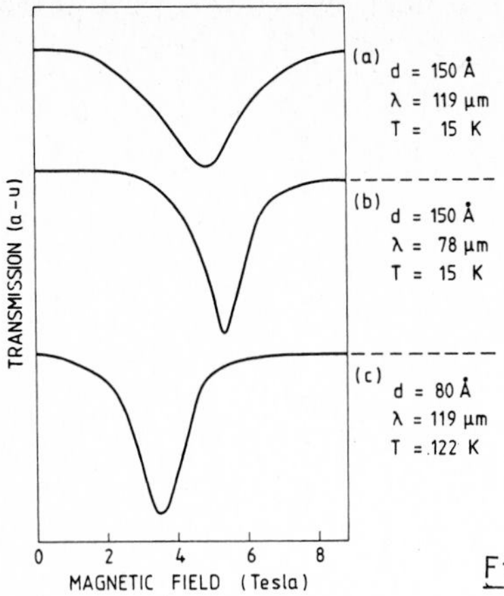

Fig. 14

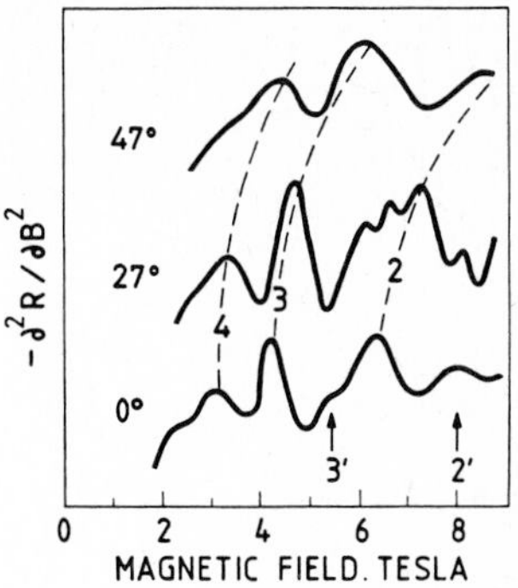

Fig. 15

Fig.14. Typical cyclotron resonance results. (a) and (b) show transmission of the 150A superlattice at 15K for 119 μm and for 79 μm radiation respectively, while (c) shows the 110 μm resonance for X the 80 A superlattice at 122K. The greater line width in this sample reflects the lower mobility

Fig.15. Magnetophonon oscillations in the 150A superlattice at 150K. Sample rotation shows that the periodicity is constant in B cos θ. The main series, 2,3,4, is due to the GaAs-like Lo mode in GaInAs, and the two weaker peaks, 2' and 3', are caused by penetration of the InP Lo phonon field into the GaInAs

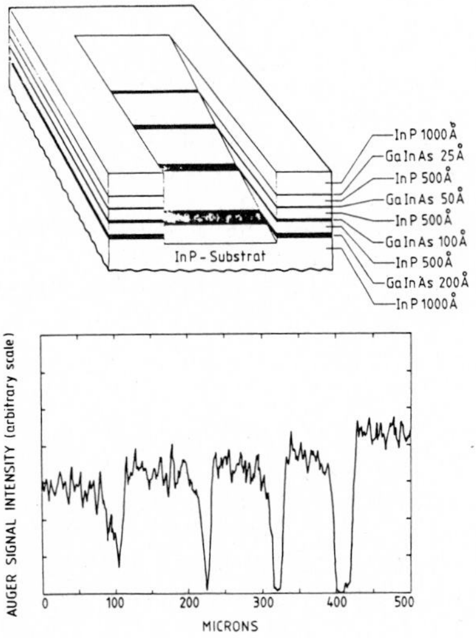

Fig.16. Auger spectrum of a chemically etched bevel with cuts of all four $Ga_{0.47}In_{0.53}As$ layers of the four well $Ga_{0.47}In_{0.53}As$ - InP sample. The abscissa indicates the distance along the bevel in microns. The zero value is determined by the start of the bevel in the outermost InP layer. Layer assignments are indicated above the trace. The bevel is schematically depicted in the inset. The actual bevel angle with respect to the outermost InP surface is between 0.02 and 0.06 of a degree

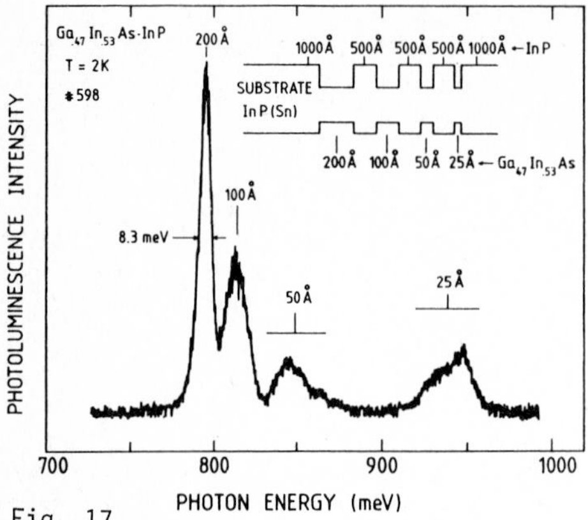

Fig. 17

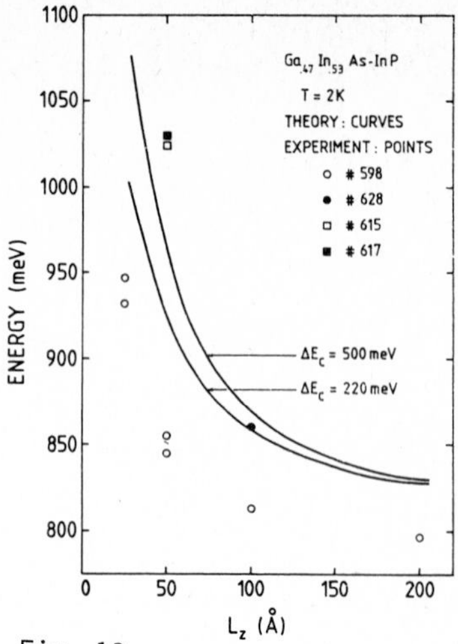

Fig. 18

Fig.17. PL spectrum of Ga$_{0.47}$In$_{0.53}$As-InP sample measured at 2K with excitation at 1170 meV (Nd YAG laser, 20mW focussed beam). Well assignments are indicated above each peak. Note that the peaks associated with the 25 and 50A wells are clearly multicomponent in nature. The FWHM of the lowest energy peak (associated with the 200A well) is 8.3 meV. The inset schematically illustrates the sample structure

Fig.18. Plot of measured PL line energies against well widths (Lz) for the four well samples and for single well samples (well width 100A) and well width 50A. Theoretical curves are shown for two values of the conduction band offset Δ_{EC}. These curves were calculated by Bastard using the band gap values quoted in (6) and the following parameters = spin-orbit splitting = $_{so}$(InP) = 100 meV, $_{so}$(Ga$_{0.47}$In$_{0.53}$As) = 360 meV, electron effective mass: m_e^* = (Ga$_{0.47}$In$_{0.53}$As) = 0.041 m_0; heavy hole effective mass m_{hh} = (Ga$_{0.47}$In$_{0.53}$As) = 0.5 m_0, m_{hh} (InP) = 0.56 m_0, where m_0 is the bare electron mass

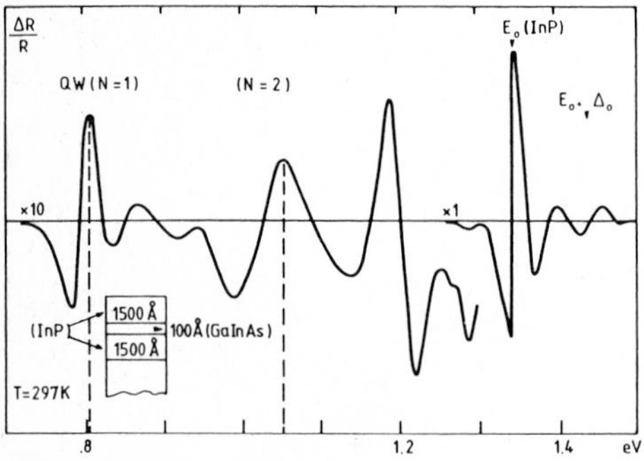

Fig.19. Electron reflectance spectrum of 100A GaInAs-InP single well at 297 K

Fig. 19 shows an electro-reflectance spectrum in the E_0, $E_0 + \Delta_0$ region of a 100 Å single well GaInAs-InP, which is well resolved, even at room temperature.

These results show that the InP-GaInAs-InP interfaces are as sharp as one monolayer.

$Ga_xIn_{1-x}As_yP_{1-y}$

Growth of GaInAsP layers was carried out at 76 torr and at a substrate temperature of 650°C, using TEI, TEG , AsH_3 and PH_3 in a $H_2 + N_2$ carrier gas:

$$X\,(C_2H_5)_3\,Ga + (1-x)\,(C_2H_5)_3\,In + Y\,AsH_3 + (1-Y)\,PH_3$$

$$\longrightarrow Ga_xIn_{1-x}As_yP_{1-y} + n\,C_2H_6\ .$$

The kinetics for $Ga_xIn_{1-x}As_yP_{1-y}$ alloys were similar. As long as the partial pressure of arsine and phosphine remained within a certain pressure range, the growth rate was proportional to the quantity of group III alkyl compounds, in the gas phase and insensitive to As and P partial pressure , which suggests that, as in the case of InP , the epitaxial growth is controlled by the mass transport of the group III species. We found that the group III elements had approximately the same ratio in the solid phase and in the gas phase. The epitaxial layer quality is sensitive to the pretreatment of the substrate and the alloy composition.

6 TWO-DIMENSIONAL ELECTRON GAS IN $Ga_{0.25}In_{0.75}As_{0.5}P_{0.5}$ - InP HETEROJUNCTIONS AND SUPERLATTICES

Heterojunction InP-GaInAsP.

Under the conditions described above InP - $Ga_{0.25}In_{0.75}As_{0.5}P_{0.5}$-INP heterojunctions were grown by LP MOCVD. The undoped GaInAsP is n type and has been shown to have a background doping level of 8 x 10^{15} cm^{-3}. A layer of GaInAsP with a thickness between 0.1 and 1 um was grown after a layer of nominally undoped InP 1000 Å thick, with an electron concentration of 10^{16} cm^{-3}, on a substrate of semi-insulating Fe-doped InP. The Hall mobility recently obtained for undoped GaInAsP (λ=1.3 um) with 0.9 um thickness was $u_H(300)$ = 4750 cm^2 $V^{-1}S^{-1}$ and u_H (77) = 13000 cm^2 $V^{-1}S^{-1}$. Shubnikov-de Haas measuremnts have been performed on $Ga_{0.25}In_{0.75}As_{0.5}$-InP heterojunctions grown by LP MOCVD. These measurments give the first evidence for the formation of a TDEG at the interface between $Ga_{0.25}In_{0.75}As_{0.5}P_{0.5}$ and InP.

Shubnikov-de Haas oscillations observed at 4-2 K with the magnetic field B perpendicular to the heterojunction interface are shown in fig. 20 which gives the reciprocal magnetic field corresponding to the magneto-oscillation maxima as a function of the Landau index for different values of Θ, which is the angle between B and the perpendicular to the interface. The oscillations are periodic in I/B, and they follow the expected ($\cos\Theta^{-1}$) dependence of a TDEG. This manifests the two-dimensionality of the electron gas under consideration. From these data, we deduce, using standard procedures, an electron density n_s = 4.16 x 10^{11} cm^{-2}; there is only one series of oscillations, showing that only the lowest bound state is occupied in sample. In this structure, as in GaInAs/InP heterojunctions (17), electrons from donors in the wide band gap material (InP) are transferred across the interface to the narrow

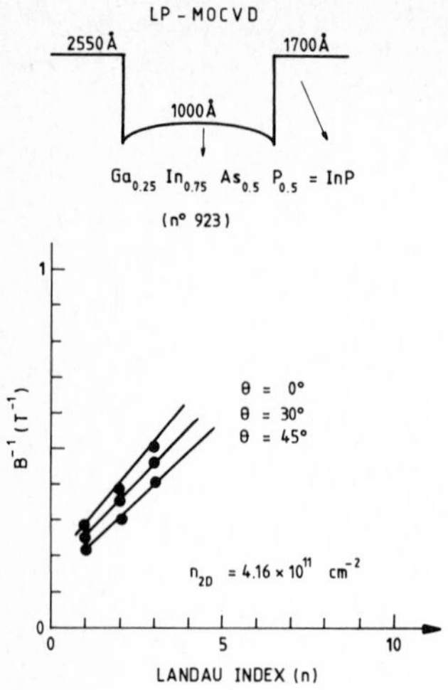

Fig.20. Reciprocal of magnetic field at maxima of magnetooscillations V_S Landau quantum number for different values of θ for a $Ga_{0.25}In_{0.75}As_{0.5}P$-InP heterojunction grown by LPMOCVD

band gap material ($In_{0.75}Ga_{0.25}As_{0.5}P_{0.5}$ to maintain a constant Fermi level in the system. This leads to the formation of the observed TDEG which is confined in the potential well resulting from the band bending occurring in the vicinity of the interface and due to the spatial separation of ionized donors and electron charges.

Superlattice InP - $Ga_{0.25}In_{0.75}As_{0.5}P_{0.5}$ - InP

In order to estimate the abruptness of InP/GaInAsP/InP heterojunctions, we have studied the growth of single and multi-quantum well layers of $Ga_{0.25}In_{0.75}As_{0.5}P_{0.5}$. Fig. 21 shows Auger spectrum of a chemically etched bevel of a GaInAsP-InP, comprising three GaInAsP wells of 100 Å thickness. The bevel angle is 0.02 degrees. It shows that InP/GaInAsP/InP interfaces are uniform and smaller than 30 Å.

21 alternate layers of n-GaInAsP and n-InP were grown on a semi-insulating InP substrate by LP MOCVD. The thickness of InP layers was 50 Å and the thickness of GaInAsP welles was 70 Å. Shubnikov-de Haas measurements carried out on superlattices and MQWS of GaInAsP-InP give the first evidence for the formation of a TDEG at the interface between InP/GaInAsP/InP in SL and MQW of GaInAsP/InP.

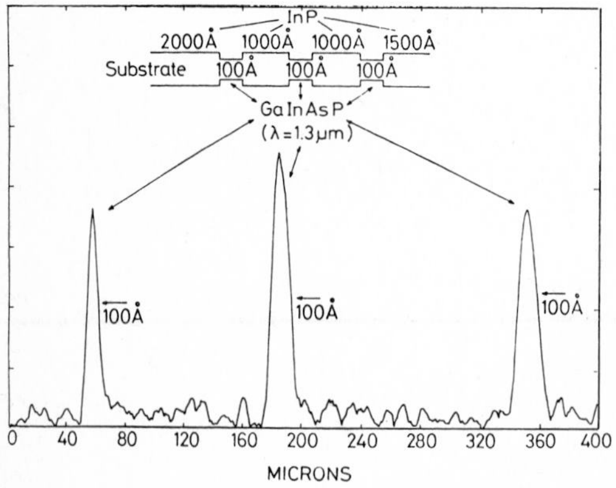

Fig.21. Auger spectrum of a chemically etched bevel of a MQW of GaInAsP-InP, comprising three GaInAsP (λ = 1.3 μm) wells of 100 Å thickness

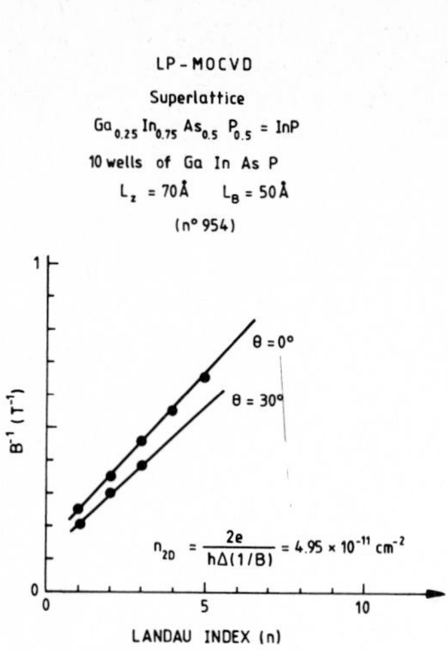

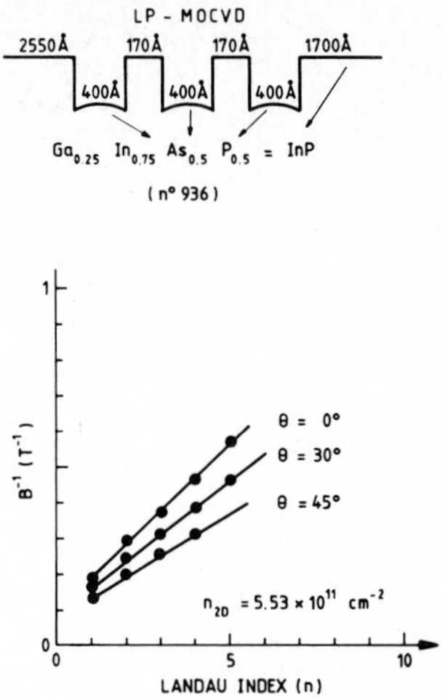

Fig.22. Reciprocal of magnetic field at maxima of magneto-oscillations vs Landau quantum number for different values of θ for a GaInAsP-InP superlattices with 21 alternate layers of ($L_z = 70$ and $L_B = 50$Å)

Fig.23. Reciprocal of magnetic field at maxima of magneto oscillations vs Landau quantum number for different values of θ for a GaInAsP-InP MQW

Fig. 22 gives the reciprocal magnetic field corresponding to the magneto-oscillation maxima as a function of the Landau index for different values for a superlattice with 21 alternate layers of $Ga_{0.25}In_{0.75}As_{0.5}P_{0.5}$ (L_z = 70 Å) and InP (L_B = 50 Å). The oscillations are periodic in 1/B, and they follow the expected $(\cos\theta^{-1})$ dependence of a TDEG. Using standard procedures, the electron density is:

$$n_{2D} = 2e/h\,\Delta(1/B) = 4.95 \times 10^{11}\ cm^{-2}.$$

Fig. 23 indicates Shubnikov-de Haas data due to a multiquantum well of $Ga_{0.25}In_{0.75}As_{0.5}P_{0.5}$-InP with $n_{2D} = 5.53 \times 10^{11}\ cm^{-2}$.

REFERENCES

1 - L. ESAKI and R. TSU I.B.M. - J. Res Develop. 14,61, 1970.

2 - M. RAZEGHI, M.A. POISSON, J.P. LARIVAIN, J.D. DUCHEMIN J. Electron. Mater., 12, pp. 371-395.

3 - M. RAZEGHI and J.P. DUCHEMIN J. of Crystal GROWTH Vol. 64, 1, (1983)

4 - M. RAZEGHI and J.P. DUCHEMIN J. Vac. Sci. et Technol. B., 1, 2, 262 (1983)

5 - M. RAZEGHI, M.A. POISSON, J.P. DUCHEMIN, B. de CREMOUX and M. VOOS. J. Appl. Phys. Lett. 43, 6, 586, (1983)

7 - M. RAZEGHI, J.C. PORTAL (unpublished)

8 - M. RAZEGHI: Light wave technology for telecommunication (to be published)

9 - J.P. DUCHEMIN , M. BONNET, G. BEUCHET and F. KOELSCH Inst. Phys. Conf. Ser. 45, Chapter 1, 10, (1979)

10 - M. RAZEGHI, Revue THOMSON-CSF Vol.15, 1. (1983)

11- M. RAZEGHI, Revue THOMSON-CSF Vol. 1, (1984)
12 - M. RAZEGHI, K.H. GOTEZ, D. BIMBERG, A.V. SOLOMONOV and G.F. GLINSKI J. Appl. Phys. 54, -(8), 4543 (1983)
13 - P. VOISIN, M. VOOS and M. RAZEGHI (unpublished)
14 - Y.S. CHEN and O.K. KIM J. Appl. Phys. 52, 7892 (1981)
15 - J.Y. MARZIN, J.L. BENCHINOL, B. SERMARGE, B. ETIENNE and M. VOOS Solid State Commun, (in Press.).
16 - R.J. NICOLAS, S.I. SESSIONS and J.C. PORTAL Appl. Phys. Lett. 37, 178 (1980)
17 - M. RAZEGHI, M.A. POISSON, Y. GULDNER, J.P. VIEREN, P. VOISIN and M. VOOS J. Appl. Phys. Lett. 40 10 , 877 (1982)
18 - A. BRIGGS, Y. GULDNER, J.P. VIEREN, M. VOOS, J.P. HIRTZ and M. RAZEGHI, Phys. Rev. B, 27 , 10 (1983)
19 - J.C. PORTAL, M.A. BRUMMELL, R.J. NICHOLAS, M. RAZEGHI and A.Y. CHO. To be published (1983)
20 - J.C. PORTAL, R.J. NICHOLAS and M. RAZEGHI J. Appl. Phys. Lett. Vol. 43, 3, 294 , (1983)

MOCVD Growth of GaAlAs/GaAs Heterostructures for Optoelectronic Devices

M. Wynne Jones

Allen Clark Research Centre, Plessey Research (Caswell) Limited
Caswell, Towceser, Northants, NN12 8EQ, United Kingdom

MOCVD offers an alternative growth technique for obtaining large abrupt multilayer III-V structures suitable for optoelectronic devices. GaAs/GaAlAs heterojunction material has been grown to meet the requirement of a large range of optoelectronic devices including lasers, LEDs, FETs, waveguides and heterojunction bipolar transistors. Assessment techniques have been established and applied to heterojunction material. MOCVD material has been processed into discrete and monolithic optoelectronic devices and the device performance evaluated.

1 INTRODUCTION

III-V materials technology has been firmly established at Plessey since the invention by the company of the chloride vapour phase epitaxy system in the early nineteen sixties [1]. The ever-expanding field of applications for fibre optic communications system has fuelled the development of a large range of devices along with development of epitaxial growth technologies. The double-heterojunction [2,3] concept provided carrier confinement with waveguiding properties and enabled continuous operation of laser devices at room temperature [4] and has impacted strongly on fibre optic systems. Thereafter understanding of heterojunction behaviour has enabled heterojunction properties to be incorporated in other optoelectronic devices and in other materials systems.

Halide VPE, HCL/hydride VPE, and liquid phase epitaxy (LPE) are used for routine growth of GaAs, InP, GaInAs and GaInAsP to meet the requirements of our optoelectronics and microwave activity. Metal organic chemical vapour deposition (MOCVD) has been developed to an advanced stage over the last five years for both GaAs/GaAlAs and InP based alloys; here work on the GaAs/GaAlAs system only is described.

2 EPITAXIAL GROWTH

MOCVD offers the capability of providing large area uniform multilayer structures with abrupt interfaces [5,6] suitable for both optoelectronic and electronic devices and will thus impact strongly on the development of optoelectronic integrated circuits.

The MOCVD epitaxial growth technique requires volatile organometallic compounds as gaseous sources of group III metals which are transported in hydrogen and react with group V hydrides in a horizontal reactor tube, Fig 1.

Fig.1 Schematic of M.O.C.V.D. system

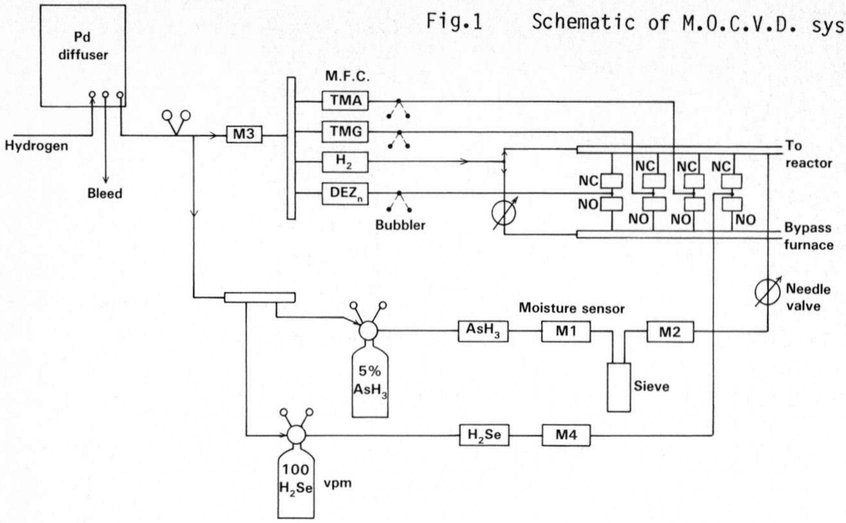

Use is made of trimethyl gallium, trimethyl aluminium as sources of gallium and aluminium and diethyl zinc and hydrogen selenide (100 vpm) as sources of acceptor and donors respectively. Control of hydrogen flow is achieved by use of mass flow controllers and both oxygen and moisture are routinely monitored. Epitaxial growth takes place at atmospheric pressure on an inductively heated graphite susceptor over the temperature range 650-800°C.

The conduction types of undoped GaAs and GaAlAs are very much a function of the ratio of group V to group III mole fraction and the incorporation of impurities is very sensitive to growth temperatures [7]. In general, the lowest carrier levels $< 10^{15}$ cm^{-3} and highest mobility $\mu_{300K} \sim 6\text{-}7000$ $cm^2V^2s^{-1}$ values are achieved at low growth temperatures (600-650°C). Si, C, Zn and others are attributed for this type change. The transition from 'p' to 'n' type background is very much a function of reactor design but more importantly of the purity of the starting organometallics.

Growth of $Ga_{1-x}Al_xAs$ layers requires careful control of both oxygen and moisture levels in the reactor in order to achieve epitaxial layers with the desired surface morphology, carrier level and photoluminescence efficiency. Various techniques have been implemented to reduce the oxygen and moisture levels to the low levels (ppm) required [8,9]. In practice a compromise is made in order to achieve acceptable background levels commensurate with reasonable photoluminescence efficiency [10,11].

The doping requirements for optoelectronic devices differs considerably, ranging from doping levels in excess of 10^{17} cm^{-3} for source devices while optoelectronic detectors, FETs and bipolar transistors require doping levels $<5x10^{16}$ cm^{-3}. Zinc and selenium allow doping levels in the range 10^{17}-10^{19} cm^{-3} and 10^{16}-10^{18} cm^{-3} to be achieved.

3 ASSESSMENT

A range of assessment techniques has been established to investigate the electrical, optical and compositional character of both uniform layers and double heterojunction material.

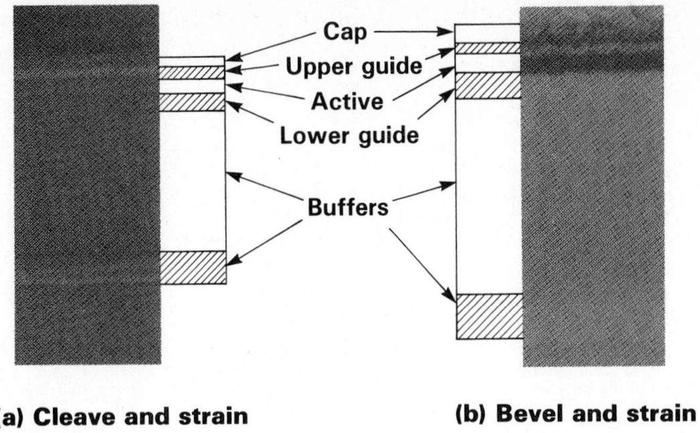

Fig.2 Thickness evaluation of LED by cleave and stain, bevel and stain

3.1. LAYER THICKNESS

Nomarski interference contrast optical microscopy was used to investigate surface morphology while layer thicknesses were evaluated by a combination of cleave and stain, bevel and stain and scanning electron microscopy.

Figure 2 gives a comparison of a cleave and stain and bevel and stain image of surface emitting diode when viewed with a microscope. The bevel angle was determined by talysurf measurement, and with knowledge of the magnification, the layer thicknesses were evaluated. The boundaries are more easily resolved in the case of Fig 2(b). The stain technique also provided contrast between doping type, where the p-type layer appeared darker, and layers containing aluminium, which appeared brighter.

3.2. AUGER ELECTRON SPECTROSCOPY

The aluminium content of $Ga_{1-x}Al_xAs$ layers was determined by Auger electron spectroscopy. Samples were bevelled at 1° and the low energy LMM Auger peak of aluminium measured as the electron beam traversed down the bevel. This signal was then compared with that from a standard sample hitherto calibrated using electron microprobe analysis. Figure 3 shows a trace from a linescan of a heterojunction bipolar transistor structure.

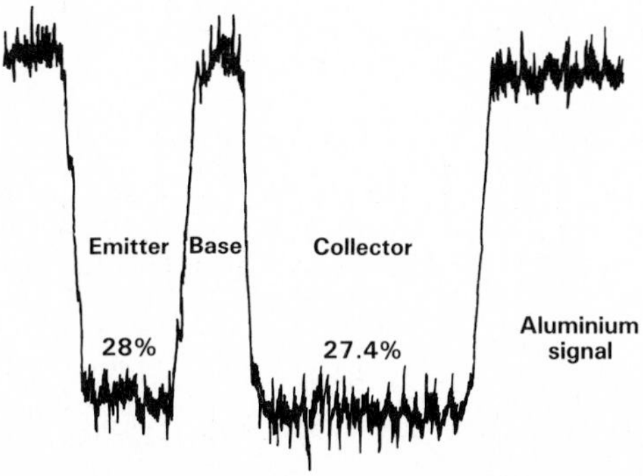

Fig.3 A.E.S. line scan of bevelled bipolar structure

3.3. X-RAY DIFFRACTION MEASUREMENTS

Lattice mismatch between GaAs and GaAlAs of the order of <.1% was routinely measured by use of a double crystal diffractometer. The 400 reflection was used for this measurement,the CuKα line wavelength being 1.5405 Å. Resolution of a few seconds of arc can be obtained using this technique. In some samples where compositional changes occurred during growth, a shoulder was observed in the GaAlAs peak corresponding to increasing lattice parameter and hence increased aluminium composition.

3.4. PLAN VIEW AND CROSS-SECTIONAL T.E.M.

The crystallographic quality of MOCVD material was assessed by plan view and cross-sectional transmission electron microscope (TEM). Plan view TEM specimens were prepared by jet chemical thinning using Cl in methanol and in some instances Br in methanol. Both dislocations and stacking faults have been seen in poor quality GaAlAs. Dislocation densities $<10^4$ were typically obtained and reflected the quality of the supplied substrates. Of further significance in investigating thin layer structures of GaAlAs has been the development of cross-sectional TEM. Samples were thinned to electron transparency by first bonding samples together in pairs using epoxy resin. The resulting structure was then lapped to around 100 μm by mechanical polishing and finally thinned by ion-milling using beams of 6 keV Ar ions. Dark field images were recorded using 200 Bragg reflections. The image obtained mirrors the local aluminium concentration and appears as a bright image. An example obtained from a bipolar structure is shown in Fig 4, and provided useful information on compositional control as well as giving further information on layer thicknesses. Furthermore, the sharp white lines at the start of both collector and

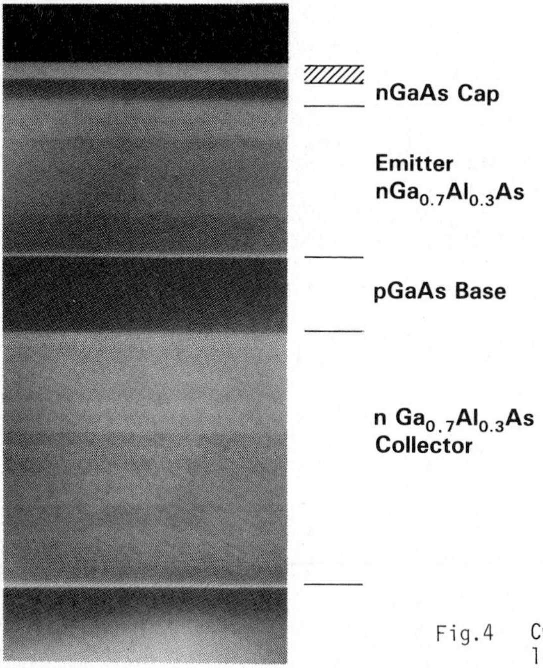

Fig.4 Cross-sectional TEM of bipolar structure

emitter growths reflected an abrupt compositional change in aluminium level on switch on of aluminium. This arose from slight inbalance of pressure between the reactor and bypass lines on first switching aluminium from the bypass to the reactor.

3.5. DOPING LEVEL STUDIES

Carrier concentration profiles have been observed using contact resistance profiling [12], mercury probe Schottky barrier CV profiling, electrolytic etching with simultaneous CV measurement (P.O.P) [13]. An example of a profile of an FET structure is shown in Fig 5. An active doping level of 10^{17} cm^{-3} was achieved with a decade fall off in carrier level occurring over an interface width of <500 Å.

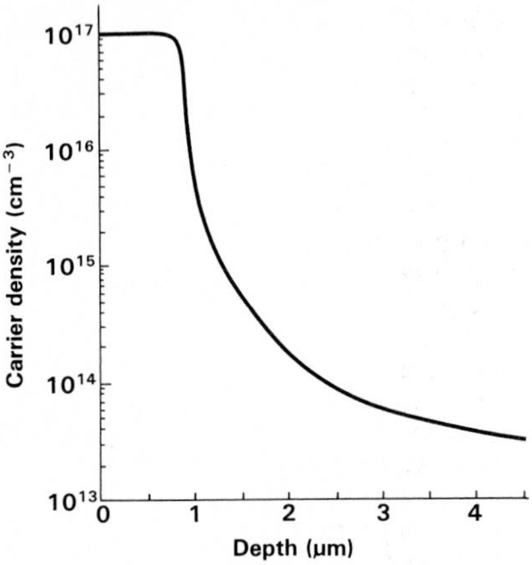

Fig.5 C-V profile of FET structure

3.6. SECONDARY ION MASS SPECTROMETRY

Elemental profiling has been made on double heterojunction material using secondary ion mass spectrometry. This work has been carried out using a Cameca IMS 3f system at Loughborough Consultants Limited. Oxygen ions with energies in the range 5.5 to 6 keV have been used for sputtering. An example of a bipolar transistor is shown in Fig 6, showing the aluminium, arsenic and Ga levels. Again in Fig 6, an increase in aluminium level on turn on of aluminium is seen as evidenced by cross-sectional TEM. Figure 6(b) shows that with due control of reactor and bypass pressure, elimination of the compositional variation at the onset of aluminium was possible.

4 DEVICE RESULTS

Epitaxial layers have been grown by MOCVD to meet the requirements of a wide range of optoelectronic structures and fabricated into discrete devices. This technology provides a firm base for the realisation of integration into monolithic optoelectronic structures. We have fabricated discrete FETs, LEDs, lasers and optical waveguide devices and the performances have been comparable with those fabricated from conventionally grown material. Laser structures exhibited threshold current densities (I_{th}) <1.5kA/cm^2, the best value being 800A/cm^2 under pulsed operation. LEDs processed into 'Burrus' type structures gave diode characteristics

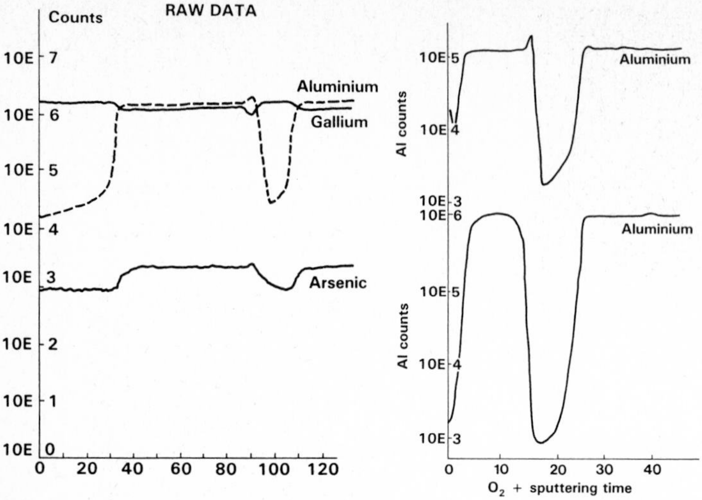

(a) Aluminium, arsenic and gallium (b) Aluminium only

Fig.6 SIMS trace of heterojunction material

with forward voltages of 1.5 V at 10 mA, and were found to be much faster than LPE devices but showed lower internal efficiency of 45% as compared to 70% for an equivalent LPE device [15]. High yields of multiple LED arrays have subsequently been fabricated from MOCVD grown material, the device results reflecting the uniformity and planarity offered by this technique.

Discrete FETs, with GaAlAs buffer layers, and an active layer of .9 μ and a doping level of around 10^{17} cm^{-3}, showed I-V curves as shown in Fig 7. Gains in excess of 100 ms/mm were attained with level saturation characteristics.

A comparison of Schottky FET prepared by MOCVD material with the specification of VPE devices is shown in Table 1 and compares favourably with existing technology.

TABLE 1

		MOCVD FET	VPE FET
I_{DSS}	mA	30	15-50
g_m	mS	16	15
V_p	Volts	-3	$-1.5 \rightarrow 3.5$

Optical waveguides have been etched from GaAs/GaAlAs multilayer structures and showed low loss behaviour [16]. Values of 1.4 db/cm for fully strip loaded waveguides at 1.3 μm and 2.5 db/cm for fully etched ridge waveguides were obtained. These results demonstrate the feasibility of incorporating waveguide-based optical devices such as modulators and

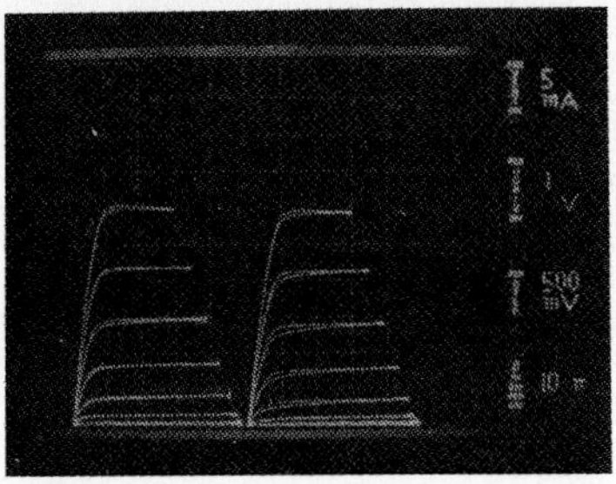

B468

N_D = 1.1 10^{17} cm^{-3}
LG = 1µm
W = 150µm
A = 0.22µm
Drift mobility 2666 $cm^2 v^{-1} s^{-1}$
g_m ~16-17 mS

Fig.7 DC characteristics of FET

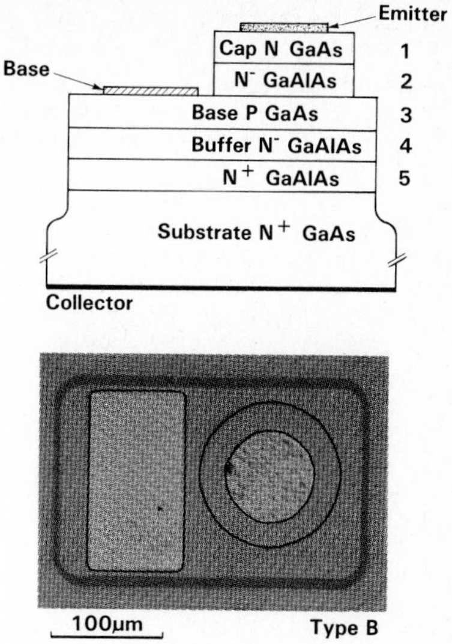

Fig.8 Structures of simple HJBT transistors

switches in monolithic structures and will impact significantly on future developments in this area.

GaAlAs heterojunction bipolar transistors offer potential advantages in terms of injection efficiency and gain from the use of heterojunctions with significant improvements in frequency response 100 GHz f_T. MOCVD material has been processed into an n-p-n transistor and low gains established ~ 100 (Fig.8). Figure 9 shows the forward characteristics of a discrete transistor.

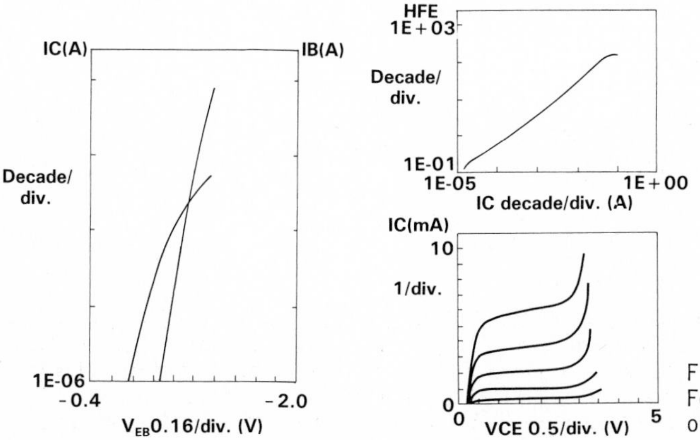

Fig.9
Forward characteristics of HJBT transistors

Local area epitaxy on both nitride and oxide coated GaAs have been made and show promise as techniques necessary for the realisation of complex optoelectronic circuit functions. Epitaxial growth occurred on the apertured regions whereas polycrystalline growth occurred on the oxide and nitride. Removal of the polycrystalline regions was achieved by etching and agitation.

The realisation of a monolithic optoelectronic device has in part been achieved by the incorporation of a Schottky FET and a double heterojunction LED. The chip featured a waveguide-linked emitter and detector, together with FET. Figure 10 shows a layout for a monolithic integration test chip along with the proposed integration scheme [12]. Suitable metallisation and etching steps provide ohmic contacts to the p and n type material as well as forming the Schottky barrier gates. Both FET D.C. transistor characteristics and LED behaviour have been investigated. Figure 11 shows doping level curves obtained by use of the POP along with the Auger line spectrum. The degree of coupling between waveguide-linked diodes is represented in Fig.12 where drive current versus detector current is shown. Coupling ratios of $.7 \times 10^{-4}$ and 1.2×10^{-4} were obtained for an emitter drive current of 62 mA. Leakage currents were roughly an order of magnitude lower than the coupled currents. These measurements compare favourably with the hydride trichloride VPE liquid phase epitaxy material previously used.

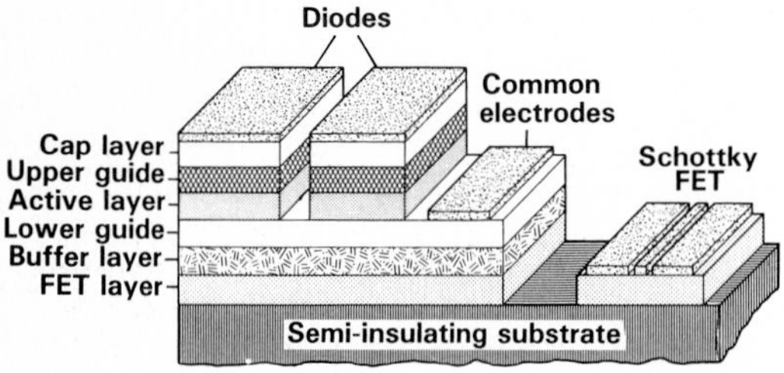

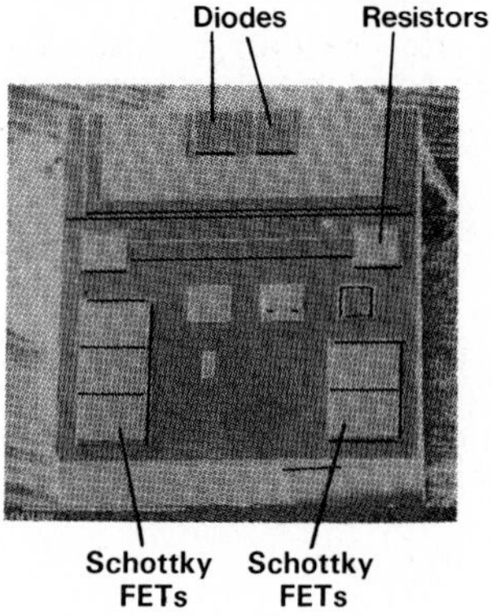

Fig.10 Schematic diagram of monolithic test structure and integration scheme

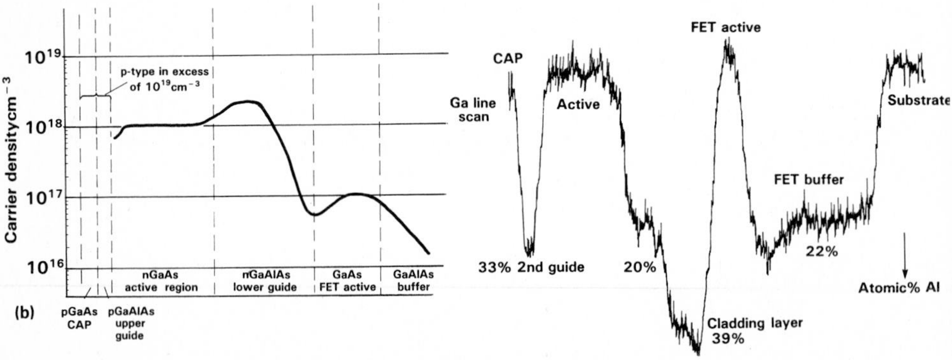

Fig.11 Doping level and A.E.S. line scan of monolithic test structure

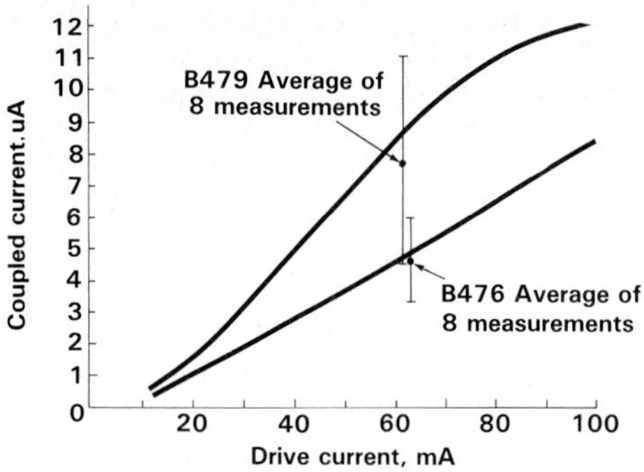

Fig.12 Optically coupled currents for varying emitter drive currents

5 CONCLUSIONS

MOCVD now offers the possibility of producing materials for a wide range of optoelectronic devices for incorporation in electronic systems as well as the more advanced optoelectronic circuits. The realisation of achieving large area thin multilayer structures by MOCVD has provided the required stimulus for its rapid development and it now offers an alternative technology well capable of meeting the material requirements of our present day devices. The possible incorporation of selective epitaxial growth on oxide or nitride surfaces offers further a necessary technology for development of monolithic optoelectronic circuits and will thus impact strongly on its future development.

ACKNOWLEDGEMENTS

The author wishes to thank R.C. Goodfellow and R.J.M. Griffiths for helpful discussions and D. Skinner for A.E.S. assessment, D.E. Sykes of Loughborough Consultants Limited for the SIMS sputter profiles and A.G. Cullis of the Royal Signals and Radar Establishment for the cross-sectional TEM work.

6 REFERENCES

1. J.R. Knight, D. Effer and P.R. Evans: Solid State Electronics. 8, 178 (1965)
2. H. Kroemer: Proc. IEEE 51, 1782 (1963)
3. Zh.I. Alferov and R.F. Kazarinov: Authors Certificate 10321551/16-15 (USSR) 1963
4. M.B. Panish, I. Hayashi and S. Sumski: IEEE J. Quantum Electronics 5, 210 (1969)
5. P.M. Frijlink and J. Maluenda: J. de Physique No.5, 185 (1982)
6. R.J.M. Griffiths, N.G. Chew, A.G. Cullis and G.C. Joyce: Electronics Lett. 19, 227 (1983)
7. K. Mohammed, J.L. Merz and D. Kasemset: Materials Lett. 2, 35 (1983)
8. D.W. Kisker, J.N. Miller and G.B. Stringfellow: Appl. Phys. Lett 40, 614 (1982)
9. J.R.Shealy and J.M. Woodall: Appl. Phys. Lett. 41, 88 (1982)
10. G.B. Stringfellow: J. Crystal Growth 53, 42 (1981)
11. A. Mircea-Roussel, A. Brière, J. Hallas, A.T. Vink and H. Veenvliet: J. Appl. Phys. 53, 4351 (1982)
12. R.C. Goodfellow, A.C. Carter, R. Davis and C. Hill: Electron Letters 14, 328 (1978)

13. T. Ambridge, C.R. Elliott and M.M. Faktor: J. Appl. Electro Chem. 3, 1 (1973)
14. C.A. Burrus and R.W. Dawson: Appl. Phys. Lett. 17, 97 (1970)
15. R.R. Bradley: J. Crystal Growth 55, 223 (1981)
16. R.G. Walker and R.C. Goodfellow: Electronics Letts. 19, 590 (1983)
17. A.C. Carter, N. Forbes and R.C. Goodfellow: Electron Lett. 18, 72 (1982)

Field-Effect Transistors with a Two-Dimensional Electron Gas as Channel

Heinrich Dämbkes and Klaus Heime

Fachgebiet Halbleitertechnik/Halbleitertechnologie
Universität-Gesamthochschule-Duisburg, Kommandantenstraße 60
D-4100 Duisburg, Fed. Rep. of Germany

The two-dimensional electron gas at an abrupt interface between highly n-doped Al(x)Ga(1-x)As and high purity undoped GaAs exhibits very high electron mobilities. Therefore it is a suitable candidate for high speed field-effect transistors (FET). It will be shown which factors determine the FET performance. Despite parasitic effects such as series resistances the (AlGa)As/GaAs heterostructure FET exhibits a pronounced improvement with respect to best GaAs homostructure MESFETs.

1. Introduction

During the last 20 years the frequency limit of high speed devices has increased drastically. Leadership at the front of high speed transistors has been passed over to GaAs MESFETs which promise cut-off frequencies up to 100GHz for quarter micron gate length /1,2/. Devices which seem to exceed these limits are the field-effect transistors with a quasi two-dimensional electron gas as channel /3,4/.

2. Structure and Characteristics of Heterostructure Field-Effect Transistors

2.1. Nomenclature of Heterostructure Field-Effect Transistors

Various names and abbreviations are used for heterostructure FETs depending on their origin. A standardization does not yet exist.

1. SDHT = selctively doped heterostructure transistors (Bell Labs)
2. MODFET = modulation doped FET (Univ. of Illinois)
3. HEMT = high electron mobility transistor (Fujitsu)
4. TEGFET = two-dimensional electron gas FET (Thomson CSF)

In this paper TEGFET will be used.

2.2. Structure and Operation of the TEGFET

The structure, band diagram and charge distribution of a TEGFET is shown in Fig.1. Electrons are confined in the potential well, thus creating a quasi two-dimensional electron gas (TEG). The current I_D through the channel between the ohmic contacts' source (S) and drain (D) is controlled by a third electrode, the gate (G). The current is defined by

$$I_D = Q_S \cdot v \cdot w \qquad (1)$$

with
- w = channel width
- v = electron drift velocity (controlled by V_{DS})
- Q_S = $q \cdot n_s$
- n_s = sheet carrier concentration per unit area (controlled by V_{GS}, V_{DS}).

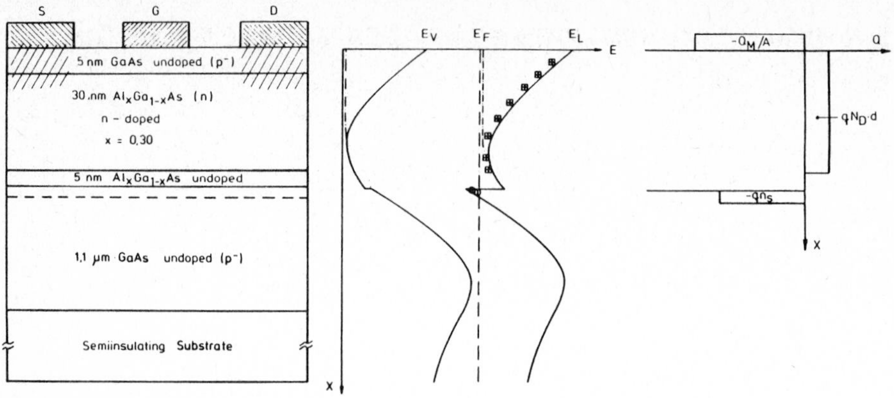

Fig.1 Structure, band diagram and charge distribution of a TEGFET

For high-speed digital and analog circuits FETs are needed with a minimum size (w small!) but simultaneously both high current (n_s large!) and short carrier transit times (v large!).

The control of the sheet carrier concentration n_s and hence the current I_D by the the gate voltage V_{GS} is explained as follows. The n^+-(AlGa)As layer below the gate is depleted a) by the transfer of electrons into the TEG, b) by the charge distribution at the metal-semiconductor contact such that an overall charge neutrality exists. An undesirable conductivity in the (AlGa)As layer is avoided by choosing its thickness such that both the contact depletion region and the hetero-interface depletion region touch. Then at $V_{GS}=0V$ the maximum value of n_s is obtained (depletion-type (normally-on) TEGFETs, D-TEGFET). By decreasing the (AlGa)As layer thickness the TEG channel is depleted, while the charge on the gate electrode remains constant. It is thus possible to deplete the channel completely at $V_{GS}=0V$ (enhancement-type (normally-off) TEGFET, E-TEGFET). A negative gate voltage ($V_{GS}<0V$) decreases n_s and I_D while a positive gate voltage increases both n_s and I_D. The drain voltage ($V_{DS}>0V$) creates an electric field in the channel which accelerates the electrons. In addition it modifies the potential drop across the heterostructure and the distribution of n_s along the channel. From these principles of operation the device characteristics can be deduced /5,6/. A more simplified consideration assumes that the space charge in the (AlGa)As layer is always constant (all donors are ionized). The derivation of the device characteristics is identical to that used for MOSFETs /7/. If it is further assumed that the electron drift velocity has reached its saturation value v_s, the current-voltage characteristics are:

$$I_{DS} = k' \, (V_{GS} - V_T) \qquad (2)$$

I_{DS} = current in the saturation regime, independent of V_{DS}

$$k' = \frac{\varepsilon_0 \varepsilon_r \, w \cdot v_s}{d_{eff}} = c \cdot v_s \cdot w \qquad \text{with} \qquad d_{eff} = d + 8\text{nm} \quad /6/$$

d = thickness of the (AlGa)As layer
c = gate capacitance per unit area
V_T = threshold voltage.

Equation (2) is a good approximation for E-TEGFETs ($L > 0.8\mu m$) /7,8/, but

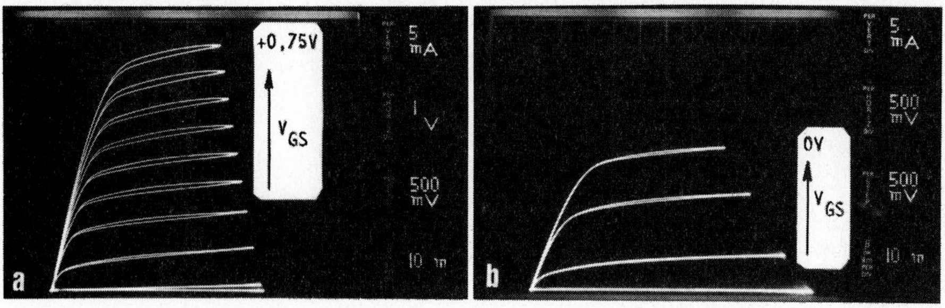

Fig.2 Characteristics of a TEGFET a) with b) without parallel conduction

less valid for D-TEGFETs. Especially if a parallel conduction in the (AlGa)As layer occurs, a significant deviation from (2) and a performance degradation results /8,9/ (Fig.2).

The transconductance g_m is

$$g_m = \frac{\partial I_{DS}}{\partial V_{GS}} = k' = c \cdot v_s \cdot w \tag{3}$$

or g_m normalized to unit gate width

$$g_m^* = \frac{g_m}{w} = \frac{\varepsilon_0 \varepsilon_r \cdot v_s}{d_{eff}} = c \cdot v_s \,. \tag{4}$$

The upper frequency limit can be described by the transit frequency f_T which is approximately given by:

$$f_T = \frac{g_m}{2\pi C_{GS}} = \frac{v_s}{2\pi L} \tag{5}$$

with $C_{GS} \approx c \cdot w \cdot L$.

Equations (2) - (5) show the important influence of the maximum drift velocity v_s, the gate length L and the carrier concentration n_s (via c) on the device performance. Experiments indicate that the effective maximum drift velocity at room temperature is $v_s = (1.8\text{-}2.5) \cdot 10^7$cm/s and increases towards lower temperatures /8,10/. The optimum thickness d for maximum n_s is around 45nm for doping of $N_D = (1\text{-}2) \cdot 10^{18}\text{cm}^{-3}$. For a 1μm gate TEGFET at 300K one obtains:

$g_m^* = (300\text{-}400)$mS/mm

$f_T = (28\text{-}40)$GHz .

These values compare favourably with experimental results /7,8,11/ (extrapolated for the intrinsic device, cf. chapter 3):

$g_m^* = (260 - 350)$mS/mm

$f_T = (25 - 35)$GHz .

Optimized GaAs-MESFETs achieve values of

$g_m^* = (200 - 350)$mS/mm

$f_T = (20 - 30)$GHz /11/.

This comparison shows the advantages of TEGFETs at room temperature for L=1µm. At lower temperatures the TEGFET performance is improved while MESFETs do not show any considerable improvement. With further reduction of the gate length L the advantage of the TEGFET over GaAs MESFET increases even more because of the more pronounced velocity overshoot in the TEG /8/ (cf. chapter 3.3).

Equations(2)-(5) lead to the conclusion that the low-field mobility does not have any influence on the device performance. Indeed, more detailed calculations /12/ show that this is approximately true for transconductance and frequency limit if the low-field mobility is higher than $3000 cm^2/Vs$. However the low-field mobility is not completely unimportant because it strongly influences parasitic (yet unavoidable) device elements such as series resistances, which in turn determine the extrinsic performance and the noise behaviour of the TEGFET. This will be dealt with in the following chapter.

3. Influence of Carrier Concentration and Mobility on TEGFET Performance

3.1. The TEGFET Small-Signal Equivalent Circuit

A real (extrinsic) FET is composed of the intrinsic FET and parasitic elements the influence of which is most easily explained by a small-signal equivalent circuit (Fig.3). The equivalent circuit elements are assumed frequency-independent, but depend on the dc bias.

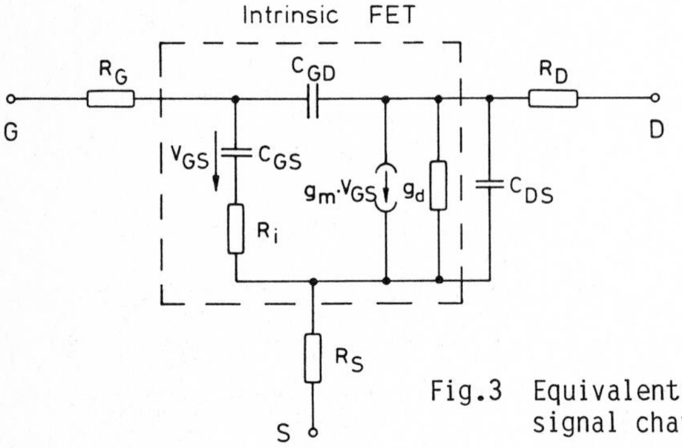

Fig.3 Equivalent circuit for small signal characterization

Since source and drain electrodes are spatially separated from the gate electrode series resistances R_S, R_D exist. In addition the "ohmic" source and drain contacts have a finite contact resistance, which is included in R_S, R_D. The gate metallization (d_{Me} = 0.1-1µm) causes the resistance R_G. Other parasitics which do not depend on the chip properties are excluded.
The voltage v_{gs} across the input capacitance C_{GS} controlling the output current source $i_d = g_m \cdot v_{gs}$ is only a part of the externally applied gate voltage v_{gsext} :

$$v_{gs} = v_{gsext} - i_d R_S \, . \tag{6}$$

The voltage drop across R_G is neglected.
The effective transconductance is thereby reduced to

$$g_{meff} = \frac{g_m}{1+R_S g_m} \tag{7}$$

and the transit frequency now is

$$f_{Teff} = \frac{g_{meff}}{2\pi C_{GS}} = \frac{v_s}{2\pi L(1+R_S g_m)} . \qquad (8)$$

The noise - an important parameter especially for microwave (analog) application - is increased by the parasitics R_S, R_D. The minimum noise figure is approximately given by /13/:

$$F_{min} = 1 + K \frac{f}{f_T} (g_m(R_S+R_G))^{1/2} \qquad (9)$$

(K is a material-dependent factor: $K_{GaAs} \approx 2.5$; $K_{TEGFET} \approx 1.5$).

The resistances R_S, R_D also increase the power dissipation. The heat has to be removed through the substrate. Any additional power dissipation reduces the integration density, increases the interconnection line lengths and reduces the speed of the circuit.
As an example the influence of n_s and μ_0 on R_S will be demonstrated. The source resistance normalized to unit width is:

$$R_S w = \frac{L_{SG}}{q\mu_0 n_s} . \qquad (10)$$

Using experimentally observed values of

$n_s = (0.5\text{-}1) \cdot 10^{12} cm^{-2}$; $\mu_0 = (5\text{-}7) \cdot 10^3 cm^2/Vs$; $L_{SG} = 2\mu m$,

one obtains

$R_S \cdot w = 2\text{-}5$ Ohm mm .

(In GaAs MESFETs with recessed gate $R_S \cdot w = 0.1\text{-}0.2$ Ohm mm is easily achieved!) Using this value g_m, f_T and F_{min} are reduced as shown in Tab.1.

Tab. 1: Influence of parasitic elements R_S, R_G on TEGFET performance

	$R_S = 0$	$R_S = 5$	$R_S = 5$	Ohm mm
	$R_G = 0$	$R_G = 0$	$R_G = 2$	Ohm mm
g_{meff}	300-400	120-135	120-135	mS/mm
f_{Teff}	28-40	12-15	12-15	GHz
F_{min} (8GHz)	1	1.6-1.8	1.7-2.0	

The table demonstrates the strong influence of the parasitics R_S, R_G, but also the influence of n_s, μ_0 and device dimensions (e.g. L_{SG}) on these parasitics. The advantage of TEGFETs over MESFETs is the higher mobility μ_0. On the contrary, the maximum product of concentration times channel thickness may be up to an order of magnitude higher in MESFETs because the layer may be thicker outside the gate region (recessed gate). Whereas the free surface potential of (AlGa)As is low (0.2-0.3V /14/) eventually resulting in an undepleted (AlGa)As channel, its influence on R_S reduction is small because of the very low mobility of (AlGa)As.

Ways towards further improvement presently known are the use of a thick additional n^{++}-GaAs top layer outside the gate region which supplies a parallel conduction path /11/ (a second TEG is excluded by growing a gradual

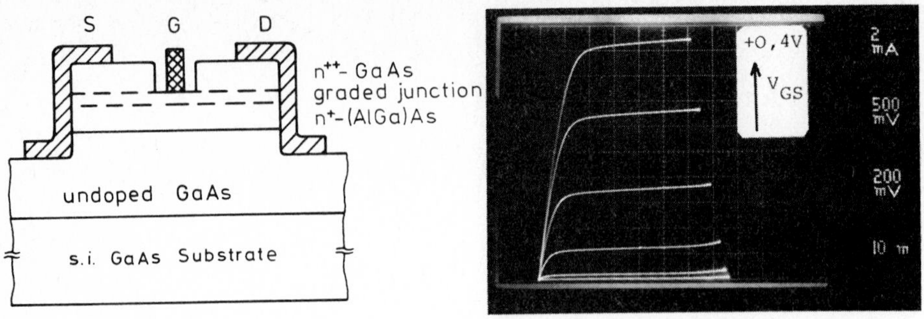

Fig.4 Cross-sectional view of improved TEGFET structure and dc characteristics (L=1.4μm, W=150μm)

transition from (AlGa)As to GaAs) (Fig.4), or an ion-implanted n^+ region outside the gate region, probably self-aligned /15,16/. Preliminary experiments by the present authors showed that R_S could be reduced from 2-5 Ohm·mm to only 0.2-0.5 Ohm·mm (increasing g_m from 60-80mS/mm to 220mS/mm (extrinsic values!) which is comparable to very good MESFETs with recessed gate.

3.2. Optimization of Sheet Carrier Concentration

The optimization of the intrinsic TEGFET requires a sheet carrier concentration n_s as high as possible. The influence of n_s on I_D (and therefore on g_m, f_T) is demonstrated in (1). It depends on the following parameters:

1. Dopant concentration N_D in the (AlGa)As layer
2. Donor activation energy
3. Compensation by deep traps
4. Height of barrier between (AlAs)As and GaAs
5. Spacer thickness.

The maximum dopant concentration N_D = (1-2) $10^{18}\,cm^{-3}$ is limited by the growth process. The activation energy of the usual Si donor is sufficiently low to allow full ionization at room temperature, but below 200K a partial freezeout of free carriers occurs /9/. From DTLS and low - frequency noise measurements several traps are known, at least one of which is an intrinsic and possibly unavoidable defect /17,18,19/. Its concentration increases with the Al concentration and may reach the donor concentration at $x \approx 0.35$.
The band bending in the (AlGa)As towards the interface is $\approx$0.2eV at $x \approx 0.35$ and depends on the Al concentration, too. In order to suppress any appreciable reinjection of electrons from the TEG, the Al content should be larger than x=0.2. These two factors limit the Al content to about 0.2-0.32. In order to reduce the residual Coulomb scattering between ionized donors and the TEG an undoped (AlGa)As spacer layer of the thickness d_i is often introduced between the undoped GaAs and the n^+-(AlGa)As. Thereby the electron transfer into the TEG is reduced but the mobility is enhanced. Electrons are transferred from a depth d_d=7-10nm above the hetero-interface, while the optimum thickness of the (AlGa)As layer is approximately 45-50nm for D-TEGFETs. Thus less than 30% of the electrons of the (AlGa)As layer are available for transfer. The maximum sheet carrier concentration in practical TEGFET structures therefore is

$$n_s = N_D^+ \, d_d = 5\cdot 10^{11}\,cm^{-2} - 2\cdot 10^{12}\,cm^{-2} . \qquad (11)$$

In Fig.5 the correlation between n_s, N_D and d_i is given.

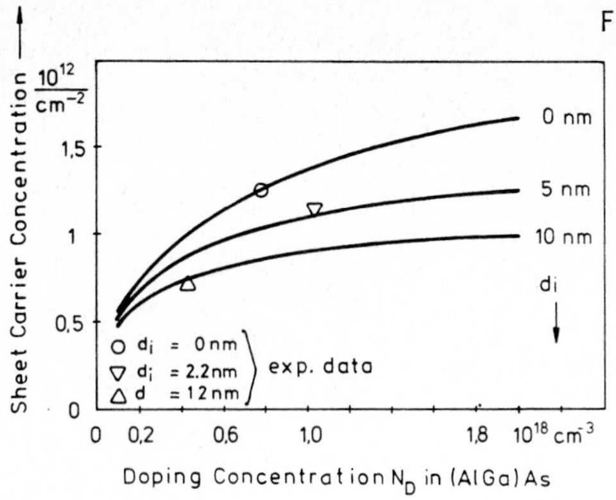

Fig.5 Dependence of n_s on d_i and N_D

3.3. Influence of Electric Field and Temperature on Electron Mobility

It has been shown in chapter 3.1. that the TEGFET performance at room temperature is superior to that of the MESFET only if the parasitics R_S and R_D are reduced considerably. At lower temperatures the low field mobility increases rapidly (Fig.6) which results in a strong reduction of the parasitics (also in comparison with GaAs MESFETs where μ is almost independent of temperature). However the high mobility decreases rapidly with increasing electric field (Fig.7), the decrease being more pronounced at lower temperatures /10,20/. Relatively small fields of 1kV/cm (=0.1V/1μm) are easily achieved in series resistances. Therefore even for low temperature operation of the TEGFET it is necessary to diminish parasitic effects. At very short gate lengths (L < 1μm) an additional effect called "velocity overshoot" becomes effective. Because of their low effective mass in the Γ minimum, electrons may initially gain very high drift velocities if they experience a sudden rise in the electric field both in space and/or in time, before they are scattered into the L-side minimum where they move with the saturation drift velocity v_s. Monte Carlo calculations have demonstrated

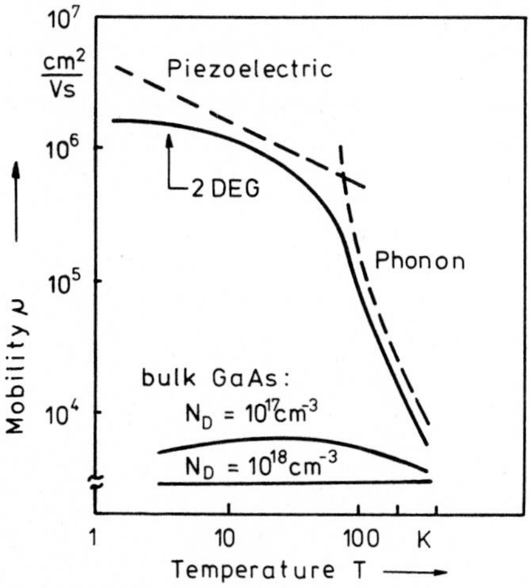

Fig.6 Dependence of mobility on temperature

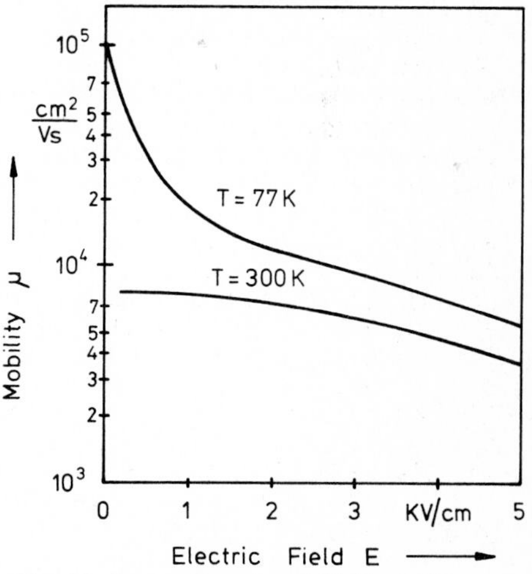

Fig.7 Decrease of mobility with increasing electric field

that this overshoot is effective for times $\leq$0.1ps and lengths of 0.1-0.3µm /21/. Maximum drift velocities of 4-8 v_s have been predicted for bulk GaAs, resulting in improvements of current, transconductance and frequency limits /21,22/. The exact correlation between low-field mobility and velocity overshoot is presently not yet known.

The simple model which was used for the calculation of the device characteristics assumed that the channel is ideally two-dimensional with a constant mobility µ at low field. In reality a mobility profile exists (Fig.8).

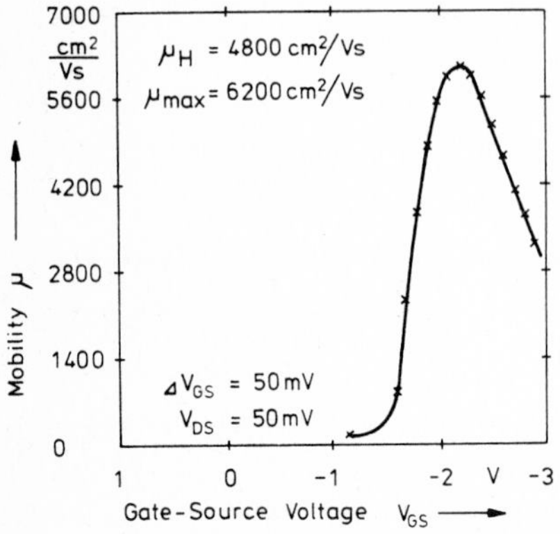

Fig.8 Mobility profile from magneto-transconductance measurements

Close to the surface ($V_{GS} \geq 0V$) the mobility is low because of increased scattering by interface roughness and Coulomb interaction with ions in the (AlGa)As layer. At the position of the (AlGa)As/GaAs interface ($V_{GS} < 0V$) µ strongly increases to high values,the peak value being clearly higher than the value of the Hall mobility, but then µ decreases again. The main reason is the decreasing carrier concentration by which the electron shielding of the Coulomb interaction with the ions in the GaAs is reduced.

4. The Performance of TEGFET Integrated Circuits

The two most important parameters for evaluating the performance of digital ICs are the switching time and the power consumed during the switching process. Also the product of these two parameters is a useful figure of merit.

As a simple example an inverter gate (Fig.9) will be discussed, which is the fundamental cell of the more complex gates /23,24/. It consists of a

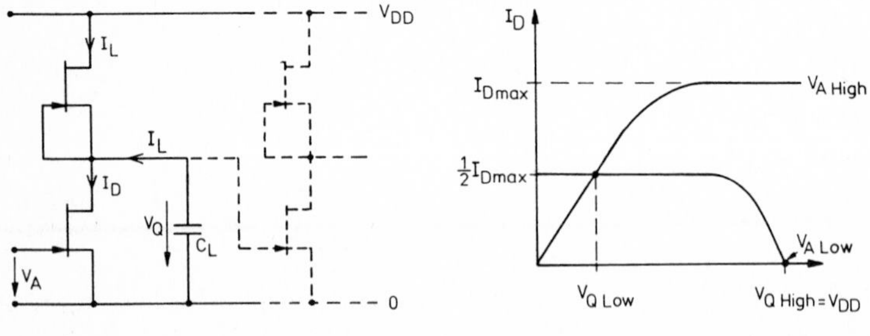

Fig.9 Simple inverter stage and principal switching diagram

switch formed by the FET and a load. If with logic high voltage at the input the switch is conducting the output voltage is logic low and vice versa. Several other gates may be coupled to the output of the inverter. They are simply represented by the load capacitance C_L. If the transistor switches from the conducting to the nonconducting state (or vice versa) a load current I_L flows.

It is now assumed that the logic swing is

$$V_{max} = V_{QHigh} - V_{QLow} = (V_{GS} - V_T)_{max} = 0.5V_{DD}.$$

Further assuming that the optimum mean load current is

$I_L = 0.5I_{Dmax}$ and C_L = const., one obtains:

$$I_L = 0.5\ k' \cdot V_{max}\ .$$

The charge to be loaded or unloaded is

$$Q_L = I_L \cdot t_d = C_L \cdot V_{max}$$

with t_d being the time interval during which I_L flows. It corresponds to the delay between the application of the input signal and the response of the output voltage. Combining the equations appropriately results in

$$t_d = \frac{2C_L}{k'}\ . \qquad (12)$$

Thus the speed of the circuit can be increased

a) by reducing the load capacitance
b) by increasing the design parameter k' which depends on the transport properties of the device. The higher k' value of TEGFETs over MESFETs especially with decreasing temperature makes them favourite candidates for high-speed circuits.

Indeed experimental results on ring oscillators (= closed chain of an odd number of inverter stages) have demonstrated the following results.

Table 2: Experimental results of TEGFETs /25,7/ and MESFETs /26/

	t_d		
T	300	77	K
TEGFET L =	12.2 1.0	17.1 2.0	ps µm
MESFET L =	20.9 1.0	16.1 1.0	ps µm

The power consumption is approximately

$$P_D = V_{DD} \cdot I_L = k' \cdot V^2_{max}\ . \qquad (13)$$

Equation (13) shows that the power increases with k' such that the power-delay-time product

$$P_D \cdot t_d = 2C_L \cdot V_{max}^2 \qquad (14)$$

does not depend on the device properties in this simple approximation. However better transport properties allow a redesign of the device, for instance a reduction of the device width if the current is to be held constant. Then k' is held constant with respect to the original value but C_L (if determined by the input capacitance of the following gates) is reduced. Thus t_d decreases and $P_D t_d$ is reduced proportional to the width of the device. A more detailed analysis is given in /27/.

5. Conclusion

The improvements of carrier transport properties in heterostructures over those in homostructures make these systems attractive candidates for high-speed devices. However at room temperature limitations exist which are mainly due to the parasitics in the extrinsic devices. It was shown that these limitations are overcome by additional technological steps, especially the introduction of a highly conductive path in parallel to the TEG but outside the gate. Then full advantages of the high mobility and higher velocity may be taken. Other advantages of TEGFETs over MESFETs are:

- reduced noise, the reason of which is not definitely clear at present,
- reduction of substrate influence by the thick undoped GaAs layer,
- reduction of surface influences as the TEG is buried below the (AlGa)As layer,
- possibility of growing the heterostructures reproducibly and very homogeneously by molecular beam or organometalic vapour epitaxy,
- possibility of defining thicknesses very accurately by reactively dry etching processes, using different "etch stoppers" for D- or E-TEGFETs, making use of differences in etch rates between (AlGa)As and GaAs layers /28/.

The authors therefore believe that both the improvements in transport properties and device performance and in device fabrication make the TEGFET a strong competitor of the conventional GaAs MESFET.

Acknowledgements: The experimental work is supported by Stiftung Volkswagenwerk in a joint project of K. Ploog and E. Schubert (Max-Planck Institut für Festkörperforschung, Stuttgart, FRG) and the present authors. An intense cooperation with G. Weimann (Forschungsinstitut der Deutschen Bundespost, Darmstadt, FRG) is gratefully acknowledged.

References

1 H.Dämbkes,W.Brockerhoff,K.Heime: IEDM 1983,Techn.Digest,pp.621-624

2 E.T.Watkins et al.: IEEE Int. MTT-S 1983,Digest,pp.145-147

3 N.T.Linh et al.: IEDM 1982, Techn. Digest

4 K.Joshin et al.: IEEE Int. MTT-S 1983,Digest,pp.563-565

5 D.Delagebeaudeuf,N.T.Linh,:IEEE ED-28 (1981) No.7,pp.790-795

6 K.Lee,M.S.Shur,T.J.Drummond,H.Morkoc: IEEE ED-30 (1983) No.3,pp.207-212

7 T.Mimura et al.:Fujitsu Scient.& Techn.Journ. 19 (1983) No.3,pp.243-278

8 A.Cappy et al.: Revue Phys.Appl. 18 (1983), pp.719-726

9 E.F.Schubert,K.Ploog,H.Dämbkes,K.Heime:Appl.Phys. A 33 (1984),pp.63-76

10 E.F.Schubert,K.Ploog,H.Dämbkes,K.Heime:to be publ. in Appl.Phys.A 33

11 H.Dämbkes, Dissertation, Universität Duisburg (FRG) 1983

12 M.S.Shur: El.Lett. 18 (1982) No.21, pp.909-910

13 H.Fukui: IEEE ED-26 (1979) No.7,pp.1032-1037

14 D.Delagebeaudeuf et al.: El. Lett. 18 (1982) No.2,pp.103-105

15 N.Nishi et al.: Jap.Journ.Appl.Phys. 22 (1983) Suppl.22-1,pp.401-404

16 N.Yokoyama et al.: IEEE ED-29 (1982) No.11,pp.1772-1777

17 K.Hikosaka,T.Mimura,S.Hiyamizu: Inst.Phys.Conf.Ser. No.63, pp.233-238

18 L.Loreck,H.Dämbkes,K.Heime: IEDM 1983, Techn.Digest,pp.107-110

19 H.Künzel,K.Ploog,K.Wünstel,B.L.Zhou:to be published in Journ.El.Mat. 13

20 H.Morkoc in: ASI on MBE & Heterojunctions, edts.: L.L.Chang,K.Ploog (Martinus Nijhoff Publishers Netherlands) 1984

21 B.Carnez et al.: Appl.Phys. 51 (1980) No.1,pp.784-790

22 T.J.Drummond et al.: Appl.Phys.Lett. 41 (1982) No.3,pp.277-279

23 K.Lehovec,R.Zuleeg: IEEE ED-27 (1980) No.6,pp.1074-1091

24 R.C.Eden et al.: IEEE Journ. SC-14 (1979) No.2,pp.221-239

25 C.P.Lee et al.: IEEE GaAs IC Symp. 1983,pp.162-165

26 R.A.Kiehl et al.: IEEE EDL-3 (1982) No.11,pp. 325-326

27 D.Delagebeaudeuf,N.T.Linh: El.Lett. 18 (1982) No.12,pp.510-512

Advantages of Multiple Quantum Wells with Abrupt Interfaces for Light-Emitting Devices

D. Bimberg and J. Christen

Institut für Festkörperphysik I, Technische Universität Berlin
Straße des 17. Juni 135, D-1000 Berlin 12

A. Steckenborn

Heinrich-Hertz-Institut für Nachrichtentechnik Berlin GmbH
Einsteinufer 37, D-1000 Berlin 10

Continuous excitation, time resolved and time delayed cathodo- and photoluminescence experiments prove that radiative recombination from narrow GaAs (and $In_{0.53}Ga_{0.47}As$) quantum wells (QWs) is of excitonic character at temperatures up to room temperature. The radiative recombination rate is strongly enhanced as compared to 3-dimensional material of the same quality and carrier capture by impurities is suppressed. Charge transfer from narrow barriers to wells occurs quasiballistically. The wavelength of emission from QWs is tunable with decreasing well thickness up 1.4 - 1.5 of the bandgap of the 3D material. All these properties lead to a novel generation of light-emitting devices having properties superior to classical ones. Examples of such devices based on GaAs are given. The importance of interface quality is emphasized in connection with InP based structures, which show much less improvement.

1. Introduction

Following the pioneering work of ESAKI and TSU [1] and DINGLE [2] and coworkers many fascinating discoveries were made in the field of physics of low-dimensional semiconducting systems during the last few years. Some of these discoveries open new insight on the fundamental structure of matter [3] and are thus of large value by their own right. Most of this work would not have been possible, had not new technologies like molecular beam epitaxy MBE, metal-organic chemical vapor deposition MOCVD and microstructuring of III-V and Si-based heterostructures would have been emerged and tremendously progressed during the past 15 years. In particular the control of interface abruptness and chemical composition turned out to be of decisive importance for all progress. G.WEIMANN, M.RAZEGHI, K. PLOOG and M.W.JONES [4] give some impression of the difficulty and importance of such work in other parts of this volume. Device physicists and electrical engineers supported strongly the development of the new technologies, anticipating that novel or at least improved devices might result. There exist indeed a number of fundamental differences of the electronic properties of three-and two-dimensional structures. Furthermore there exists now experimental evidence that a number of devices based on, e.g., III-V heterostructures or multiple quantum wells showing a two-dimensional gas of charge carriers have properties different or superior to devices of a similar type based on three-dimensional structures. These new properties comprise:

- Wavelength tunability as a function of the thickness L_z of the quantum well
- simpler technology, since quaternary materials containing phosphorus for optoelectronic applications in the 1.2-1.6 µm range can be avoided
- larger gain and differential gain of lasers
- lower laser threshold currents
- larger injection efficiency and thus larger total efficiency of the devices
- larger modulation band width of light-emitting diodes.

After a brief survey of basic electronic properties in quantum wells, which is given in section 2 of this paper, we will concentrate in section 3 on recent results of photo- and cathodoluminescence experiments in quantum wells, which nicely illuminate a number of fundamental differences between two- and three-dimensional materials, relevant for light-emitting devices. In section 4 we will briefly discuss the speed and efficiency by which transfer of charge carriers occurs from barriers to wells. Most recent results of time-resolved luminescence experiments are summarized in section 5. A few novel QW devices which indeed show properties superior to classical three-dimensional ones are presented in section 6.

2. Basic Electronic Properties of Quantum Wells: Band Structure, Excitons, Shallow Impurities

Let us assume that the thickness L_z of the quantum well is smaller than the de Broglie wavelength and that the thickness of the potential barrier is much larger than the tunnel length of the particle involved. Then any multiple quantum well problem reduces to the problem of a single quantum well or quantum mechanically speaking to the problem of a particle in the box. If we assume that the potential is infinitely high, it is particularly easy to solve the 1-dimensional Schrödinger equation [1,2]. The eigenenergies are

$$E_n = (\hbar^2/2\,m^*) \cdot (\pi n/L_z)^2 \qquad (1)$$

for n = 1, 2, 3, ... , m* is the effective mass.

Thus the energies of the electrons and holes depend on a geometrical factor L_z. The energy dispersion is split into subbands forming series for the different types of quasiparticles. Furthermore the degeneracy of the Γ_8 valence band is lifted, light holes and heavy holes form series which are, however, not completely independent of each other. The $n \geq 2$ states interact [5]. In reality the height of the potential barrier is rather small (typically 400 - 500 meV for a $GaAs/Ga_{0.6}Al_{0.4}As$ quantum well) and a transcendental equation [1, 2] has to be solved to determine the energy eigenvalues. Figure 1 shows solutions of this equation for a typical MQW structure with parameters identical to those of a sample used for the experiments presented later. There exist three bound states for the electrons and two for the light and heavy holes, respectively.

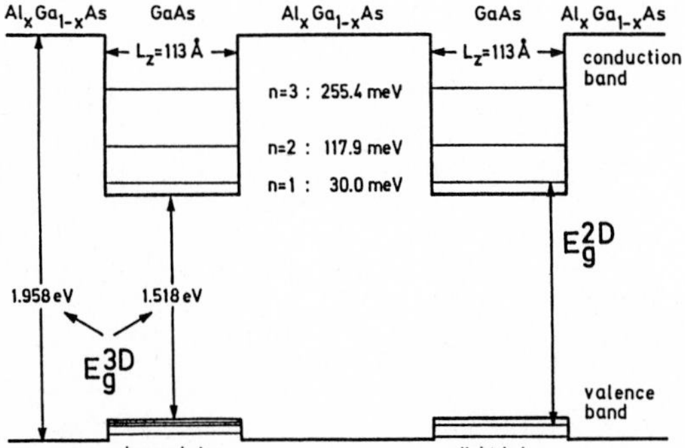

Fig. 1
Energy levels in GaAs quantum well for a $Al_xGa_{1-x}As$ barrier with $x \cong 0.4$. It is assumed that 85 % of the band gap difference occurs in the conduction band

The energy difference between the n = 1 electron subband minimum and the n = 1 (heavy) hole subband maximum define a new L_z-dependent and barrier height dependent band gap $E_g(2D)$, which is larger than $E_g(3D)$. Using finite barrier heights we determined E_g as a function of L_z for GaAs quantum wells and $Al_{0.4}Ga_{0.6}As$ barriers, assuming that 85% of the band gap discontinuity occurs in a conduction band (see Fig. 2). Similarly E_g for $In_{0.53}Ga_{0.47}As$ quantum wells with InP barriers is depicted in Fig. 2, assuming that 50% of the discontinuity occurs in the conduction band. All values are given for 4 K. The energy shifts become very large for L_z < 150 Å and the band gap of, e.g., $In_{0.53}Ga_{0.47}As/InP$ almost doubles. Technological progress has indeed made it possible to verify these predictions. The GaAs band gap has been reported to shift at RT from 860 nm to 650 nm in the red [6] and most recently Welch et al. [7] observed luminescence of large intensity at 4 K at 967 nm from a 15 Å wide SQW of $In_{0.53}Ga_{0.47}As$ surrounded by $Al_{0.48}In_{0.52}As$ barriers. A particularly pretty experimental verification of the shift of the luminescence is shown in Fig. 3 [8]. $In_{0.53}Ga_{0.47}As$ quantum wells of width 25 Å, 50 Å, 100 Å and 200 Å were grown in one sample isolated by 500 Å InP barriers.

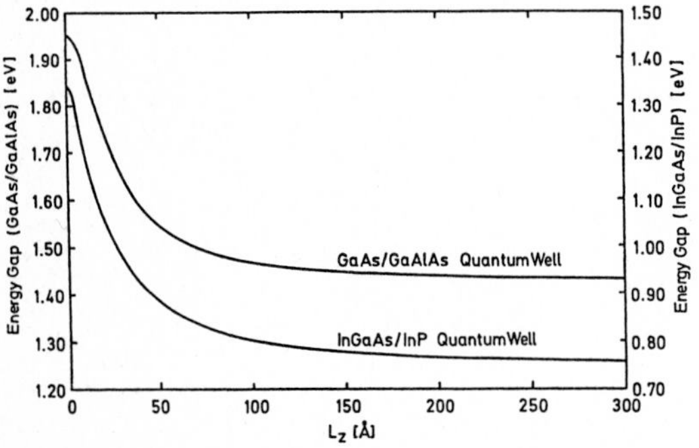

Fig. 2
L_z dependence of the 2D band gap for $GaAs/Ga_{0.6}Al_{0.4}As$ (left vertical scale) $In_{0.53}Ga_{0.47}As/InP$ (right vertical scale) quantum wells

Thus structural localisation of charge carriers leads indeed to such large shifts of the band gap that the luminescence of the binary material GaAs extends far into the visible red part of the spectrum without that any Al has to be introduced into the active layer, which usually reduces the efficiency. The luminescence of $In_{0.53}Ga_{0.47}As$ quantum wells cover the whole range of the spectrum from 1.65 µm up to 1 µm, which is the range where optical fibers have the highest transmission and the lowest dispersion. The addition of phosphorus, the use of quaternary $In_xGa_{1-x}As_yP_{1-y}$ crystals difficult to grow by MBE because of the [P]/ [As] partial pressure ratio which is very critical to control, can be avoided.

The luminescence of quantum wells is still at room temperature excitonic of nature [9, 10]. The Bohr radius a_B of the free exciton is 145 Å in GaAs [11]. Thus it is intuitively clear that starting at L_z values of the order 300-400 Å the three-dimensional exciton will not fit in the quantum well any more. By further reducing L_z the shape of the exciton will become pancake like. The radius in the L_z direction is forced to decrease strongly followed by a smaller decrease in the perpendicular direction. This situation resembles very much the situation of the exciton in high magnetic fields [11] where the magnetic field dependent cyclotron radius presents an upper limit for the extension of the excitonic wave function perpendicular to the magnetic field. There the exciton deforms to a prolate (baseball-like) shape.

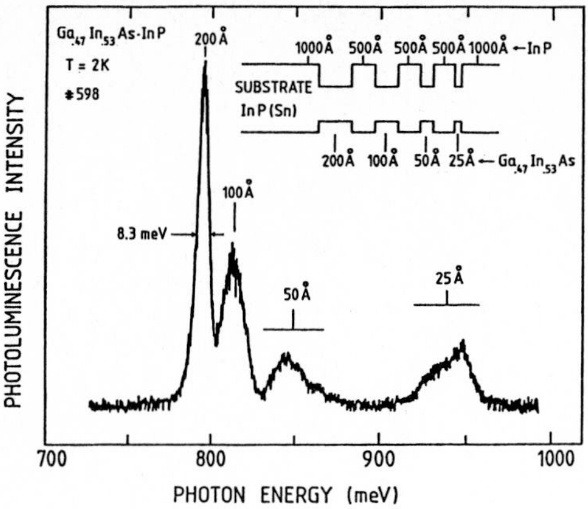

Fig. 3
Luminescence spectrum of $Ga_{0.47}In_{0.53}As$-InP sample shown on the top right. The well widths are 25 Å - 200 Å

Decreasing L_z the binding energy increases, approaching theoretically four times the 3D effective Rydberg in the limiting case of the double interface [12]. The transition probability increases and the lifetime [13, 14] decreases with the square of b, the "radius" of the oblate exciton. Bastard [12] and Miller and coworkers [15] calculate explicitly the anisotropic extension of the exciton wave function as a function of L_z/a_B. Using their results we estimate a decrease of the lifetime by $\approx$ 12 between the 3D case and a monolayer.

The wave functions and binding energies of impurities are also strongly influenced by increased localisation [16, 17] . The binding energy of a shallow impurity in the center of the well increases up to a factor 4, similar to the exciton. Beyond that, however, the binding energy also depends on the position of the impurity inside the well. The binding energy decreases for impurities which are closer to the potential barrier, since the potential mixes in higher order wavefunctions. Thus there is a pronounced difference in binding energy of an interface impurity and a center well impurity. This difference increases with well width. Consequently the impurities form an energy band. The density of states function shows two pronounced maxima at the impurity positions z_i = 0 (center of well) and $z_i = \pm L_z/2$ (interface) which converge to δ functions for $L_z \rightarrow 0, \infty$. Bastard [16] calculated the theoretical lineshape of the free electron- neutral acceptor luminescence assuming equal occupation probability of the different possible sites throughout the well. For small Fermi energies he finds a strongly dominating central acceptor peak and a much weaker interface acceptor peak, in contrast to experiment as we shall see.

3. Radiative Recombination Processes in Quantum Wells

The radiative recombination in direct gap semiconductors is strongly dominated by impurity related processes like pair recombination, free to bound ((e, A°) or (h, D°)) recombination and bound excitons after moderate to low excitation or injection. Intrinsic excitonic recombination contributes very little to the emission. Figure 4a visualizes this situation nicely for an ultrapure bulk GaAs sample ([p] = $1 \times 10^{15} cm^{-3}$) grown by LPE. The 5 K cathodoluminescence spectrum is shown on a logarithmic scale with moderate spectral resolution.

The spectrum of a nominally undoped ([n] = 2×10^{14} cm^{-3}) 52 Å GaAs MQW system surrounded by 176 Å $Al_{0.42}Ga_{0.58}As$ barriers taken under similar con-

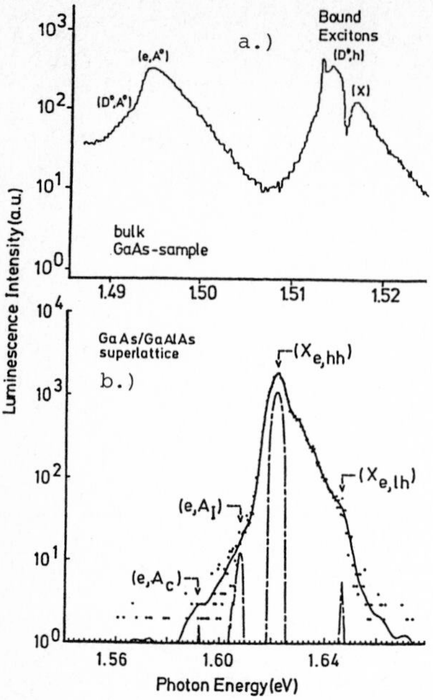

Fig. 4
(a) 5K cathodoluminescence of a pure GaAs LPE layer. (b) 5K cathodoluminescence of $L_z = 52$ Å GaAs quantum wells taken under similar excitation conditions as a)

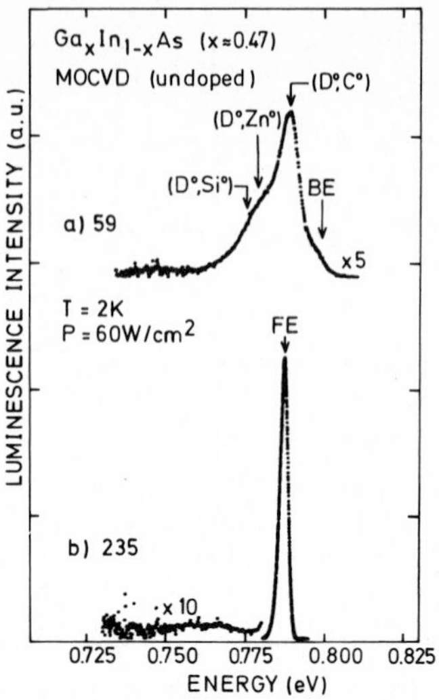

Fig. 5
(a) 2K photoluminescence from a pure $In_{0.53}Ga_{0.47}As/InP$ sample showing no 2 DEG. (b) 2K photoluminescence from a sample of similar thickness and composition but more perfect interface exhibiting a 2 DEG. The luminescence is dominated here by the (e,hh) exciton [18]

ditions as the spectrum of the bulk sample is shown in Fig. 4b. The much more intense recombination is strongly dominated by the heavy hole exciton. At ≈ 24 meV higher energy the light hole exciton is identified. The recombination radiation of free electrons with the interface acceptor A_I^o and the central acceptor A_C^o is approximately 2.5 orders of magnitude less intense than the dominating excitonic process. Apparently acceptors are preferentially accumulated at the interface in high purity material, since the (e, A_I^o) intensity is by far larger than the (e, A_C^o) intensity. Theory - assuming equidistribution of impurities on the available sites - predicts the opposite [16]. Similar results are obtained for $In_{0.53}Ga_{0.47}As/InP$ heterojunctions: Fig. 5a shows a typical luminescence spectrum of a pure MOCVD sample having a mobility $\mu = 8600\ cm^2/Vs$ at RT and $\mu = 33000 cm^2/Vs$ at 77 K. The different impurity related features are identified in the figure. In complete contrast Fig. 5b shows only one extremely intense recombination ascribed to the (e, hh) exciton [18]. The MOCVD sample has a slightly smaller impurity content ($\mu_{300K} = 12000\ cm^2/Vs$, $\mu_{77K} = 55000\ cm^2/Vs$, $\mu_{2K} = 90000 cm^2/Vs$) but a much better interface. The mobility results and de Haas-Shubnikov experiments show clearly the presence of a 2D electron gas at the interface. Apparently all electrons are dragged so rapidly to the interface that the recombination takes place only there and no recombination occurs in the bulk of the material. Capture of carriers by impurities is apparently bypassed by the more rapid excitonic recombination process. Figure 6 shows that these conclusions also hold for moderately doped samples and room temperature. Cathodoluminescence spectra

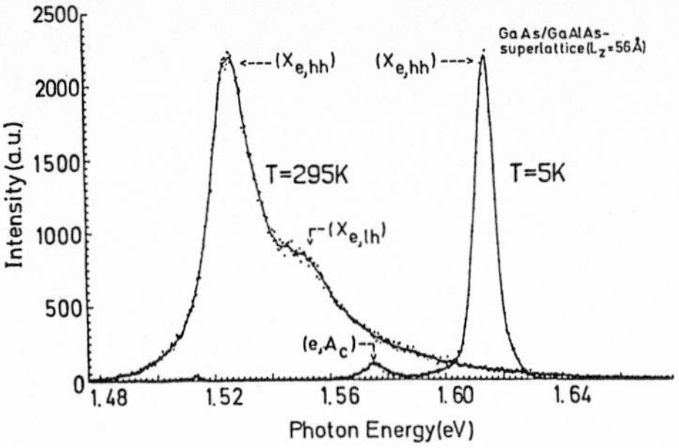

Fig. 6
Comparison of the 5 K and 295 K cathodoluminescence spectra of a Be-doped [p] = 1 x 10^{16} cm^{-3}) L_z = 54.6 Å GaAs MQW sample. Both spectra are dominated by excitons

from Be-doped, [p] = $10^{16}cm^{-3}$, L_z = 54.6 Å GaAs QW's taken at 5 K and 295 K are compared with each other. Excitonic recombination indeed dominates also at RT.

4. Charge Carrier Transfer from the Barriers to the Well

Charge carriers of either type can transfer from barriers to wells. Without such a transfer modulation doping would have no effect and the two-dimensional electron gas FET would not exist. The velocity of charge carrier transfer, the time in which it occurs and the transfer efficiency are of fundamental importance but remained unknown until recently [13]. Results like that displayed in Figure 5b indicate that still in this 0.7 µm thick layer the transfer must occur with high speed and efficiency, otherwise impurity induced recombination lines from the bulk of the sample would be observed. To become quantitative we shall discuss now Fig. 7,which shows a typical luminescence spectrum of a MQW system in the energy interval 1.5 - 2.1 eV. The MQW system consists of a GaAs s.i. substrate on top of which 0.5 µm thick GaAs and 0.13 µm thick $Ga_{0.58}Al_{0.42}As$ buffer layers are grown. On top of that there are 60 alternating $Ga_{0.58}Al_{0.42}As$/GaAs barriers and wells, 176 Å and 54.8 Å thick, respectively. 30 keV electrons having a projected range of 3.5 µm are used for excitation. The intensity of the luminescence lines stemming from GaAlAs ((N°, X) recombination) is here 2 x 10^3 lower than the intensity of the lines from the GaAs QWs, although the number of generated e-h pairs in the GaAlAs barriers is a factor of 4.0 larger than in the GaAs layers, assuming a composition and depth-independent generation rate and taking into account the ratio of layer thickness d^{GaAlAs}/d^{GaAs}. The intensity ratio of well and barrier luminescence is:

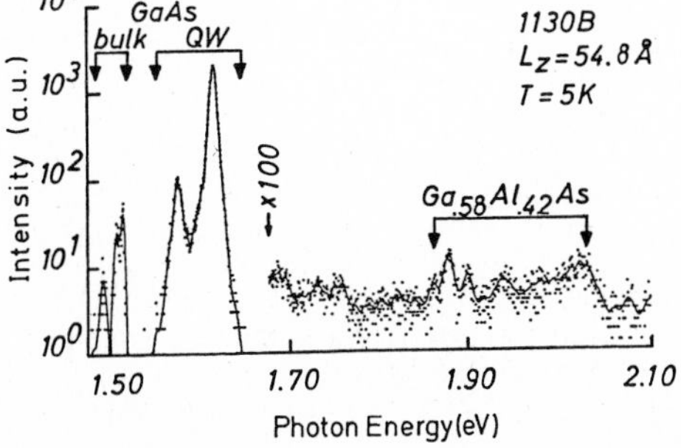

Fig. 7
Low temperature luminescence spectrum of a Be-doped MQW sample (|p|≈ 10^{16} cm^{-3}) with L_z = 54.6 Å in a large photon energy range. Different groups of lines emitted from bulk GaAs, the GaAs MQWs and the GaAlAs are depicted [13]

$$I^{GaAs}/I^{GaAlAs} = (\tau_r^{GaAlAs}/\tau_{tot}^{GaAlAs}) \cdot (d^{GaAs}/d^{GaAlAs}) \qquad (2)$$

assuming a unity quantum efficiency in the GaAs layer. τ_r and τ_{tot} are the radiative and effective lifetimes respectively. Assuming $\tau_r^{GaAlAs} = 10^{-9}$ s a $\tau_{tot}^{GaAlAs} = 1.5 \times 10^{-13}$ s is derived. This value represents the average sweep out time of the layer. The average LO-phonon emission time of a hot electron (hole) in GaAs is exactly of the same magnitude: 1.3×10^{-13} s (1.8×10^{-13} s) [19]. For $Ga_{0.58}Al_{0.42}As$ a very similar value is expected. Thus the charge carriers are swept out of such narrow layers ballistically or quasiballistically which is probably not too surprising given the short distance. One might expect that the sweep out from thicker layers is at least much slower or that no sweep out at all occurs. Surprisingly, recent results by Göbel et al. [14] indicate again a strong carrier sweep out from very thick layers with a speed of $\approx 1.5 \times 10^6$ cm/s. Here a 50 nm SQW of GaAs is sandwiched in between two 1 µm thick AlGaAs layers. Excitation source is a mode-locked dye laser emitting 4.7 ps pulses at 575 nm which are completely absorbed in the first 0.5 µm of the cap layer. Naively one would expect no luminescence from the QW since no charge carriers are directly excited there and its thickness is comparatively negligible. Nevertheless $I^{GaAs}/I^{GaAlAs} \approx 0.2$! Streak camera images of the different luminescence bands show that the carriers are swept out of the top layers within 50 ps. The results of Christen et al. [13] indicate a transfer or injection efficiency from thin barriers into QWs of 100 %. The physical mechanism behind the rapid transfer is not yet clear. Driving forces might be built-in electric fields and/or rapid plasma expansion.

5. Time-Resolved Cathodoluminescence

30 keV electron pulses with rise and decay times $\leq$ 200 ps are used as a source of excitation for the experiments described below. Time-delayed spectra of the luminescence and life times are measured after a long pulse (100 ns - 1 µs) is abruptly switched off. Thus the decay from a clearly defined quasi-equilibrium state is monitored in contrast to most time resolved photoluminescence experiments, where the excited state never reaches a quasi-equilibrium because of the shortness of the pulses used.

Fig. 8 shows a typical luminescence spectrum of the same pure sample of bulk GaAs as in Fig. 4: 0, 20 and 50 ns after excitation. The originally dominat-

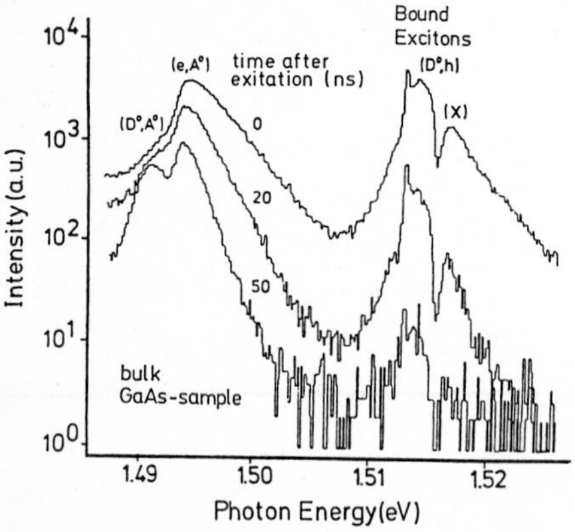

Fig. 8
Time delayed spectra of a high putity $[p] = 1 \times 10^{15}$ cm $^{-3}$ bulk GaAs sample [20]

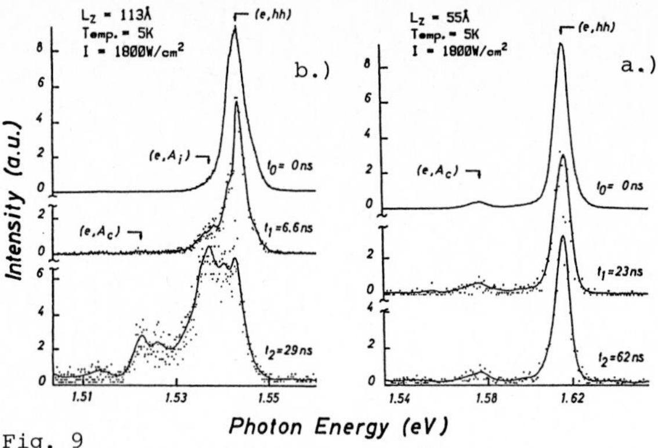

Fig. 9
Time delayed spectra from a 113 Å undoped (a) and a 54.8 Å Be-doped (b) ([p] ≈ 1 x 10^{16} cm^{-3}) MQW system

ing free and bound excitons disappear rapidly, whereas the pair band (D°, A°) and the free to bound (e, A°) transition sustains for rather long times [20]. Thus the luminescence changes character completely with increasing time and shifts to lower energy, displaying nicely the sequence of different steps of carrier capture and recombination. This result has to be contrasted with the time-delayed spectra taken from a 113 Å undoped GaAs MQW system and a Be-doped (p ≈ 1 x 10^{16} cm^{-3}) 54.8 Å GaAs MQW system shown in Fig. 9. Although the 54.8 Å layer is doped the spectrum does not change character as long as luminescence light can be observed. Electrons and holes apparently have the same lifetime, no preferential trapping of holes by impurities occurs. Excitonic recombination is by far the dominating recombination channel not only at quasiequilibrium (Fig. 4) but at any time after the excitation is switched off; the impurities are bypassed. The time delayed spectra from the 113 Å MQW system show a behavior which is in between the narrow QW and the bulk material. At zero and at small delay times excitonic recombination dominates, whereas at long times acceptor-induced features dominate.

Apparently the lifetime of the excitons becomes shorter with decreasing L_z, whereas the carrier capture times do not shorten similarly. The decay times of the X (e, hh) peak shown in Fig. 10 for two undoped and one doped samples at RT and at 5 K support this conclusion. The initial fast decay in the L_z = 112.8 Å QW occurs at 5 K in 0.8 ns. This value is lower than the value of ≈ 2 ns found for similarly doped bulk p-type GaAs. The initial decay at 5 K in a L_z = 51.6 Å QW occurs still faster in 0.5 ns. The room temperature decay times are all longer by the same factor 8. The strictly exponential decay at RT presents clear support of the conclusion that the luminescence at RT is of excitonic origin. The ratio τ (113 Å) / τ(52 Å) = 1.8 is in excellent agreement with a theoretical value of 1.8 we predict on the basis of numerical calculations of the L_z-dependent excitonic radii of Bastard et al. [12]. These and other experiments prove unamboguously that the excitonic recombination rate is strongly enhanced due to carrier confinement in the quantum wells. The excitonic recombination rate of narrow QWs is more than one order of magnitude larger than the rate of similarly doped bulk material. The 300 K decay times show also that p doping has an indirect influence on it. The decay time of the doped L_z = 56 Å sample is shorter than the decay time of the undoped L_z = 52 Å sample. More excitonic channels participate in the recombination process at RT since more hole subbands are occupied. A detailed analysis of the decay times gives also information on intersubband relaxation times [10].

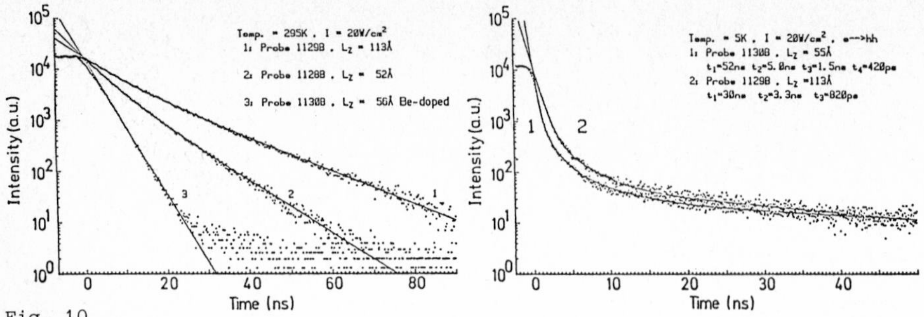

Fig. 10
Decay times of 3 different MQW systems with L_z = 113 Å (undoped), 52 Å (undoped) and 55 Å (Be doped) at 5 K and 295 K

6. Advantages of Light-Emitting Devices Based on Quantum Wells

The shift of the photon energy of the emission from QWs to much higher energies than given by the band gap of the 3D material presents for the GaAs and the InP based groups of materials a breakthrough. The wavelength range of GaAs-based light-emitting devices which have the highest energy conversion efficiency of all known such devices can be extended into the visible red part of the spectrum below 700 nm. Thus GaAs QW LEDs might one day replace the GaP: (Zn, O) or the GaAlAs LEDs. An immediate application of QW lasers might be the optical disc or optical data storage. The wavelength range of InGaAs/InP or InGaAs/InAlAs/InP based QW light-emitting devices might be extended to or below 1 μm, thus avoiding a number of technological problems connected with the use of phosphorus in the growth of quaternary materials like $In_xGa_{1-x}As_yP_{1-y}$ by MBE.

Holonyak et al. [21] were the first to report drastically lowered threshold current densities of GaAs MQW lasers as compared to classical double heterostructure ones following the pioneering work by v.d. Ziel and Dingle et al. [22]. Fig. 11 compares schematically a conventional laser and a MQW laser [23]. The lowest j_{th} values of a DH laser reported up to now are ≈ 500 A/cm² whereas Tsang found 160 A/cm² for his MQW structure. Hersee et al. [24] found a still lower value of j_{th} = 121 A/cm² for a GaAs-GRIN-SCH-SQW laser (graded refractive index separate confinement heterostructure single quantum well)

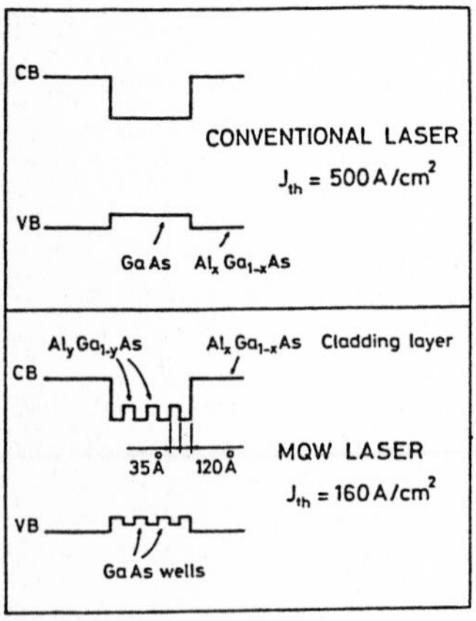

Fig. 11
Schematic band diagram of a MQW laser (after Tsang [23]) and of a DH laser. Note the improvement in threshold current density of the MQW laser

shown in Fig. 12. The threshold current is only 3 mA for a 250 μm long cavity. These authors [21 - 24] concluded that MQW systems must be from a fundamentally physical point of view superior to 3-dimensional systems. We believe that the enhanced radiative transition probability reported above and the subsequently reduced capture of carriers by impurities and defects in QWs present the basic explanation of the greatly improved device performances. A rough theoretical estimate of j_{th} (MQW) / j_{th} (DH) for GaAs gives a reduction to 28 % of the 3-dimensional value. A rate equation model [25] including bimolecular recombination, known GaAs materials parameters and the assumption of an order of magnitude increase of the rate constant of the lasing mode,but no other change of the input parameters,leads to this number which is rather close to the experimentally observed one. Chin, Holonyak et al. [26] also reported a much decreased sensitivity of j_{th} to temperature changes (T_o = 400 K instead of 240 K for DH lasers [27] . Recent theories seem to be able to explain also this observation as a result of carrier localisation [28].

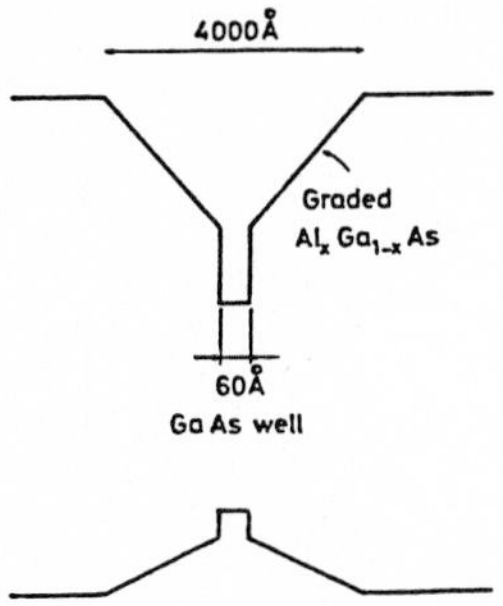

Fig. 12
Schematic band diagram of a GRINCH-SCH-SQW laser (after Hersee et al. [24]) j_{th} = 121 A/cm^2 was reported for this laser type

The gain g of a laser is inversely proportional to the spontaneous radiative lifetime τ_s. Thus it should improve by the same amount by which τ_s decreases. The results and discussion of the transfer of charge carriers from barriers to wells presented above also show that it is as close to ideal as it can be. Finally it should be mentioned that the high-frequency cut off f_c of a light-emitting diode is given by $1/(2\pi\tau_s)$.Thus an order of magnitude improvement of the maximum modulation frequency of a QW LED as compared to a normal LED seems to be feasible.

These final remarks are certainly not exhaustive, but they show that there exists evidence based on the properties of actually existing QW light-emitting devices and on the fundamental physical understanding of devices and QWs which seems to indicate that QWs have advantages from many different points of view.

Much remains still to be done in this rather young field. In particular, as far as InP based lasers are concerned no such dramatic improvements have been reported as yet in the scarce literature which exists. The InP technology, however, has not yet reached the same maturity as the GaAs technology and a basic understanding of some fundamental parameters of these lasers like the temperature dependence of the threshold current (T_o problem) is still missing. Further, judging from luminescence experiments [8] the interface quality of these structures seems to be still much inferior to the GaAs-based structures and this might be the main field where technological improvements could cause an impact on device properties.

1 L. Esaki and R. Tsu: IBM J.Res. Dev. 14, 61 (1970)
2 R. Dingle: Advances in Solid State Physics 15 (ed.: H.J.Queisser), Pergamon-Vieweg, Braunschweig, 1975, p. 21
3 D.C. Tsui, H.L. Störmer and A.C. Gossard: Phys.Rev.Lett. 48, 1559 (1982)
4 see the papers by G.Weimann, M.Razeghi, K.Ploog and M.W.Jones in this volume
5 M.Altarelli: this volume
6 M.D.Camras, N.Holonyak Jr., K.Hess, J.J.Coleman, R.D.Burnham and D.R.Scifres: Appl.Phys.Lett. 41, 317 (1982)
7 D.F.Welch, G.W.Wicks and C.F.Eastman: Appl.Phys.Lett. 43, 762 (1983)
8 M.Razeghi, J.P.Hirtz, K.O.Ziemelis, C.Delalande, B.Etienne and M.Voos: Appl.Phys.Lett. 43, 762 (1983)
9 P.Dawson, G.Duggan, H.I.Ralph and K.Woddbridge: Phys.Rev.B 28,7381 (1983)
10 J.Christen, D.Bimberg, A.Steckenborn, G.Weimann: Verhandlg. DPG (VI)19, 214 (1984) and to be published
11 D.Bimberg: Advances in Solid State Physics (ed.: J.Treusch), Pergamon-Vieweg, Braunschweig, 1977, p. 195
12 G.Bastard, E.E.Mendez, L.L.Chang and L.Esaki: Phys.Rev. B 26, 1974 (1982)
13 J.Christen, A.Steckenborn, D.Bimberg and G.Weimann: Verhandlg. DPG VI,18, 668 (1983) and J.Christen, D.Bimberg, A.Steckenborn and G.Weimann: Appl. Phys. Lett. 44, 84 (1984)
14 E.O.Göbel, H.Jung, J.Kuhl and K.Ploog: Phys.Rev.Lett. 51, 1588 (1983)
15 R.C.Miller, D.A.Kleinmann, A.C.Gossard, O.Munteanu: Phys.Rev. B 25, 6545 (1982)
16 G.Bastard: Phys.Rev. B 24, 4714 (1981)
17 C.Mailhiot, Y.C.Chang, T.C.McGill: Phys.Rev. B 26, 4449 (1982)
18 D.Bimberg, K.H.Goetz and M.Razeghi: to be published
19 D. von der Linde and R.Lambrich: Phys.Rev.Lett. 42, 1090 (1979)
20 H.Münzel, D.Bimberg and A.Steckenborn: Physica 117 + 118 B, 214 (1983)
21 N.Holonyak, R.M.Kolbas, W.D. Laidig, B.A.Vojak, R.D.Dupuis and P.D. Dapkus: Appl. Phys. Lett. 33, 737 (1978)
22 J.P. Van der Ziel, R.Dingle, R.C.Miller, W.Wiegmann and W.A.Nordland: Appl.Phys. Lett. 26, 464 (1975)
23 W.T.Tsang: Appl. Phys. Lett. 39, 786 (1981)
24 S.D.Hersee, M.Baldy, P.Assenat, B.de Cremoux and J.P.Duchemin: Electron. Lett. 18, 870 (1982)
25 H.E.Schöll, D.Bimberg, H.Schumacher and P.T.Landsberg: IEEE J.Quantum Elect. in print
26 R.Chin, N.Holonyak, B.A.Vojak, K.Hess, R.D.Dupuis and P.D.Dapkus: Appl. Phys. Lett. 36, 19 (1980)
27 H.D.Wolff, K.Mettler, K.H.Zschauer: Jap. J. Appl.Phys. 20,L 693 (1981)
28 N.K.Dutta: J. Appl. Phys. 54, 1236 (1983)

MBE-A Tool for Fabricating IV-VI Compound Diode Lasers

K.-H. Bachem, P. Norton, and H. Preier

Fraunhofer-Institut für Physikalische Messtechnik, Heidenhofstraße 8
D-7800 Freiburg, Fed. Rep. of Germany

The molecular beam epitaxy (MBE) and the closely related hot wall epitaxy (HWE) techniques are excellent tools to fabricate lead chalcogenide heterostructures for device applications. Whenever exact compositions of a ternary or quaternary film are required, like in lasers, the MBE method is best suited. Lattice-matched PbTe-based and non-lattice-matched PbSe-based laser structures have been realized for the 4 to 12 μm wavelength range. While no significant difference exists between lattice-matched and non-lattice-matched devices, both kinds exhibit much larger operation temperatures than diffused homojunction diodes. The properties of an MBE system specially designed for the growth of lead chalcogenides are described in detail. Characteristic data of PbTe- and PbSe-based DH lasers are presented and compared with those of homojunction devices.

1. Introduction

Quite a variety of lead chalcogenide heterostructures have been fabricated by molecular beam epitaxy and the related technique of hot wall epitaxy. The main purpose for growing these heterostructures was to prepare double heterostructure laser diodes /1 - 4/ with properties superior to those of diffused homojunction lasers. In addition, also detectors of the kind PbTe-PbSnTe /5/ and PbS-PbSSe /6/ have been fabricated using evoporation techniques.

Most of the laser structures realized to date have lattice-mismatched heterojunctions, like PbTe-PbSnTe-PbTe /1/ or PbS-PbSSe-PbS /2/. The properties of these devices were already superior to homojunction lasers in spite of the lattice mismatch. The most significant improvement was the maximum operation temperature which was raised up to 120 K from a value of about 80 K for homojunction lasers.

Recently lattice-matched double heterostructures with PbEuSeTe active and confinement layers as well as PbSnTe active layers and PbSnYbTe confinement layers were prepared successfully at GENERAL MOTORS (GM). The PbEuSeTe diodes have been operated at 4.06 μm up to 147 K cw, the highest operation temperature observed so far in lead chalcogenide lasers. With the lattice-matched PbSnTe diodes a cw tuning range from 10.3 μm (at 10 K) to 7.1 μm (at 128 K) has been realized. However, if the pulse properties are considered, no large differences between lattice-matched lead telluride and non-lattice-matched lead sulfide selenide lasers has been noticed. Apparentely lattice matching is less important in the second system. The higher cw operation temperatures obtained for telluride-based structures are not significant because the cw temperature is strongly affected by the properties of the contacts and bounding technique. This leads to the conclusion that not inherent laser properties but rather a superior contacting procedure leads to the high cw operation temperatures of the GM lasers. In this paper the MBE system used in our laboratory for the fabrication of lead chalcogenide heterostructures is described. Design and performance of the evaporators are discussed in detail, since both affect most critically the control of layer composition. The properties of PbSe-PbSnSe-PbSe and PbS-PbSSe-PbS lasers are discussed and compared with published data.

2. Evaporation Techniques

Deposition of lead chalcogenides including Sn-containing ternary or quaternary compounds is most straightforwardly accomplished by evaporation techniques. All lead and tin chalcogenides vaporize already at moderate temperatures /7/. Practical beam flux densities in the order of 10^{13} to 10^{15} $cm^{-2}s^{-1}$ are obtained at temperatures below 1100 K. The lead chalcogenides vaporize almost undissociated. Besides PbX(X = S, Se, Te) no other compounds are found in the gas phase and congruent subliming compositions exist within the stability region of the solids. The vaporization behavior of the tin chalcogenides is more complex, their vapor consists of X_2, SnX, SnX_2, Sn_2X_3 and Sn_3X_4, but SnX is the major constituent. SnTe, and SnSe, evaporate incongruently. Therefore, the source composition shifts into the two-phase region. Complications do not arise from this complexity because a two-phase source provides vapor pressures just as well defined as a congruently subliming one-phase source. Alternatively to compound source materials, elemental source materials have been successfully employed /8/. However, this approach has been abandoned since it was found that closer stoichiometry control and therefore lower carrier concentrations are achievable by growth from sources of the binary compounds /8/. Only in cases where elemental source materials are inevitable are these used in combination with binary compound source materials. For example, in order to deposit ternary lead chalcogenides containing rare earth or earth alkali elements, it is neccessary to use elemental source materials because the chalcogenides of these elements cannot be evaporated at practical temperatures due to their extreme thermal stability.

Assuming binary source materials are used and these materials are evaporated from thermocouple-controlled Knudsen cells, the beam flux stability can be derived from the temperature dependence of the vapor pressures of the binaries /7/. Considering that the evaporators are operated at a temperature of 1000 K and taking into account that a state-of-the-art thermocouple controlled feedback system provides a relative temperature accuracy of 0.2 to 0.5 K, it can be estimated that the vapor pressures of lead and tin chalcogenides are accurate within 0.2 to 0.5 percent inside the Knudsen cells. Thus, the composition of a molecular beam combined from the beams of Knudsen cells would be accurate within approximately 0.4 to 1 percent. The same kind of precision is expectable for the composition of ternary films for constant deposition temperature. Considering the most critical case of a $PbSe_{1-x}Te_x$ film to be lattice matched to a given substrate, it follows that lattice matching down to a precision of 5×10^{4} should be achievable. An even better composition control is expected if the beam is not combined from two individually temperature-controlled Knudsen cells but the two cells are combined inside one temperature-controlled heater /5/. In this case temperature fluctuations affect the vapor pressures in both cells in the same direction. However, one has to consider that the higher stability of the beam composition is gained at the expense of flexibility.

The above presented estimations are based on vapor pressure calculations, the technical limitations resulting from the relative accuracy of the temperature control system, and assuming ideal Knudsen cells. In reality, the evaporation cell is a complex device whose behavior is critically dependent on its design and operation conditions. Differently from III-V compound deposition systems where open evaporators have to be used for obvious reasons, in lead salt deposition systems, Knudsen cell like evaporators are usually preferred. The vapor pressures of the lead and tin chalcogenides approach the mbar region already at convenient temperatures around 800 °C. Therefore, evaporation cells with millimeter sized orifices can be used. Such a cell provides a beam flux rate of the order of 10^{15} cm^{-2} s^{-1} at a distance of 15 cm, which is sufficient to obtain growth rates of a few µm per hour. Figure 1 shows a schematic view of the cells we are using for evaporation of binary compounds. The orifice has a diameter of about

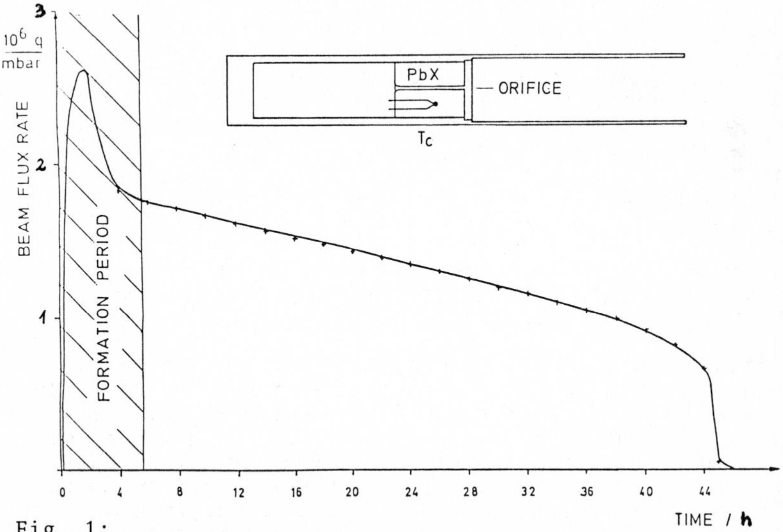

Fig. 1:
Dependence of beam flux rate on operation time for an evaporator shown schematically in the inset

1.5 mm. The temperature inside the cell is not uniform, but decreases from the bottom toward the open end. Therefore the source material, loaded with as crushed pieces, sublimes towards the cover where it forms a solid block with a channel in its axis. The thermocouple is fed through the bottom of the cell and its tip is placed where the block is formed. The vapor pressure in the cell corresponds to the temperature of the open surface of the block because this surface is the hottest part of the block. Fig. 1 shows the beam flux rate obtained from such a cell in dependence of time for uninterrupted operation and unchanged set point of the temperature controller. The beam flux rate has been measured at the substrate position with standard ion gauge. The conversion factor between pressure reading and beam flux rate has not been determined. Therefore the beam flux rates are given in pressure units. For orientation, a PbSe beam flux rate corresponding to 3×10^{-6} mbar measured at the substrate position yields a growth rate of 11 Å per second for a deposition temperature of 350 °C, and for a combined beam of PbSe and SnSe with rates corresponding to 3×10^{-6} mbar and 2×10^{-7} mbar, respectively, yields a ternary film with a Sn content of approximately 10 % (atomic fraction). The dependence of beam flux rate on time as shown in Fig. 1 is typical. For the first few hours of operation of a freshly loaded cell the beam flux is not controllable. In this period the material sublimes into the colder part of the cell thereby forming a solid block. When sublimation is completed the flux stabilizes and becomes precisely controllable. The steady decrease of the beam flux rate in dependence on operation time results from the axial temperature gradient in the cell. Slightly modified cells are employed for evaporation of Se and Bi selenide. These species are used for control of stoichiometry and n-type doping. For these purposes only small beam flux rates, typically in the range of 10^{-9} to 10^{-8} mbar, have to be provided. Thus only small sized low temperature operated cells are required. The low operation temperatures, typically 300 and 600 °C, request for a special heater design because radiation coupling between heater, cell and thermocouple is weak in the low temperature regime. With appropriately designed cells, a beam flux stability at the above-mentioned level of better than 20 % is obtained. This accuracy is sufficient for the desired purposes.

The SIMS profile of a four-layer PbSnSe structure is shown in Fig. 2. The PbSe flux rate was $1{,}5 \times 10^{-6}$ mbar, the SnSe flux rate was increased in steps. The ratio of the flux rates q(SnSe)/q(PbSe) is indicated in the diagram. The deposition was interrupted for 15 minutes between adjacent films in order to reach steady-state source conditions for the SnSe cell. The data have not been smoothed, but even without sophisticated data processing procedures it can be seen that the composition in each film is reasonably constant. The total thickness of the multilayer is approximately 1 µm. Therefore, the interface regions are broadened due to sputter effects. Nevertheless, the transition between the first and second film is still abrupt although the deposition temperature is 400 °C. The laser structures are usually grown at lower temperatures (350 - 375 °C). Multiquantum well structures should be obtainable with relative ease. The ion rate axis and the sputter time

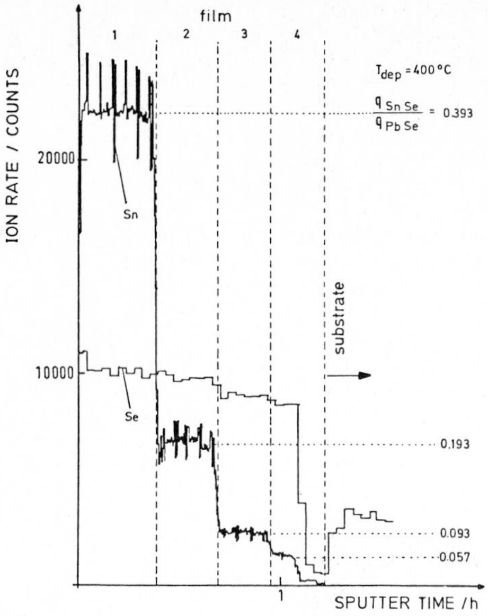

Fig. 2:
SIMS profile of a layer structure consisting of $Pb_{1-x}Sn_xSe$ films with 4 different compositions deposited at a temperature of 400 °C

axis do not correspond linearly to composition and depth, respectively, because of SIMS typical matrix effects. For orientation, the third layer has a tin content of 6.8 ± 0.1 %. Its count rates agreed precisely with those of a calibration standard of the composition $Pb_{0.932}Sn_{0.068}Se$.

The substrate holder is uncritical in lead salt deposition systems because of the typical low deposition temperatures (350 - 400 °C) and small substrate areas (1 - 2 cm^2). The commercially available substrate station from VARIAN, originally designed for depositon of III-V compounds, is well suited.

The PbSnSe and PbSSe laser structures, being described later, have been grown under background pressures ranging from 5 x 10^{-10} mbar up to 3 x 10^{-9} mbar. PARTIN /3, 4/ has deposited PbSnTe laser structures under apparently identical vacuum conditions. These vacuum conditions are standard for an UHV system obtained after backout at medium temperatures (≃ 150 °C). It is important to notice that, up to now, no one has proven that such low background pressures are neccesary for deposition of laser active materials in an MBE system. In HWE systems PbSSe laser structures with excellent performance /2/ have been grown under vacuum conditions of about 10^{-7} mbar, far beyond UHV standard. It has been claimed that, in contrast to the HWE technique, the MBE technique is less tolerable as far as background pressure is concerned. However, experimental proof to support this judgement has not been presented yet as far as we know. Independently of the outcome of appropriate experiments, an MBE system for deposition of lead chalcogenide based compounds should meet UHV standards. Lead chalcogenide compounds containing rare earth and earth alkali elements will most likely require UHV conditions and these compounds are the ones to close the gap in the laser emission spectra between the more classical lead tin chalcogenide lasers and the III-V compound lasers.

PbSe and PbS films grown from binary source materials which have been prepared by the procedure described above are always n-conducting (3×10^{17} - 1×10^{18} cm^{-3}). The stoichiometry of the starting material has no significant effect on the carrier concentration. Carrier concentration and type of conductivity of the deposited films can be controlled by adjusting the deviation of stoichiometry via coevaporation of Se or by intentionally doping with foreign elements.For telluride rich systems, doping with Tl and Bi is the common procedure for growing p- and n-type films, respectively /9/. We prefer a combination of both techniques. Bi is employed for growing n-type films and p-type conductivity is achieved by deposition under Se-rich conditions. Doping experiments have shown that Tl is not incorporated as acceptor if coevaporated with PbS and PbSe under typical metal-rich deposition conditions in the MBE system. Films covered with Tl droplets formed during deposition in high Tl flux remained n-conducting. This finding is contrary to results of doping experiments in PbTe /9/. In this material Tl forms an acceptor-like center in chalcogen-rich as well as metal-rich matrix. Contrary to Tl, Bi is a well-behaving efficient n-type dopant in PbS, PbSe, $PbS_{1-x}Se_x$ and $Pb_{1-x}Sn_xSe$. Electron concentrations up to 10^{20} cm^{-3} have been obtained in PbSe at a deposition temperature of 400 °C and a PbSe flux rate of 3×10^{-6} mbar, a Bi rate of 10^{-7} mbar and a Se rate of 10^{-7} mbar. Carrier concentrations in the range of 2 - 5×10^{18} cm^{-3} require Bi fluxes of 10^{-9} mbar. This is the typical doping level of the active regions of our laser structures.

3. Laser structures

The work on lead salt laser structures is currently concentrated on lead telluride and lead selenide based systems.

At GENERAL MOTORS several attempts have been made to realize short-wavelength lasers selecting quaternary telluride systems whose band gaps are larger than the lead telluride band gap. The experimental results on lead germanium and lead ytterbium tellurides showed that these systems are less attractive for laser fabrication. Ge-containing laser structures could not be prepared, because of the extraordinary small sticking coefficient of Ge. For Yb-containing materials a complex doping behavior has been reported. A DH laser structure emitting at about 10 µm with an active PbSnTe layer and PbSnYbTe confinement layers operated up to 128 K (cw). The maximum temperature for pulsed operation of this particular laser has not been reported. Although the reported cw temperature is the highest achieved for a 10 µm laser, the temperature exceeds the best values for diffused lasers operating at the same wavelength only slightly. For diffused PbSnSe lasers with Sn contents of 3 % and 7 % we have obtained maximum operation temperatures of 170 and 160 K, respectively, for pulsed operation and of 115 K for continuous operation.

Excellent results have been obtained with lead europium telluride lattice matched DH laser structures /4/. High operation temperatures up to 190 K for pulsed operation and 147 K for continuous operation have been reported.

Compared with the more sophisticated lattice-matched structures realized by GENERAL MOTORS, the DH laser fabricated by our group is rather simple. We have prepared three types of structures. Long-wavelength laser structures consisting of PbSe confinement layers and an active PbSnSe film. The structure is deposited on a p-type PbSe substrate. The first confinement layer and the active film is Se-doped (p-type) and the upper confinement layer is Bi-doped. The structure with the thinnest active film (0,5 μm) operated up to 190 K for pulsed operation (I_{th} = 5 A) and up to 125 K for continuous operation (I_{th} = 0,5 A). The tuning range was 9.1 to 11.9 μm for cw operation. In a series of structures with different thicknesses of the active layers (0,5, 1, 2, 3 μm) the maximum operation temperature drops steadily down to the temperature typically achieved with diffused lasers (150 K). In Fig. 3 the dependence of threshold current on operation temperature for two of these lasers is shown for pulsed operation. One laser has an active film thickness of 0.5 μm and the other has a 1 μm thick active film. For high temperatures both lasers show almost identical behavior, but for low temperatures the temperature dependence is markedly different, indicating that different loss mechanisms are dominating at low temperatures. This behavior is typical for these two batches.

The second type is a $PbS/PbS_{1-x}Se_x/PbS$ structure (x ≃ 0.9) deposited on a p-type PbSe substrate. This laser structure operates up to 210 K in pulsed operation (I_{th} = 5 A) and up to 110 K in continuous operation (I_{th} = 0.35 A). The thickness of the active film was chosen to be 0.5 μm. The dependence of optical output power on laser current and different operation temperatures for one of these lasers is shown in Fig. 4.

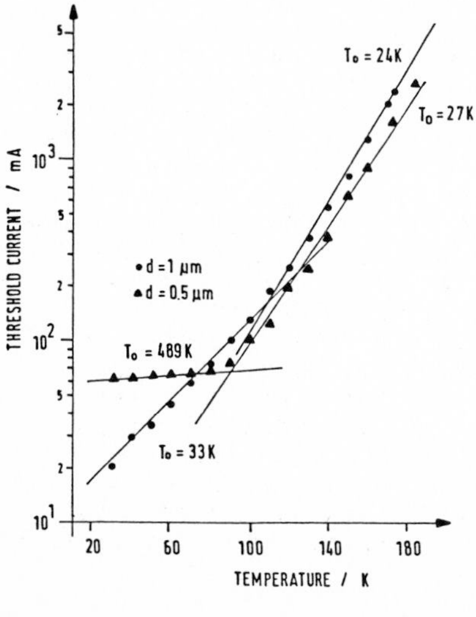

Fig. 3:
Threshold current versus temperature for two $Pb_{1-x}Sn_xSe$ DH lasers with an active film thickness of 0.5 and 1 μm

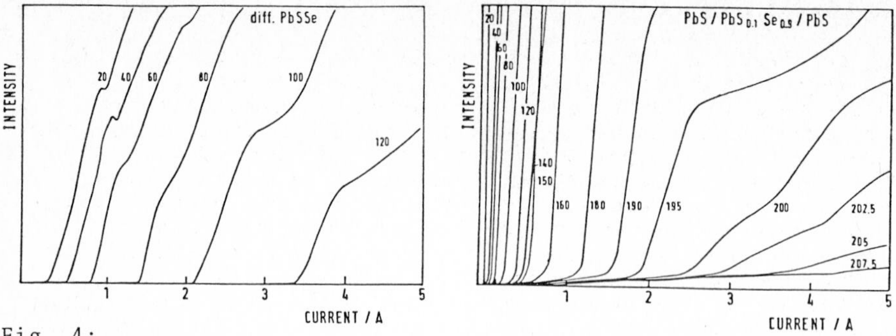

Fig. 4:
Dependence of optical output power of a DH laser and of a diffused homojunction laser on current for different temperatures indicated in the graphs in degrees K

The full scale intensity corresponds to an output power of approximately 1 mW. The power scale is not linear because of non-linear wavelength-dependent detector response. In the upper part of Fig. 4 output power versus current curves for a diffused laser are presented.

The third structure consists of PbS confinement layers and a quasi-ternary active layer of $(PbS)_{1-x}(SnSe)_x$. The x value has not been determined explicitly, but from the emission frequency of the laser it can be deduced that the band gap of the active film is approximately 35 meV smaller than the bandgap of the PbS confinement layers. These lasers operate up to 135 K in pulsed and up to 75 K in continuous operation (I_{th} = 0.4 A) and tune from 5.3 to 4.75 µm in the temperature range of 20 to 75 K. The maximum operation temperatures of the short wavelength lasers fall clearly behind the values obtained for the other type of lasers. The step in the refractive index between active layer and confinement layer is most likely too small for sufficient optical confinement.

The results are summarized in table 1.

Table 1: Properties of MBE grown DH lasers

laser structure	maximum operating temperature (K) pulse	cw	cw tuning range (µm)
PbEuSeTe-PbEuSeTe-PbEuSeTe	190	147	4.93 - 4.06
PbYbSnTe-PbSnTe-PbYbSnTe	?	128	10.7 - 7.1
PbSe-PbSnSe-PbSe	190	125	11.9 - 9.1
PbS-PbSSe-PbS	210	110	8.95 - 5.48
PbS-PbSnSeS-PbS	135	75	5.3 - 4.75

4. Conclusions

The fabrication of lead chalcogenide laser structures by the MBE technique is relatively easy. The major constituents and doping materials evaporate at convenient temperatures and the beam fluxes can be controlled. Lattice-matched laser structures of the PbTe type and non-lattice-matched lasers of the PbSe type have been realized. Concerning the highest operation temperatures obtained for the various types of lasers, a particular feature of extreme practical importance has to be noticed. Non-lattice-matched DH lasers based on PbSnSSe reach higher operation temperatures under pulsed conditions than lattice-matched laser structures based on telluride-rich systems prepared by the same deposition technique. It should be mentioned that the highest operation temperature (230 K) has been reported for a PbS/PbSe/PbS structure /2/. The lattice mismatch is almost 3 % for this combination. From this fact it certainly cannot be deduced that lattice mismatch is uneffective in the lead sulfide selenide system. However, there is evidence that lattice matching is of minor importance at least in PbSe/PbSnSe/PbSe structures. In the latter case, the maximum operation temperature increases with decreasing thickness of the active film in the investigated range from 3 to 0.5 µm. This indicates that a non-radiative recombination process at the interfaces is most likely not the dominating loss mechanism. However, more fundamental studies are necessary to clarify this point. For example, it is not known whether the mismatch dislocations themselves or only the combination of mismatch dislocations and dopants create an effective non-radiative recombination center. In this context it is worthwhile to notice that in all the laser structures we have fabricated so far it was avoided to dope the interfaces. The pn junctions have been placed after the first 2000 Å of the confinement film have been deposited undoped. Concerning fabrication of short-wavelength lasers, rare earth element containing lead chalcogenide systems are apparently better suited than Ge, Mn or Cd containing systems.

Acknowledgement

This work is sponsored by the German Ministry for Research and Technology (Contract 13 N 5281). The films and DH lasers have been prepared with the help of H. BÖTTNER, M. KÖHNE, G. KNOLL, O. PRASSE and N. SCHÄL, which is highly appreciated.

References

1. J.N. WALPOLE, A.R. CALAWA, T.C. HARMAN, and S.H. GROVES: Appl. Phys. Lett. 28, 552 (1976).

2. H. PREIER, M. BLEICHER, W. RIEDEL and H. MAIER: Appl. Phys. Lett. 28, 669 (1976).

3. D.L. PARTIN: J. Vac. Sci. Technol. B1 (2), 174 (1983).

4. D.L. PARTIN: Proceedings of the SPIE conference on "Tunable Diode Laser Development and Spectroscopy Applications", San Diego, August 25 - 26, 1983, Vol. 438, p. 17.

5. H. HOLLOWAY and J.N. WALPOLE, Prog. Cryst. Growth Charac. 2, 49 (1979).

6. R.B. SCHOOLAR, J.D. JENSEN, and G.M. BLACK: Appl. Phys. Lett. 31, 620 (1977).

7. A.V. NOVOSELOVA, V.P. ZLOMANOV, S.G. KARBANOV, O.V. MATVEYEV and A.M. GAS'KOV: Progress in Solid State Chemistry, Vol. 7, eds. H. REISS and J.O. McCALDIN; p. 85 (1972), Pergamon Press.

8. D.L. SMITH and V.Y. PICKARDT: J. Electr. Mat. 5, 247 (1976).

9. D.L. SMITH and V.Y. PICKHARDT: J. Electrochem. Soc., 125, 2042 (1978).

Hot-Wall Epitaxy of IV-VI Compounds

H. Sitter
Institut für Experimentalphysik, Universität Linz, A-4040 Linz, Austria

In the past few years the Hot-Wall Epitaxy technique has rapidly become a recognized method to prepare thin films of IV-VI and II-VI compound semiconductors. The technique is capable of producing epitaxial films with controlled stoichiometry of simple binary compounds as well as of mixed crystals. It was established that the most important factors affecting the structure of the films were the nucleation and growth process of the first monolayers.

1. Introduction

Nowadays it is possible to prepare excellent semiconductor materials by means of evaporation methods such as molecular beam epitaxy (MBE) [1] and hot-wall epitaxy (HWE) [2]. Samples with different electrical properties can be grown in the same system. This capability makes these methods attractive for the fabrication of devices and the preparation of materials for use in many different types of investigations. HWE and MBE are in many respects closely related. The difference between HWE and MBE is mainly the insertion of a heated wall between source and substrate. In this manner, a) loss of evaporating material is strongly reduced, b) a clean environment is kept within the growth tube, compared to the rest of the vacuum system and therefore no UHV is needed, and c) a relative high pressure of the evaporating materials can be maintained inside the tube [4]. In principle, a HWE apparatus which is contained in a vacuum chamber consists of a quartz tube sealed at one end which contains the sources and open at the other end where the substrate is placed to close the tube. A schematic diagram of the apparatus used for the growth of PbTe is shown in Fig. 1. The ovens can be heated independently and

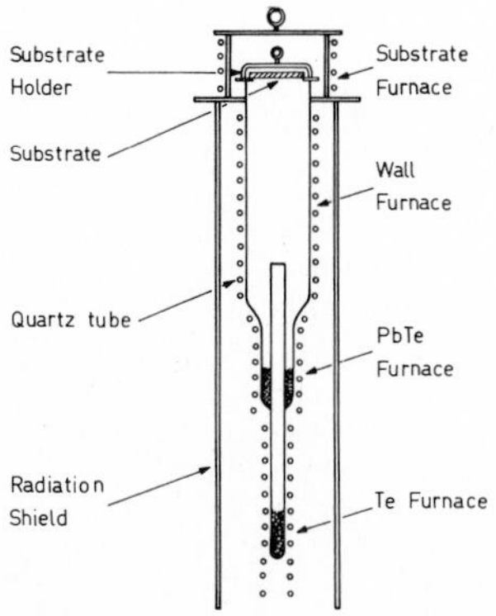

Fig. 1: Schematic diagram of the Hot-Wall evaporation system [4]

contain PbTe and pure Te. The tellurium oven serves to maintain a required Te_2 pressure during the process of growth to adjust the stoichiometry of the layers. Two important growth conditions, high substrate temperature and low supersaturation, can be obtained by the HWE. This capability can be used to control the nucleation stage and consequently high crystalline quality of the epitaxial layers can be obtained. The HWE satisfies all requirements for the growth of II-VI and IV-VI compounds and represents the advantages of high flexibility, little loss of material and low costs.

2. Theoretical Aspects of the Nucleation Stage

To our knowledge a theoretical description of the heteroepitaxy of compound materials does not exist. Since the crystalline structure and the lattice parameters of the substrate and the deposited materials matched reasonably we used a theoretical model designed for homoepitaxy as a first approach to describe the nucleation process.

2.1 Nucleation of PbTe on KCl Substrates

An important parameter in crystalline growth from the vapour phase is the supersaturation. For growth to be possible the supersaturation must be greater than one. In the case of nuclear growth an effective supersaturation can be used, defined as the ratio of the total incident flux of material to be deposited by direct impingement and by surface diffusion to the loss by evaporation and by surface diffusion from the periphery of the nucleus. The area from which material is collected by diffusion is determined by the lifetime and the mobility of the molecules on the substrate. The condition for growth of a nucleus is that the incoming flux must be larger than the loss.

In the following considerations it is assumed that the Te and Pb bonds have equal strength Φ. In a model for the growth of a nucleus we have to distinguish only between three different kinds of positions at which a molecule can be adsorbed. A schematic drawing of the possible positions is shown in Fig. 2.

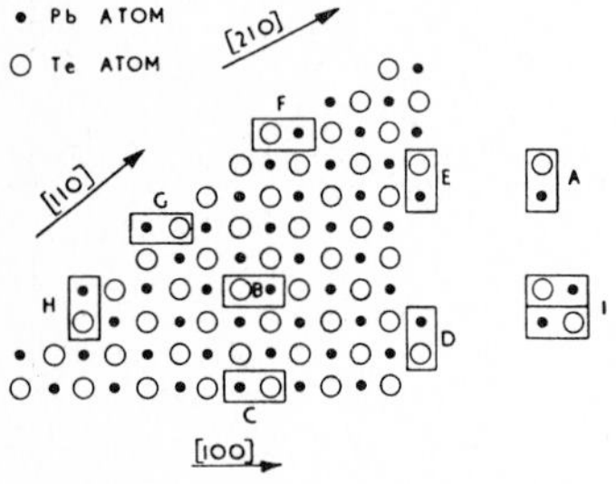

Fig. 2: Structure and intermolecular bonding of (001) plane of PbTe (after Ref. [5])

A molecule A can join the nucleus in positions B and C which do not increase the lateral growth of the nucleus. At peripheral sites of islands there are positions with an attachment energy 2 Φ such as D on an <100> edge or at I, forming a pair. In addition a molecule can find kink sites such as those at E, F, G and H with an attachment energy 3 Φ. Due to the larger binding energy the sites E, F, G and H are more stable resulting in a higher growth rate in <110> directions than in <100> directions. As soon as bridging or touching of islands has occurred, the composite island is believed to undergo a series of shape changes, as indicated schematically in Fig. 3. These stages can all

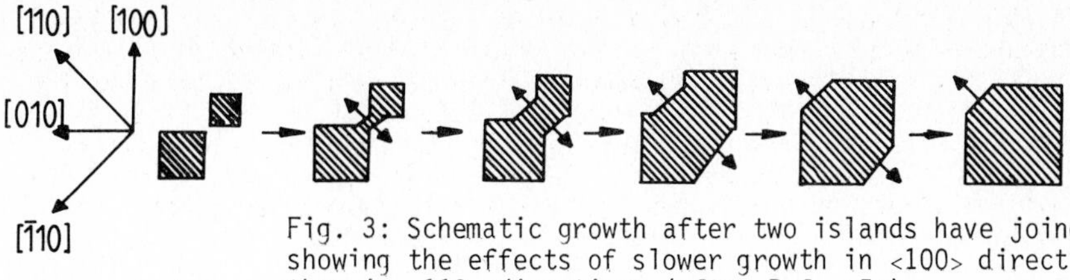

Fig. 3: Schematic growth after two islands have joined, showing the effects of slower growth in <100> directions than in <110> directions (after Ref. [5])

be accounted for by the assumption that growth is faster in <110> directions than in <100> directions.

2.2 Growth of II-VI Compounds

One of the basic problems of the epitaxial growth of II-VI compounds is the high probability of twin formation due to stacking faults. The critical temperature T_{CRIT} for twin formation as a function of supersaturation, neglecting surface diffusion, defined by p/p_0 (p and p_0 are the vapour pressures of the source and the substrate, respectively) is shown in Fig. 4 for Si, ZnS, CdSe and CdTe [6]. The probability of twin formation increases in the sequence CdS-CdSe-ZnTe-CdTe for the II-VI compounds. These results are based on a calculation of the energy of twin formation (W_{TW}) [6]. The energy for twin formation as a function of the substrate temperature (T_{SUB}) is drawn in Fig. 5 for two different supersaturations of $p/p_0 = 10$ and $p/p_0 = 100$. The lowest twin density is to be expected at low supersaturation and at low substrate temperatures.

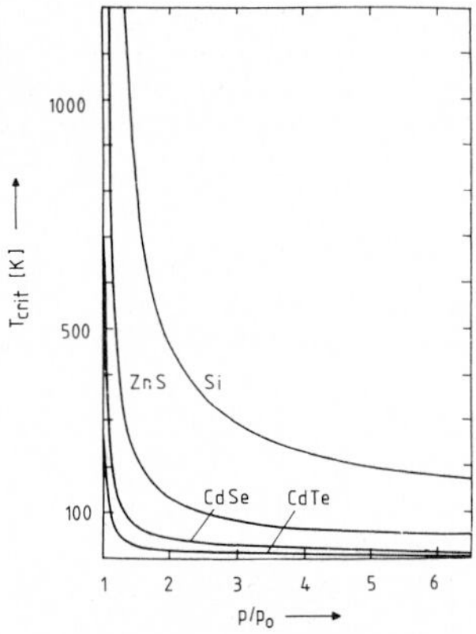

Fig. 4: Critical temperature for twin formation as a function of supersaturation p/p_0 (after Ref. [6])

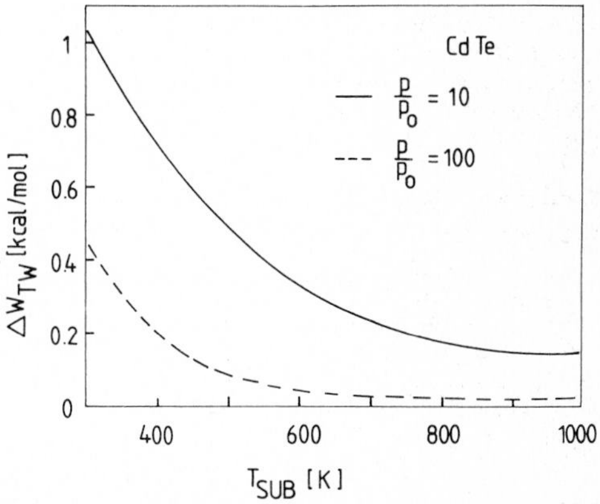

Fig. 5: Calculated energy of twin formation in CdTe vs. substrate temperature for two different values of supersaturation p/p_0 (after Ref. [6])

3. Experimental Procedures

The PbTe and CdTe layers were grown in a HWE system described in section 1. In order to exercise some control over the carrier concentration of the p and n layers of PbTe we made use of a p-x (pressure-composition) diagram for

PbTe films [2]. In the case of II-VI materials a third concentric quartz tube was inserted into the growth apparatus containing the doping material [7]. The doping of n-type material was achieved by In for CdTe and CdS and by Sb for p-type CdTe. As in bulk material CdS could not be inverted into p-type material. As substrate materials KCl and BaF_2 served for PbTe, BaF_2 for CdTe and SrF_2 for CdS. The substrate crystals were cleaved in air immediately before deposition.

Two different temperature programs were used for the growth of CdTe on BaF_2 as depicted in Fig. 6.

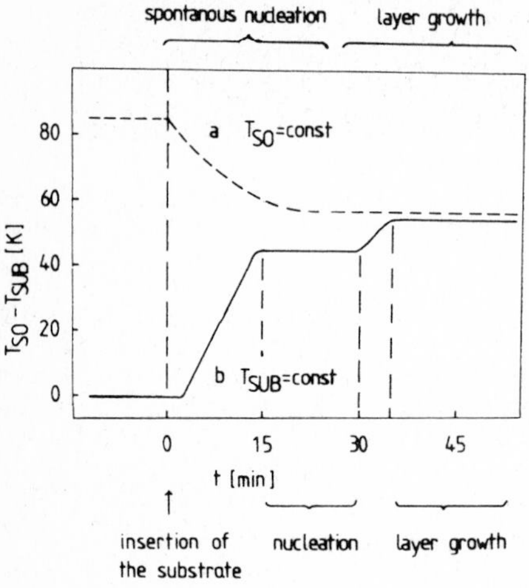

Fig. 6: Temperature time programs for the growth of epitaxial layers. a) Substrate insertion at high supersaturation. b) Substrate insertion into an undersaturated system

For the program a) the source temperature T_{SO} was kept constant and at the time of the insertion of the substrate at much lower temperature due to high supersaturation spontaneous nucleation takes place. In the temperature program b) T_{SUB} was kept constant. The substrate was inserted into the growth system at undersaturated conditions so that no spontaneous nucleation could take place. Then T_{SO} was raised until the critical supersaturation for nucleation was reached. The samples were kept under those conditions for 15 minutes. Then T_{SUB} was further increased in order to get practicable growth rates for the subsequent layer growth.

As a test for the monocrystalline perfection of the layers X-ray diffraction and electron-diffraction measurements were performed. For the investigation in the transmission-electron microscope the layers were lifted off from the substrates by dissolving the substrate in distilled water [8]. In the case of BaF_2 the substrate was polished to a thickness of 50 μm and dissolved in a 3% HNO_3 solution. Following several water rinses the films were picked up on standard electron microscope grids.

4. Experimental Results and Discussion

4.1 Transmission Electron Microscopy

4.1.1 PbTe Layers

Transmission electron microscopy (TEM) studies were performed on PbTe layers grown on KCl and BaF_2 in the nucleation stage. All samples described in this section were grown by temperature program a). To follow the different steps

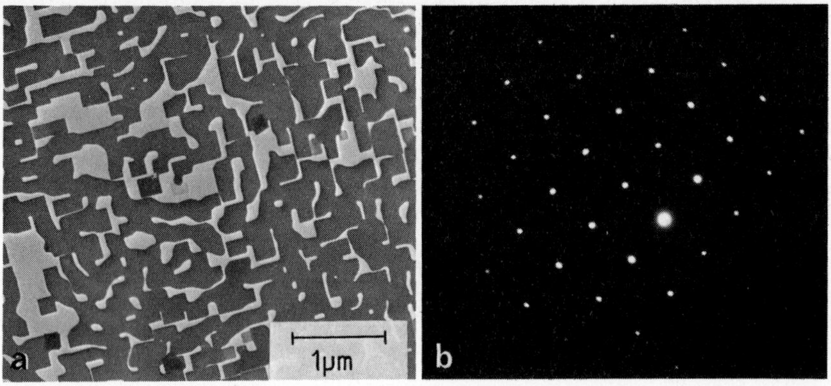

Fig. 7: a) TEM micrograph of a 40 nm thick PbTe film grown on KCl. b) Electron diffraction pattern of the PbTe film depicted in a)

of nucleation and coalescence stage the growth was interrupted after a period of 0.5 - 2 minutes. For the early stage of growth on (100) surfaces of KCl, (100) oriented PbTe square shaped islands were found with edges lying along (100). The islands were connected by bridges along (100) directions. Fig. 7a shows a TEM micrograph of islands with an average size of 200 x 200 nm^2 and a thickness of 40 nm. The shape of the islands found is in good agreement with the model presented in section 2. The islands are free of crystalline imperfections and well oriented in the (100) direction. The diffraction patterns gave a very symmetric picture (see Fig. 7b) without any evidence for misorientation of the islands.

As long as the coalescence is not complete dislocations are very rare but are generated in films thicker than approximately 120 nm. The dislocations are very mobile and glide in the (100) plane parallel to the substrate surface until they are pinned by a hole in the film as illustrated in Fig. 8. In the single crystalline layers with few grain boundaries a dislocation density of 10^8 cm^2 was found. Since the dislocations were all parallel to the growth surface the influence on the further growth must be weak. So we conclude that a great deal of the lattice mismatch is absorbed by the network of dislocations in the first 150 nm of the film and the subsequent growth of

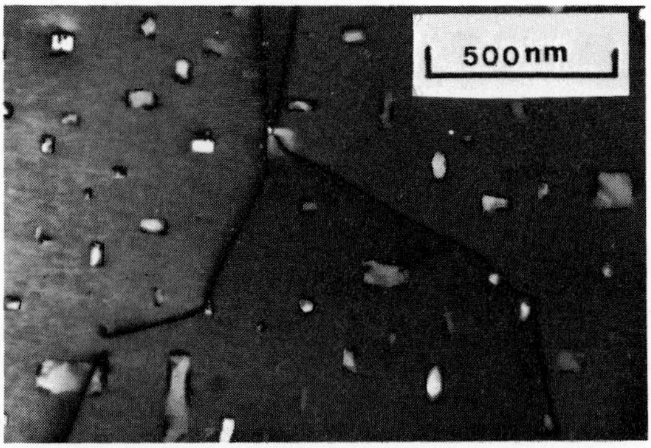

Fig. 8: TEM micrograph of dislocations pinned at holes of a 150 nm thick PbTe film

the layers is not disturbed by the crystalline imperfections of the buffer layer. That is in agreement with the high mobilities obtained in 5 μm thick layers described below.

In the case of PbTe grown on (111) surfaces of BaF_2 islands with (111) and (100) orientations were detected. The (100) islands cover less than 10% of the well-oriented (111) matrix but are more frequently found near cleavage steps. Fig. 9a shows a transmission micrograph of the matrix with triangular holes and the dark spots correspond to the (100) oriented grains. The diffraction pattern of the same film (see Fig. 9b) illustrates the [111] poles of the matrix and [100] poles of several islands. The (111) oriented matrix is represented by the hexagonal arranged bright spots. The (100) islands are oriented in such a way that their (110) directions coincide with the (110) directions of the matrix. Therefore the (100) grains give a twelve-fold symmetry in the diffraction patterns according to the three <110> directions, which can be seen clearly inside the first hexagon of the bright spots.

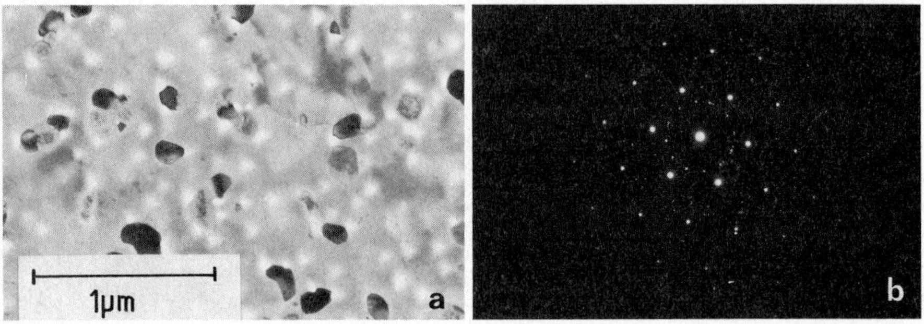

Fig. 9: a) TEM micrograph of a PbTe film grown on BaF_2. b) Electron diffraction pattern of the film depicted in a)

The grain boundaries between the (111) oriented matrix and the (100) oriented islands contained a network of dislocations absorbing the misorientation of the lattices (see Fig. 10). Thus the higher density of dislocations (10^{10} cm^{-2}) can be understood by this effective source of dislocations which is present in this buffer layer of the epitaxial films. The (100) oriented is-

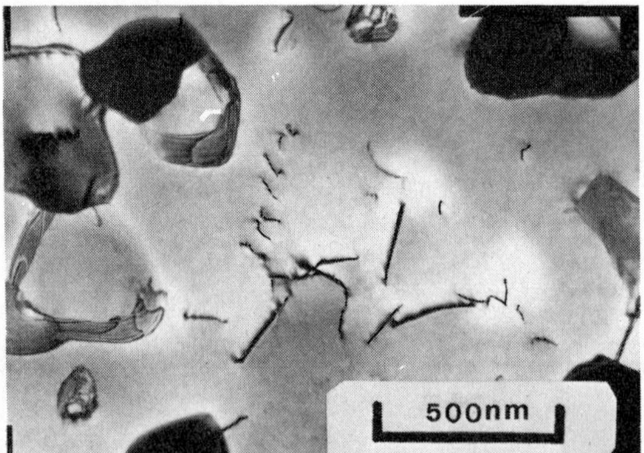

Fig. 10: (100) oriented islands surrounded by grain boundary dislocations acting as an effective source for matrix dislocations

lands are overgrown by the (111) oriented matrix since there is no evidence of (100) orientation in films thicker than 1 µm as shown by X-ray diffraction.

4.1.2 CdTe Layers

The CdTe layers grown with temperature program a) (see Fig. 6) showed a three-dimensional growth with many pyramids and hillocks on the surface. In contrast, the temperature program b) gave mirror-like surfaces with an average roughness of 36 Å measured with the surface-plasmon technique [9]. This roughness is not much higher than that due to cleaving steps of the substrate (around 14 Å).

To investigate the influence of the preheating temperature of the substrates on the nucleation stage the growth was interrupted after 15 minutes of nucleation. The resulting TEM pictures are shown in Fig. 11. On the surface preheated to 430°C a much higher density of nucleation was found compared to the substrate preheated to 550°C. We conclude that the preheating treatment reduced the adsorbed impurities and gave a thermally cleaned surface with less nucleation centres.

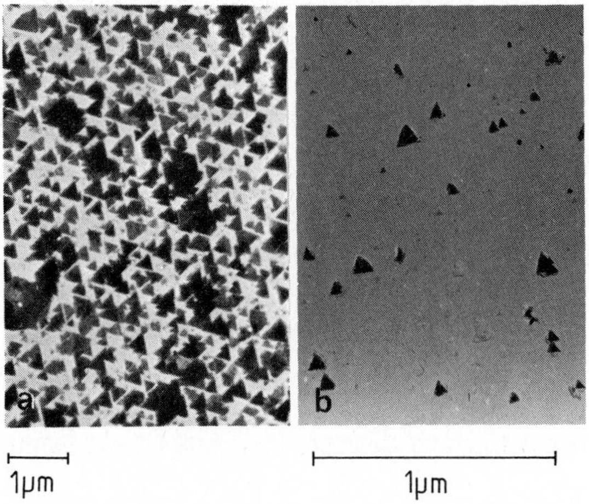

Fig. 11: TEM micrograph of the CdTe film after the nucleation time of 15 minutes a) preheating temperature T_p = 430°C, b) T_p = 550°C

In order to test the theoretical predictions concerning the twin density some films were grown at a reduced substrate temperature. The result was a drastic reduction of the density of twins, which could be clearly demonstrated by TEM [10]. As a test of the crystalline quality of the layers, the halfwidth of the peaks in the X-ray diffraction measurements were investigated. As shown in Fig. 12 an increase of the halfwidth was detected with increasing difference between source and substrate temperature, which is a measure for the supersaturation, in good agreement with the theory of twin formation.

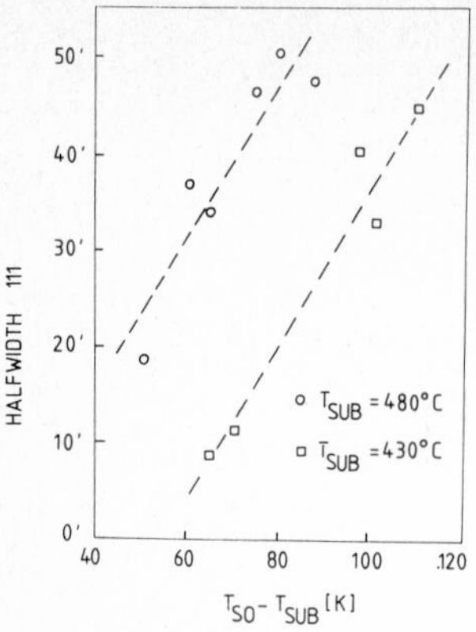

Fig. 12: Halfwidth of the 111 X-ray diffraction peak as a function of the difference between source and substrate temperature

4.2 Electrical Characterization of the Epitaxial Layers

Our results indicate that it is possible to grow consistently good films by HWE using the same source for approximately 100 films with a thickness of 5 μm. A concise summary of the film properties is given in Table I. In PbTe films grown on BaF_2 substrate mobilities were obtained as high as 3×10^4 cm^2/Vs at 77 K and above 10^6 at 4.2 K for n-type samples and 1×10^4 cm^2/Vs at 77 K for p-type samples.

Table I: Characteristic parameters of some IV-VI and II-VI compounds

Material	Mobility (300 K) [cm^2/Vs] a)		Carrier concentration (300 K) [cm^{-3}]	
	n-type	p-type	n-type	p-type
PbTe	1500	800	$3\text{-}4 \times 10^{17}$	$2\text{-}3 \times 10^{17}$
PbSe	1000	800	$5\text{-}7 \times 10^{17}$	$2\text{-}3 \times 10^{17}$
$Pb_{1-x}Sn_xTe$ b)	13000 (77K)	20000 (77K)	2×10^{17}	2×10^{17}
$Pb_{1-x}Ge_xTe$	1300	700	1×10^{17}	$3\text{-}4 \times 10^{17}$
$Pb_{1-x}Mn_xTe$	1100	700	$3\text{-}6 \times 10^{17}$	$2\text{-}6 \times 10^{17}$
CdTe	600	40	$0.5\text{-}2 \times 10^{17}$	5×10^{18}
CdS	230	--	$0.2\text{-}3 \times 10^{18}$	---

a) Maximum values obtained in as-grown layers.
b) Because of the small energy gap of $Pb_{1-x}Sn_xTe$ only values at low temperatures are significant.

5. Conclusion and Applications of HWE Films

The crystalline quality of the CdTe layers grown by HWE could be increased by the use of a special temperature-time program which allowed a controlled nucleation stage followed by layer growth. So CdTe layers with mirror-like surfaces could be grown which can be used for further investigations.

In the case of PbTe we have shown that the lattice mismatch is absorbed in the buffer layer by a network of dislocations which does not influence the subsequent growth of the film. Some experiments in fundamental solid state physics could be performed only because of the high quality of the films

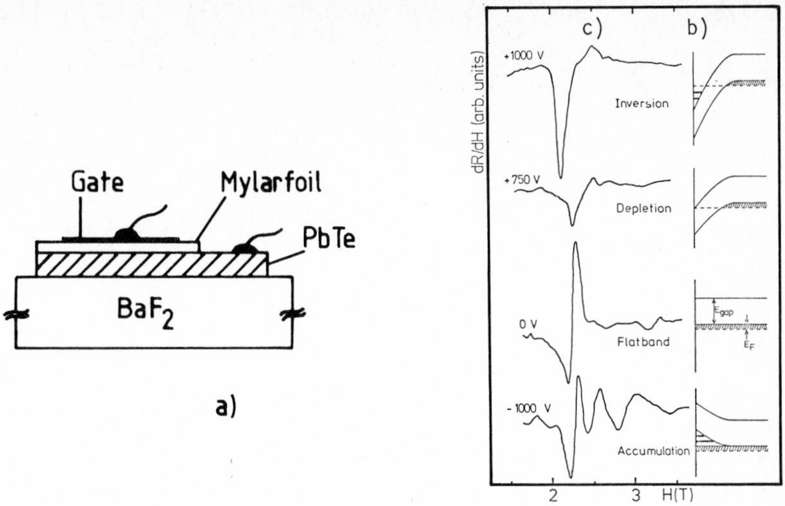

Fig. 13: a) Schematic cross section of the samples. b) Band-bending and sub-band edges at the surface for inversion, depletion, flat band conditions and accumulation. c) Field derivative of the magneto-reflectivity of PbTe for various values of gate voltage at a wavelength of 119 μm and a temperature of 4.2 K (after Ref. [13])

grown by HWE. For example,two observations should be mentioned: Hall effect of hot electrons in n-type PbTe [11] and free carrier magnetooptical effects in the Reststrahlen region could be observed [12].

Since p-type PbTe layers with hole concentrations of the order of 10^{16} cm^{-3} were available, cyclotron resonance in the far infrared of bound carriers in inversion and accumulation layers could be performed [13]. The experimental results are summarized in Fig. 13.

The samples used in these experiments were (111) oriented 5 μm thick PbTe layers grown on BaF_2 substrates. A schematic cross section of the samples is shown in Fig. 13a. The gate voltage was applied between the PbTe film and a semitransparent Ni-Cr gate electrode across a 3.5 μm thick mylar foil in a MIS capacitor arrangement. Because of the low carrier density the accumulation and depletion were induced by moderate positive and negative voltage for inversion. So it was possible to achieve inversion before the dielectric breakdown strength of the mylar foil was reached.

The different surface conditions of the sample gave different field derivatives of the magneto-reflectivity as shown in Fig. 13c. The flatband curve (OV) represents the bulk-hole cyclotron resonance. Applying a positive voltage of 750 V,the depletion layer acts as a mask to the volume resonance, thus reducing its observed amplitude. Under inversion and accumulation conditions new resonances appeared,reflecting the bound carriers in the subbands.

Acknowledgements

I would like to thank Prof. A. Lopez-Otero, who developed the HWE technique, for all his help and suggestions. I also wish to thank Dr. D. Schikora for growing the CdTe layers and Dr. P. Pongratz for the TEM investigations. I gratefully acknowledge the valuable discussions with Prof. H. Heinrich and his critical reading of the manuscript. Last, but not least, my thanks to Ch. Leitner, C. Hrdlicka and O. Fuchs for technical assistance.
The work was supported by the "Fonds zur Förderung der wissenschaftlichen Forschung in Österreich".

References

[1] L.L. Chang and R. Ludeke, in Epitaxial Growth, ed. by J.W. Mattews (Academic, New York, 1975), Part A
[2] A. Lopez-Otero, Appl. Phys. Lett. 26, 470 (1975)
[3] S.A. Semiletov, Sov. Phys. Crystallogr. 9, 65 (1964)
[4] A. Lopez-Otero, Thin Solid Films 49, 3 (1978)
[5] B. Lewis and D.J. Stirland, J. of Crystal Growth 3, 200 (1968)
[6] L. Däweritz, Kristall u. Technik 7, 167 (1972)
[7] A. Lopez-Otero and W. Huber, J. of Crystal Growth 45, 214 (1978)
[8] The TEM studies were done by P. Pongratz, TU Wien
[9] G. Harbeke, E.F. Stiegmeier, A.E. Widmer, H.F. Kappert and G. Neugebauer, Appl. Phys. Lett. 42, 249 (1982)
[10] H. Sitter and D. Schikora, Thin Solid Films, to be published
[11] J. Rozenbergs, H. Heinrich and W. Jantsch, Sol. State Comm. 22, 439 (1977)
[12] H. Burkhard, G. Bauer, P. Grosse and A. Lopez-Otero, Phys. Stat. Sol. (b) 76, 259 (1976)
[13] H. Schaber, R.E. Doezema, P.J. Stiles and A. Lopez-Otero, Sol. State Comm. 23, 405 (1977)

Part III

Multi Quantum Wells and Superlattices

Selected Topics in Semiconductor Quantum Wells

G. Bastard

Groupe de Physique des Solides de l'Ecole Normale Supérieure, 24 rue Lhomond
F-75231 Paris Cedex 05, France

This paper presents some theoretical results on the electronic properties of semiconductor quantum wells. They include the influence of band non-parabolicity on the discrete and continuous energy spectra ; the evaluation of the quantum well-projected density of states and a discussion of the binding energies of delocalized (moving) excitons. Exciton trapping on interface defects is briefly outlined.

1 Introduction

Advanced growth techniques (Molecular Beam Epitaxy or Metalorganic-Chemical Vapour Deposition) have made the realization of semiconductor quantum wells (Q.W.) or superlattices (S.L) possible [1,2]. In these narrow structures (characteristic length in the range of few hundreds Angströms), quantum size effects become a dominant feature changing the bulk, three-dimensional behaviour into a quasi two-dimensional one. Linked to the lower dimensionality, one finds confinement energies and thus level shifts from the bulk energy diagram. For instance, bulk GaAs emits in the near infra-red whereas GaAs Q.W. can be taylored to emit in the visible (red) region [3]. The Q.W.'s and S.L.'s have increasing technological implications in micro-electronics (high speed transistors) and opto-electronics (Q.W. lasers). Device applications have prompted basic research on these new materials. Here we wish to present a brief survey of some of the electronic properties of semiconductor quantum wells.

2 Electronic states in quantum wells

. Boundary conditions

In idealized Q.W.'s, the current continuity imposes that the wavefunctions and their derivatives be continuous at the interfaces [4]. In semiconductor Q.W.'s or S.L.'s, there is a difficult matching problem for the wavefunction at the interface between two materials A and B. This is because each of the A and B layers have already a natural periodicity. Therefore, within each layer, the wavefunction is a combination of Bloch waves and not of plane waves. However if
i) we restrict our attention to a narrow portion of the host Brillouin zone (to make sure that a limited number of host bands contribute to the Q.W. wavefunction)
ii) one assumes a perfect lattice-matching between A and B layers without alteration of the host crystallographic structures,
iii) one assumes that the periodic parts of the Bloch functions of relevance at a given point of the Brillouin zone (the Γ point in direct gap of III-V and II-VI based Q.W.'s) are identical in A and B layers ;
then we may adopt the following boundary conditions [5,6,7] :

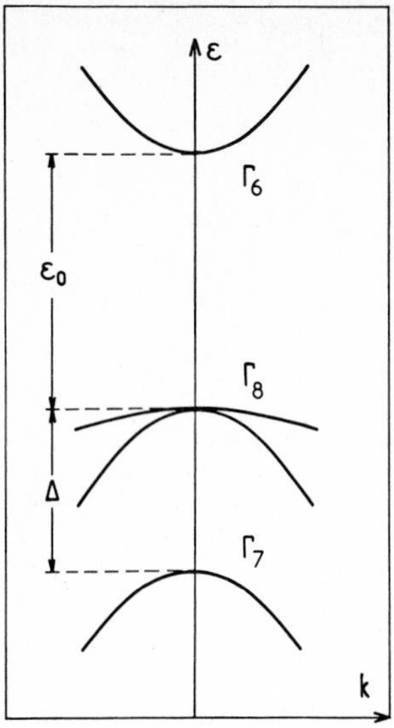

Fig.1 Band structure of direct gap III-V and II-VI semiconductors near the Γ point

i) the envelope function F(z) is continuous at the A-B interfaces,
ii) $\mu^{-1}(\varepsilon,z)(dF/dz)$ is continuous at the A-B interfaces,
where F(z) is the solution of an effective mass-like Hamiltonian projected on one of the band edges of interest and $\mu(\varepsilon,z)$ a quantity which has the dimension of an effective mass and which depends on the energy ε of the states to be built if the hosts' band structures are non-parabolic. In III-V based and II-VI based materials, the relevant host band structure (Fig.1) is accurately described by the KANE model [8]. Eight bands (Γ_6, Γ_7, Γ_8) are assumed to contribute to the Q.W. wavefunctions. For zero wavevector in the layer plane, there exists an exact decoupling between Γ_8 heavy hole states and the coupled Γ_6, Γ_7, Γ_8 light particle states. The light particles' effective mass Hamiltonian, projected on one of the S-like edges (Γ_6), reads :

$$\tilde{\mathcal{H}} F_s = \{\frac{P^2}{3} p_z \left(\frac{2}{\varepsilon+\varepsilon_A-V_p(z)} + \frac{1}{\varepsilon+\varepsilon_A+\Delta_A-V_\delta(z)}\right) p_z + V_s(z)\} F_s = \varepsilon F_s \qquad (1)$$

where ε_A, Δ_A are the band gap and Γ spin-orbit coupling of the A material. $V_s(z)$, $V_p(z)$, $V_\delta(z)$ are step-like functions which are zero in A layers and equal to V_s, V_p, V_δ in B layers where V_s, V_p, V_δ are the algebraic energy shifts for the Γ_6, Γ_7, Γ_8 edges when going from the A to the B layers. In Eq.(1), P is the KANE matrix element :

$$P = \frac{1}{m_0} \langle S|p_x|X\rangle . \qquad (2)$$

Integrating (1) across an interface leads to

$$\mu^{-1}(\varepsilon,z) = \frac{2P^2}{3}\left(\frac{2}{\varepsilon+\varepsilon_A-V_p(z)} + \frac{1}{\varepsilon+\varepsilon_A+\Delta_A-V_\delta(z)}\right) . \qquad (3)$$

For parabolic materials (i.e. $\varepsilon \ll \varepsilon_A$, $|\varepsilon-V_s| \ll \varepsilon_B$ for Γ_6-related Q.W. bound states), $\mu(\varepsilon,z)$ reduces to

$$\mu(\varepsilon,z) = \begin{cases} m_A & \text{in A layers} \\ m_B & \text{in B layers} \end{cases} \qquad (4)$$

where m_A, m_B are the Γ_6 band edge masses in A and B layers. Except in the HgTe-CdTe system [9] (light particles at the Γ_8 edge) m_A and m_B have the same sign.

. Quantum Well bound states

Once we know the boundary conditions, we may search for the Q.W. bound states. Suppose A confines the carrier for the band edge of interest (i.e. A is GaAs for Γ_6, Γ_7, Γ_8 edges in GaAs in the GaAs-$Ga_{1-x}Al_xAs$ system). Then the bound state equation for non-parabolic Q.W. is

$$\cos k_A L - \frac{1}{2}\left(\tilde{\xi} - \frac{1}{\tilde{\xi}}\right)\sin k_A L = 0 \quad ; \quad \tilde{\xi} = \frac{k_A(\varepsilon)}{\mu_A(\varepsilon)}\frac{\mu_B(\varepsilon)}{\kappa_B(\varepsilon)} \tag{5}$$

where L is the well thickness. $k_A(\varepsilon)$, $\kappa_B(\varepsilon)$ are the wavevectors of the propagating and evanescent waves in A and B layers respectively. They are found from the KANE three-band equation :

$$\varepsilon(\varepsilon+\varepsilon_A)(\varepsilon+\varepsilon_A+\Delta_A) = \frac{2P^2}{3}\hbar^2 k_A^2(\varepsilon+\varepsilon_A+2\Delta_A/3) \tag{6}$$

$$(V_s-\varepsilon)(\varepsilon-V_s+\varepsilon_B)(\varepsilon-V_s+\varepsilon_B+\Delta_B) = \frac{2P^2}{3}\hbar^2\kappa_B^2(\varepsilon-V_s+\varepsilon_B+2\Delta_B/3)\,. \tag{7}$$

In (6,7), ε_B, Δ_B are the band gap and Γ spin-orbit coupling of B material and the energy origin has been taken at the bottom of the Γ_6 band of the A material. Like idealized Q.W.'s,semiconductor Q.W.'s always admit one bound state (per edge). By increasing the well thickness L from zero to infinity more and more states are bound. On a chart ε versus L, the onset of bound Γ_6 levels occur when

$$\varepsilon = V_s \quad \text{i.e.} \quad \tilde{\xi} = \infty \quad \text{or} \quad k_A(V_s)L = n\,\pi\,. \tag{8}$$

The influence of the barrier height on the confinement energy of the (E_1) Γ_6-related ground bound state of GaAs-$Ga_{1-x}Al_xAs$ Q.W.'s is shown on Fig.2.

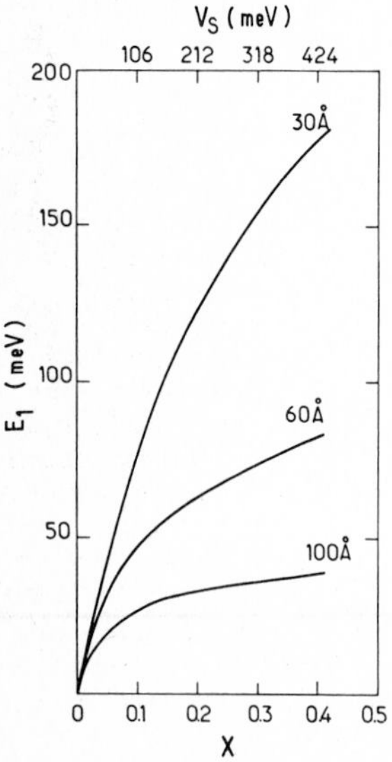

Fig.2 Influence of the barrier height on the energy E_1 of the ground conduction states in GaAs-$Ga_{1-x}Al_xAs$ single Q.W. for three different well thicknesses

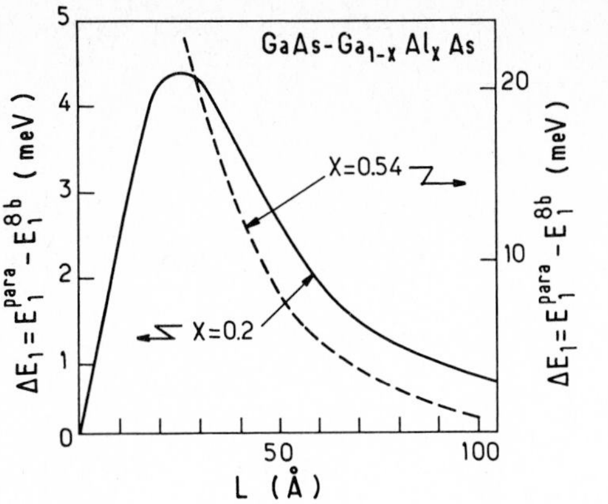

Fig.3 Calculated energy difference ΔE_1 between the confinement energies of the ground conduction state E_1 of GaAs-Ga$_{1-x}$Al$_x$As single Q.W.'s assuming parabolic and non-parabolic dispersion relations

On Fig.3, we illustrate the influence of host band non-parabolicity on the E_1 energy in the same system (x = 0.2 and x = 0.54). The band non-parabolicity has a weak effect on E_1 since the GaAs bandgap is large. Larger confining barriers increase E_1 and in turn the non-parabolicity corrections. Non-parabolicity effects are more pronounced in InP-In(Ga)As or AlSb-GaSb Q.W. energy levels since In(Ga)As and GaSb have smaller band gaps than GaAs.

. Quantum well delocalized states

Beyond V_s the Q.W. energy spectrum is continuous and each level is twice degenerate. The transmission coefficient of an electron across a non-parabolic, symmetric, double Q.W. is given by [10] :

$$T(\varepsilon) = \{1+\frac{1}{4}(\xi-\frac{1}{\xi})^2\sin^2 k_A L_2 [2\cos k_A L_2 \cos k_B L_3 - (\xi+\frac{1}{\xi}) \sin k_A L_2 \sin k_B L_3]^2\}^{-1} \quad (9)$$

where L_2 is the thickness of both wells and L_3 the middle barrier thickness. Single well (thickness $2L_2$) results are simply obtained by letting $L_3 = 0$ in (9). Transmission resonances occur whenever

$$k_A L_2 = p\pi \quad ; \quad p > 0 \quad (10)$$

$$\cos k_A L_2 \cos k_B L_3 - \frac{1}{2}(\xi+\frac{1}{\xi}) \sin k_A L_2 \sin k_B L_3 = 0 \quad (11)$$

where k_B differs from κ_B in (7) only by changing $V_s - \varepsilon$ into $\varepsilon - V_s$ (real propagation in B layers). The first kind of resonances is characteristic of single well structures [4,11] whereas the second kind (11) arises from the double well configuration. We see from (8) and (10) that single well resonances are actually the continuation of the bound levels when their confinement energies have come to exceed the height of the cladding barriers. These resonances are quite pronounced when ε weakly exceeds V_s (or $|V_s-\varepsilon_B|$ or $|V_s-\varepsilon_B-\Delta_B|$) in such a way that $k_B/k_A \ll 1$. They are increasingly smeared out when going deeper in the continuum. With each resonance, one may associate [4,11] a virtual bound state whose physical image is that of a classical particle which is trapped inside the Q.W. The trapping time is much longer than the classical time needed to cross the Q.W. if the resonance is narrow ($k_B/k_A \ll 1$). The Fig.4 shows the L dependence of the light hole bound and virtual bound states in GaAs-Ga(Al)As symmetric double well structure. When ε approaches the onset of the continuum, the single well resonances (10) match with the symmetric bound states S_n of the double well whereas the resonances (11) match with the antisymmetric bound states AS_n.

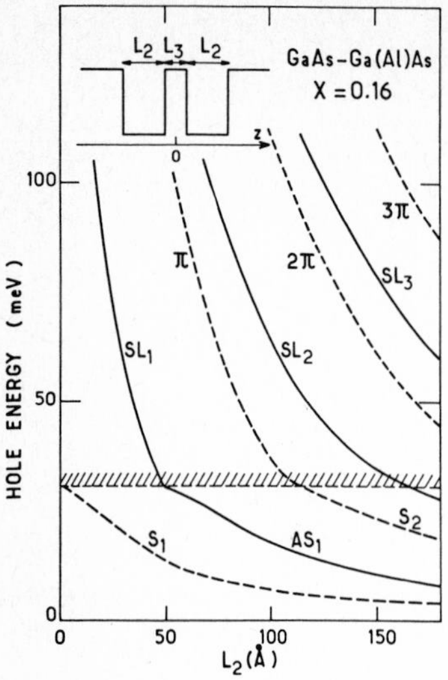

Fig.4 Calculated thickness dependence of the energy of the bound and virtual bound light hole state in symmetric double $GaAs\text{-}Ga_{1-x}Al_xAs$ Q.W.'s. L_3= 12 Å; x= 0.16; after [10]

. Density of states

For Q.W.'s with parabolic host band structure, the density of states (D.O.S.) per unit energy is given by

$$\tilde{\rho}(\varepsilon) = \rho_0 \sum_n Y(\varepsilon-\varepsilon_n) \qquad ; \qquad \rho_0 = \frac{m^* S}{\pi \hbar^2} \tag{12}$$

where m^* is the carrier effective mass, S the sample area and the summation runs over both bound and extended states. For bound states, the D.O.S. is staircase-like, in contrast with the bulk behaviour (D.O.S. $\sim \sqrt{\varepsilon}$). This two-dimensional behaviour was clearly revealed in optical absorption by DINGLE in GaAs-Ga(Al)As [2], VOISIN et al. in GaSb-AlSb [12] and CHANG et al. in InAs-GaSb (S.L.'s) [13].

Another D.O.S. illustrates more directly carrier localization in the Q.W. : the Q.W.-projected D.O.S. which is defined as

$$\rho(\varepsilon) = \rho_0 \sum_n Y(\varepsilon-\varepsilon_n) P_n \qquad ; \qquad P_n = \int_{well} |\chi_n(z)|^2 dz \, . \tag{13}$$

When a carrier is injected in the Q.W. continuum, it quickly relaxes towards the vicinity of the continuum edge ($\varepsilon = V_b$) by emitting optical phonons. These are intracontinuum transitions. To become captured by the Q.W. during a subsequent phonon emission, it should shrink its delocalized wavefunction to match Q.W. bound χ_n's. Hence $\rho(\varepsilon)$ appears better suited than $\tilde{\rho}(\varepsilon)$ to have a qualitative idea of which continuum states are important for the capture process. A typical plot of $\rho(\varepsilon)$ for low-lying continuum states is shown on Fig.5 for an idealized separate confinement heterostructure which consists of a Q.W. of variable thickness w flanked by two cladding barriers of height V_b terminated by impenetrable walls. The total thickness L of the structure is kept constant (3000 Å) as well as V_b (195 meV). Due to the terminating walls, all the states are discrete. The states of positive energy are delocalized over L whereas the states of negative energies, which would reduce to Q.W. bound states at infinite L, are essentially localized over w. The

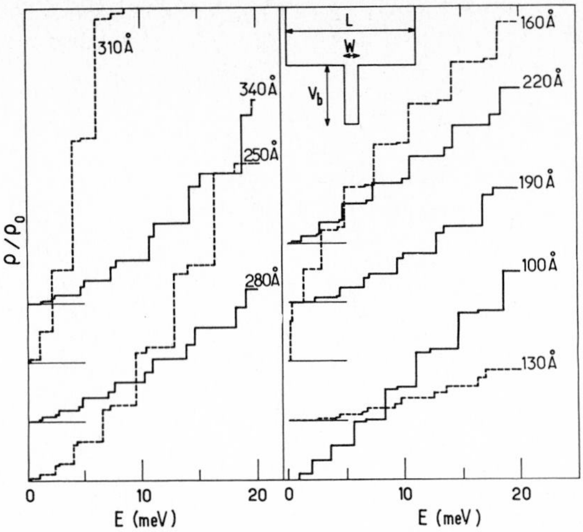

Fig.5 Quantum-well-projected density of states (in units of ρ_0) for an idealized separate confinement heterostructure. V_b = 195 meV ; L = 3000 Å Each horizontal division is 5 %. Curves corresponding to different w are displaced vertically for clarity and are alternatively drawn in dashed and solid lines

various curves shown on Fig.5, displaced vertically for clarity, show the energy dependence of $\rho(\varepsilon)$ for several w. Two features are apparent :
i) $\rho(\varepsilon)$ is very small near the onset of the continuum. This is consistent with the general consideration that a particle of vanishing energy in the barrier is actually repelled by the Q.W., evidencing the wave-like behaviour of the particle near a sharp interface;
ii) for some w, $\rho(\varepsilon)$ increases much more steeply with ε than it does for neighbouring w's. This occurs when a virtual bound state has just popped outside the Q.W.

3 Excitons in Quantum Wells [14-19]

When a Q.W. is radiated with light of energy $\hbar\omega$ larger than $\varepsilon_A+E_1+HH_1$, where E_1 (HH_1) are the ground electron (heavy hole) Q.W. bound states, the photo-excited electron-hole pair forms a bound state for the electron-hole reduced motion : the exciton. The center of mass of the pair moves freely in the layer plane. Hence the exciton is quasi bi-dimensional. Due to the Q.W. confinement, the exciton binding energy is increased over the bulk value [14-16]. If the barrier heights are infinite for both electrons and holes and if both kinds of particles are confined within the same layer, the exciton binding energy increases monotonically with decreasing layer thickness (Fig.6) reaching four times the three-dimensional effective Rydberg at L = 0 [14,15]. For Q.W.'s, where electrons and holes are spatially separated (GaSb-InAs Q.W.'s with $L_{InAs} \lesssim 100$ Å), the increase in binding energy is very small [15]. Finite barrier effects lower the exciton binding energy with respect to the infinite barrier situation. GREENE and BAJAJ [16] have shown that exciton binding energy first increases with decreasing L and then decreases in very narrow Q.W.'s, the barrier heights being kept constant. The low temperature radiative recombination of high quality GaAs Q.W.'s is excitonic [17], in contrast with bulk GaAs where extrinsic photoluminescence dominates. The luminescence line is often Stokes-shifted by few meV with respect to free exciton (HH_1-E_1) absorption peak. WEISBUCH et al. [17]

Fig. 7

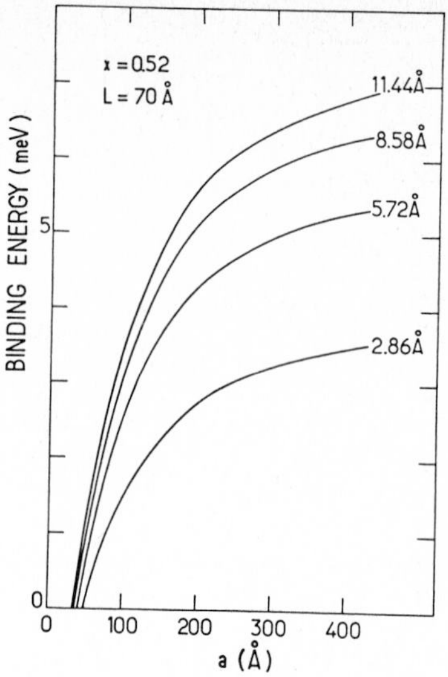

Fig. 6

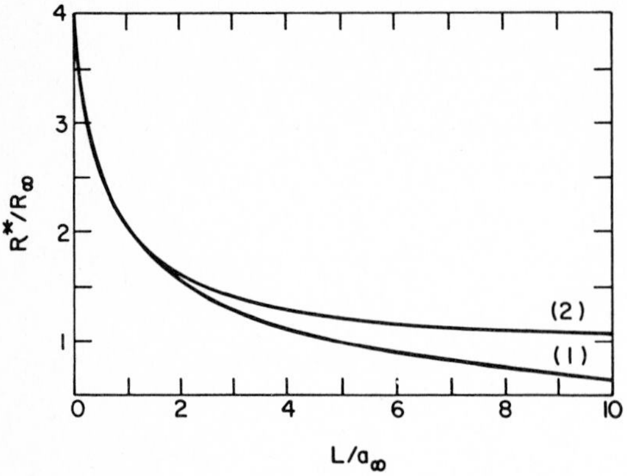

Fig.6 Exciton binding energy versus well thickness in the infinite barriers approximation. Energy and length are scaled to the bulk exciton effective Rydberg R_∞ and Bohr radius a_∞ respectively. Curves (1) and (2) correspond to two different trial wavefunctions. After [15]

Fig.7 The exciton binding energy on a semi-Gaussian interface defect is plotted versus the lateral size a of the defect for several values of the defect depth b in a GaAs-$Ga_{1-x}Al_xAs$ single quantum well. L = 70 Å ; x = 0.52. After [18]

have interpreted these features in terms of exciton trapping on interface defects (island-like) whose depth is $\sim$ one monolayer (2.86 Å) and whose lateral dimension is $\sim$ few hundreds Å. Recently [18] we have investigated exciton binding energy on model interface defects (semi-Gaussian), defects whose lateral and along-the-axis parameters are denoted by a and b respectively . Fig.7 shows the a dependence of the exciton binding energy for several values of the parameters b in a GaAs Q.W. (L = 70 Å) clad between $Ga_{0.48}Al_{0.52}As$ barriers. From the results shown on Fig.7, we may conclude that the luminescence line can hardly be separated from exciton absorption peak by more $\sim$ 10 meV. Larger Stokes shifts have however been observed on "poorer" structures. To explain these shifts coulombic impurities, possibly trapped at the interfaces during the growth process, appear to be likely candidates. Once an exciton gets trapped in an interface defect , its only possible motion at low temperature is phonon-assisted hopping between traps. A calculation of such a process shows that for reasonable defect concentrations ($\lesssim 10^{10}\,cm^{-2}$), the hopping process by acoustical phonon emission lasts sensitively longer than the exciton radiative lifetime ($\sim 3\times10^{-10}$s [19]). Thus the trapped exciton actually does not move. This implies that the luminescence line involves unthermalized trapped excitons, a feature which is consistent with our experimental findings.

Acknowledgments

I am pleased to thank Drs. L. Esaki, L.L. Chang, E.E. Mendez, C.A. Chang from IBM Laboratories and Drs. M. Voos, Y. Guldner, P. Voisin, C. Delalande, J.A. Brum from E.N.S. Laboratory for their participation in the works described here.

References

1. L. Esaki, R. Tsu, IBM J. Res. Develop. 14, 61 (1970).
2. R. Dingle in Festkörperprobleme XV (Advance in Solid State Physics) edited by H.J. Queisser (Pergamon Vieweg 1975).
3. B.A. Vojak, W.D. Laidig, N. Holonyak Jr., M.D. Camras, J.J. Coleman, P.D. Dapkus, J. App. Phys. 52, 621 (1981).
4. A. Messiah, Mécanique Quantique Vol.1 (Dunod, Paris 1959).
5. G. Bastard, Phys. Rev. B 24, 5693 (1981) and Phys. Rev. B 25, 7584 (1982).
6. S. White, L.J. Sham, Phys. Rev. Lett. 47, 879 (1981).
7. M. Altarelli, Physica 118 B, 747 (1983).
8. E.O. Kane, J. Phys. Chem. Solids 1, 249 (1957).
9. Y. Guldner, this volume.
10. G. Bastard, U.O. Ziemelis, C. Delalande, M. Voos, A.C. Gossard, W. Wiegmann, Solid State Commun. (1984), in the press.
11. D. Bohm, Quantum Theory (Prentice Hall, New York 1951).
12. P. Voisin, G. Bastard, M. Voos, E.E. Mendez, C.A. Chang, L.L. Chang, L. Esaki, J. Vac. Technol. B 1 (2), 409 (1983). P.Voisin, this volume.
13. L.L. Chang, G.A. Sai-Halasz, L. Esaki, R.L. Aggarwal, J. Vac. Sci. Technol. 19, 589 (1981).
14. R.C. Miller, D.A. Kleinman, W.T. Tsang, A.C. Gossard, Phys. Rev. B 24, 1134 (1981).
15. G. Bastard, E.E. Mendez, L.L. Chang, L. Esaki, Phys. Rev. B 26, 1974 (1982).
16. R.L. Greene, K.K. Bajaj, Solid State Commun. 45, 831 (1983).
17. C. Weisbuch, R. Dingle, A.C. Gossard, W. Wiegmann, J. Vac. Sci. Technol. 17, 1128 (1980).
18. G. Bastard, C. Delalande, M.H. Meynadier, P.M. Frijlink, M. Voos, to be published.
19. E.O. Göbel, H. Jung, J. Kuhl, K. Ploog, Phys. Rev. Lett. 51, 1588 (1983).

Subband Structure and Landau Levels in Heterostructures

A. Fasolino* and M. Altarelli

Max-Planck-Institut für Festkörperforschung, Hochfeld-Magnetlabor, 166X
F-38042 Grenoble Cedex, France

* Present adress: SISSA, Strada Costiera 11, I-34100 Trieste, Italy

The motion of two-dimensional carriers in quantum wells and superlattices is discussed, with emphasis on subband dispersion parallel to the interfaces and on quantization in a perpendicular magnetic field. For coupled bands (valence bands, coupled s-p bands in narrow-gap semiconductors, etc.) striking non-parabolicities of the subband dispersion and non-linearities of the Landau levels versus magnetic field occur. Results for GaAs-GaAlAs and InAs-GaSb systems are compared to analytical solutions for simple models and to experiments.

1 Introduction

The most interesting aspect of the physics of heterostructures is the quantization of the electron and hole motion in the direction perpendicular to the interfaces, which we shall from now on denote as z. It is therefore natural that the majority of theoretical work has concentrated on the motion along z, and the problem of subband dispersion in the (k_x, k_y) plane has been addressed in detail less frequently [1-3]. There are at least two compelling reasons, however, that make this problem very interesting. On the one hand, the dispersion of subbands in this plane is highly non-trivial for states derived from coupled bands, such as the p-like valence band of all semiconductors, or the k·p coupled valence and conduction bands of narrow-gap materials. On the otherhand, the dispersion of subbands along the layers is essential in determining the density of states, the properties of exciton and impurity states, transport properties and, of most interest to us, the Landau level quantization in a perpendicular magnetic field.

In the present contribution, we shall first discuss the strongly non-parabolic dispersion of subbands in the (k_x, k_y) plane, in terms of analytical results for a simple but instructive model, and of realistic calculations in the envelope-function approximation for GaAs-GaAlAs and InAs-GaSb systems. We shall then discuss the extension of the many-band envelope-function approximation to the calculation of Landau levels in a field perpendicular to the layers. The striking non-linear dependence of the levels on the magnetic field will be discussed and compared to magneto-optical experiments.

2 Motion of Holes Along a Quantum Well

To demonstrate the highly non-trivial behavior of subband dispersion along a quantum well for coupled band structures, we shall first consider the motion of holes in an infinitely deep quantum well. The upper valence band edge (J=3/2) of cubic semiconductors is described by LUTTINGER's Hamiltonian [4], in terms of the material parameters γ_1, γ_2 and γ_3. The subband structure of a Luttinger hole confined in an infinitely deep quantum well can be calculated analytically as shown by NEDOREZOV [5]. For the materials of interest here, a further simplification with little loss of accuracy is obtained by neglecting the warping of the bands, i.e., by assuming $\gamma_2=\gamma_3\equiv\bar{\gamma}$ in Luttinger's Hamiltonian, with $\bar{\gamma}=(3\gamma_3+2\gamma_2)/5$ (spherical model [6]). The bulk band structure is then spherically symmetric. If we take the $\vec{k}$ vector in the (y,z) plane, i.e., normal to the direction x of quantization of angular momenta, the 4x4 Luttinger matrix decouples into two identical 2x2 blocks (Kramers degeneracy), each with eigenvalues $E_{l,h}=-(\gamma_1\pm2\bar{\gamma})k^2/2$ for light (l) and heavy (h) holes respectively. If the quantum well is perpendicular to z and has width L, we find that for $k_y=0$ light and heavy holes decouple and give rise to independent quantum levels according to the familiar prescription (atomic units are used throughout the paper)

$$E_{(l,h)n} = -(\gamma_1 \pm 2\bar{\gamma})k_z^2/2 \tag{1}$$

with $\quad k_z = n\frac{\pi}{L} \qquad n = 1,2,\ldots$.

For $k_y \neq 0$ (motion along the quantum well) this is no longer possible. Light- and heavy-hole waves must be mixed to satisfy the boundary conditions. The subband energy $E(k_y)$ is determined by the dispersion relation :

$$[4\, k_{lz}^2\, k_{hz}^2 + k_y^2\,(k_{hz}^2 + k_{lz}^2) + 4k_y^4]\,\sin k_{lz}L\,\sin k_{hz}L + 6k_y^2\, k_{lz}\, k_{hz}$$
$$(1-\cos k_{lz}L\,\cos k_{hz}L) = 0 \tag{2}$$

where (remember that E<0)

$$k_{(l,h)z} = [-2E/(\gamma_1\pm2\bar{\gamma}) - k_y^2]^{1/2}.$$

Since (2) is in analytic form, one can derive from it effective masses in the layer plane, defined as:

$$\frac{1}{m_{(l,h)n}} = 2(\partial E_{(l,h)n}/\partial k_y^2)_{k_y=0}\ . \tag{3}$$

One finds:

$$\frac{1}{m_{(l,h)n}} = \frac{-1}{(\gamma_1\pm2\bar{\gamma})}\cdot\left(1+3\,\frac{\cos\theta_n+(-1)^{n+1}}{\theta_n\sin\theta_n}\right) \tag{4}$$

with

$$\theta_n = n\pi\sqrt{(\gamma_1+2\bar{\gamma})/(\gamma_1-2\bar{\gamma})} \qquad \text{for light holes, l}$$

$$\theta_n = n\pi\sqrt{(\gamma_1-2\bar{\gamma})/(\gamma_1+2\bar{\gamma})} \qquad \text{for heavy holes, h.} \tag{5}$$

For GaAs parameters (γ_1=6.85, $\bar{\gamma}$=2.58 [7]) this gives positive (i.e., electron-like!) masses for some of the heavy-hole subbands (n=2, n=5, etc.). The masses are independent of the width L of the well, as the subband energies scale like $1/L^2$ and the momenta k_y like 1/L. The hole subbands derived from (2) in a L=200Å well of GaAs are shown in Fig.1. The strong non-parabolicities and the positive mass of the second subband, in agreement with (4), are quite evident.

These results for the infinite well are interesting because all their salient features are retained, or even emphasized, by the subband dispersion in the realisitic case of finite potential wells. In Fig.2 we show indeed results for a 120Å quantum well of GaAs between $Ga_{1-x}Al_xAs$ barriers with

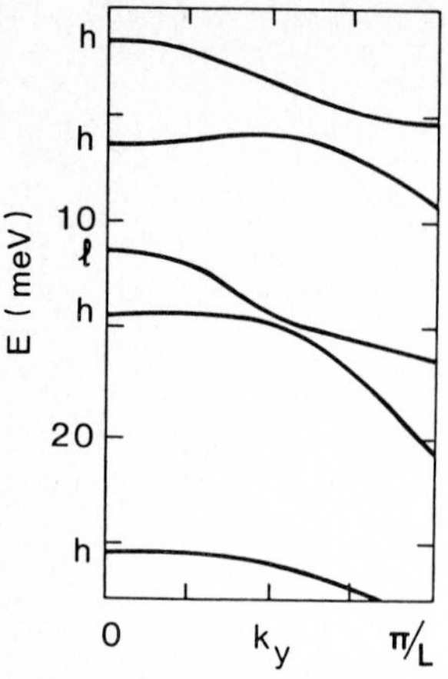

Fig.1 Hole subbands in a L=200Å GaAs infinite quantum well. The direction y is in the plane of the well. The labels l and h denote light- or heavy-hole character at k_y=0

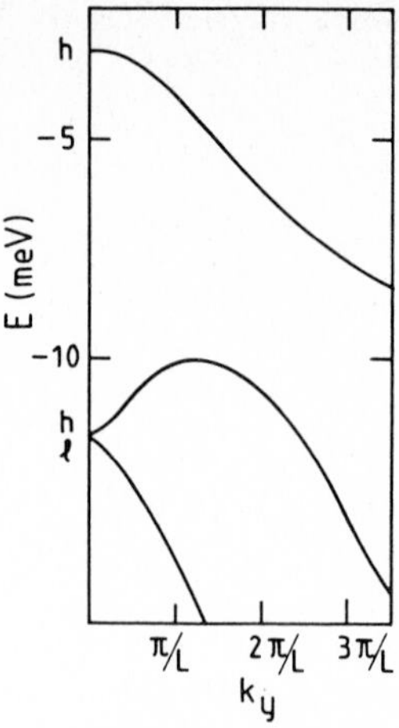

Fig.2 Hole subband dispersion versus k_y for L=120Å GaAs quantum well between $Ga_{1-x}Al_xAs$ barriers with x=0.24

x=0.24. Although the barrier heights are only 41 meV, the second subband with heavy-hole character at k_y=0 has indeed an upward dispersion in a region of size ∿ π/L in momentum space and ∿ 2meV in energy. It is interesting to notice that these are the relevant momentum and energy regions in the formation of exciton states.

The results of Fig.2 were obtained following the many-band envelope-function method described in Ref. [3] and [2] for the superlattice geometry. When the GaAlAs thickness is increased, the isolated well limit is easily recovered. For thinner GaAlAs layers, the small dispersion along k_z does not alter the features of the subbands along k_y, as also shown in Ref. [3]. Considering also our previous results on InAs-GaSb superlattices [2,8], we can conclude that striking non-parabolicities in the subband dispersion along the interfaces occur for all coupled band systems in heterostructures [9].

3 Landau Levels in Heterostructures

There are good experimental reasons to investigate the electronic states of superlattices and quantum wells in an external magnetic field, and there is also the interesting theoretical question of the Landau quantization for the strongly non-parabolic two-dimensional subbands just described. In fact whenever energy vs.k_y dispersion curves are non-parabolic we can anticipate energy vs.magnetic field curves to deviate from linearity. For high oscillator quantum numbers this can be seen from the semiclassical limit, according to which Landau levels of a subband $E(k_y)$ are given by the correspondence

$$E_n(B) \simeq E\left(k_y=\sqrt{\frac{2eB\,n}{c}}\right) \quad \text{for } n>>1. \tag{6}$$

Some insight into the behavior for small n, on the other hand, can be gained through the analytical solution of the problem for a 2x2 model [10]. Consider indeed the subband structure given, for $\vec{k}$ in the (x,y) plane, by the following matrix

$$H(\vec{k}) = \begin{pmatrix} \Delta/2-\alpha k^2 & P(k_x+ik_y) \\ P(k_x-ik_y) & -\Delta/2+\beta k^2 \end{pmatrix} \tag{7}$$

in which a hole-like subband is an energy Δ above a conduction-like one, the two being coupled by a k·p term with matrix element P, which produces an anticrossing behavior. Adding a field B in the z direction, one obtains solutions in terms of harmonic oscillator eigenfunctions of the form $(c_1\phi_n, c_2\phi_{n-1})$ where $c_2=0$ for n=0 and $c_1,c_2\neq 0$ for n=1,2... . The eigenvalues as function of the field are sketched in Fig.3. The band coupling introduces strong deviations from linearity for all Landau levels except the n=0, which is purely hole-like. This simple model provides good qualitative insight for systems with anticrossing subbands, such as the InAs-GaSb superlattices.

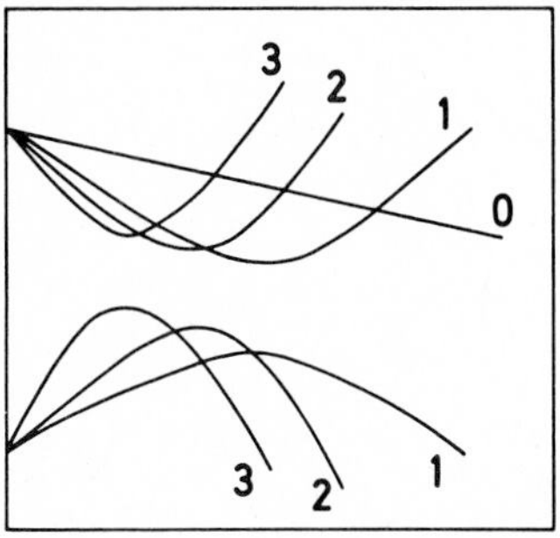

Fig.3 Landau levels versus magnetic field (arbitrary units) for a two-subband model with anticrossing. The quantum number n (see text) is indicated for each level

For the interpretation of experiments, however, we do need realistic calculations. We treat the general case of full k·p coupling between an s-like conduction band and the upper (J=3/2) edge of a spin-orbit split p-like valence band. Although the split-off band is not explicitly included, its effect on the effective mass m^* and g factor g^* of the conduction band is introduced following ROTH et al.[11]. The k·p Hamiltonian for each material in the heterostructure, when the conduction band spin is included, is a 6x6 matrix. In the absence of the magnetic field, it can always be decoupled into two 3x3 blocks [8] by choosing the spin quantization axis along x, say, while $\vec{k}$ lies in the (y,z) plane. However, when a field is present in the z direction, we must always retain a 6x6 formalism (one reason is that extra couplings between spin-up and spin-down arise if z is not the quantization axis), so that we conform to tradition and quantize angular momenta along z.

The many-band envelope function method as sketched in [8] and [2] is modified to include the external field along z. This is accomplished by :

(i) Replacing $\vec{k}$ by $\vec{k}-(e/c)\vec{A}$ in the Hamiltonian of each constituent
(ii) Adding new terms arising from electron or hole-spin coupling to the field. For the conduction band this means that the diagonal terms become:

$$H_{cc}=E_c+\frac{1}{2m^*}(\vec{k}-(e/c)\vec{A})^2+\frac{e}{2m_0c}g^*s_zB \tag{8}$$

where $\vec{s}$ is the electron spin and :

$$\frac{1}{m^*}=\frac{1}{m_0}\left(1+\frac{2P^2}{3(E_c-E_v+\Delta)}\right), \quad g^*=\frac{2m_0}{m^*} \tag{9}$$

Δ being the spin-orbit splitting of the valence band, and P the basic momentum matrix element. Eq.(9) embodies the split-off band contribution mentioned above [11]. For the valence band, this means adding the term [4]

$$\frac{e}{c}\,\kappa\, J_z B + \frac{e}{c}\, q\, J_z^3 B \tag{10}$$

where J_z is the spin 3/2 matrix and κ and q are material parameters. Actually, q is very small for the semiconductors of interest here and the second term can be neglected.

(iii) Modifying the boundary conditions at the interfaces which determine the eigenvalues. Indeed, the new terms introduced above modify the probability current operator and therefore the conditions enforcing its conservation at the interfaces. Before writing the new matching conditions, we must specify the structure of the solutions in each material. Introducing the harmonic oscillator representation for motion in the (x,y) plane based upon the commutation rules for the x and y components of $\vec{k}-(e/c)\vec{A}$ [12], the envelope wavefunction can be written :

$$\Psi_n=(c_1(z)\phi_n, c_2(z)\phi_{n-1}, c_3(z)\phi_{n+1}, c_4(z)\phi_{n+1}, c_5(z)\phi_n, c_6(z)\phi_{n+2}) \tag{11}$$

where ϕ_n are harmonic oscillator wavefunctions, n=-2,-1,0,1,... and the c coefficients are automatically vanishing for those components which have a negative oscillator index, if n<1. For the superlattice case all six $c_j(z)$ functions have the common prefactor exp $(ik_s z)$ where k_s is a quasi-momentum in the superlattice Brillouin zone, parallel to the external field. In terms of the Ψ_n states (11) the effective Hamiltonian can be written as a 6x6 matrix with the structure (see the Appendix) :

$$H_{jj'}=(D_{jj}^{zz}\, k_z^2 + E_j)\delta_{jj'} + (\Pi_{jj'} + A_{jj'}(n,B))k_z + C_{jj'}(n,B) \tag{12}$$

where j and j' label the 6 basis states, $\Pi_{jj'}$ is the coefficient of k_z in the terms present at zero field via the k·p interaction, and the A and C matrices represent the "new" magnetic field induced terms. With the replacement $k_z \to -i\,\partial/\partial z$, (12) gives a differential system for the $c_j(z)$ functions. Deriving from (12) an expression for the current operator and imposing its divergence-free character at the interfaces [8] we obtain, besides the continuity of the c functions, the following conditions :

$$-i\, D_{jj}\frac{\partial}{\partial z}\, c_j(z) + \frac{1}{2}\sum_{j'} A_{jj'}\, c_{j'}(z) \text{ continuous} \qquad j=1,\ldots 6. \tag{13}$$

We then apply to (12) and (13) the numerical method of solution explained in Ref.[8] to obtain the Landau levels as a function of the field B and, in a superlattice geometry, of the quasi-momentum k_s. This is shown in Fig.4 at $k_s=0$ for a GaAs-$Ga_{1-x}Al_xAs$ 120 Å - 120 Å superlattice with x=0.24, in the region corresponding to the valence subbands. The deviations from linearity of the Landau levels are quite striking, in particular the nearby field-independent behavior of some n=±1 levels over a very large range of B values. Although partly masked by the much steeper behavior of the conduction band Landau levels, these peculiar features should be observable in interband magneto-optical experiments.

The results of this type of calculations are in fact directly comparable to magneto-optical experiments, once the selection rules $\Delta k_s=0$, $\Delta n=\pm 1$ and the

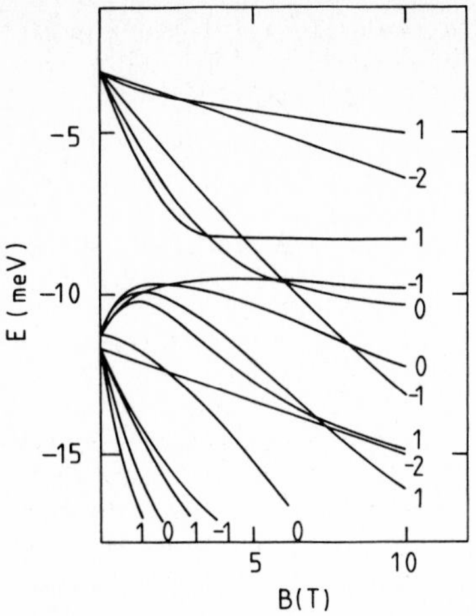

Fig.4 Landau levels of the holes for a 120Å-120Å GaAs-$Ga_{1-x}Al_xAs$ superlattice with x=0.24 at k_s=0. The quantum number n (see text) is indicated

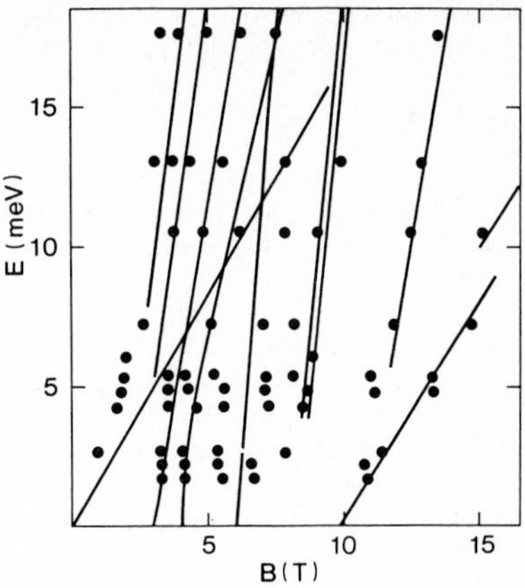

Fig.5 Far-infrared absorption maxima as a function of photon energy and magnetic field. Dots: experimental results [13, 14]. Solid lines: present theory

position of the Fermi level are taken into account to identify the allowed transitions between Landau levels. In the case of InAs-GaSb, the comparison with far-infrared results is satisfactory. In Ref.[10] results for a 120 Å - 80 Å sample were presented; in Fig.5 we show a comparison for a 200 Å -100 Å sample [13, 14]. Here the only disposable parameter is the energy difference between the top of the GaSb valence band and the bottom of the InAs conduction band, which was taken as Δ_{CV}=158 meV, in agreement with the current estimate Δ_{CV}=0.15 eV [15]. Of particular interest is the fact that although the band structure at zero field has a small energy gap [2], magneto-optical transition with arbitrarily small energies are observed [10]. This can be understood by considering the simple 2x2 model of (7), with a finite gap at B=0, in which we assume the Fermi level to lie. At the finite field where (see Fig.3) the n=0 and n=1 levels in the upper family cross, the Fermi level is at the crossing point, so that the gap vanishes at that particular field value, as well as at lower fields, at the crossing points between n=1 and n=2, n=2 and n=3, etc.

Acknowledgement

The numerical results were obtained with the computer facilities of the Centre de Calcul Vectoriel pour la Recherche, Palaiseau. We are grateful to G. Aubert for his interest.

References

1 G.Bastard: Phys.Rev. B25, 7584 (1982)
2 M.Altarelli: Phys.Rev. B28, 842 (1983)
3 Y.C.Chang and J.N.Schulman: Appl.Phys.Letters 43, 536 (1983)
4 J.M.Luttinger: Phys.Rev. 102, 1030 (1956)
5 S.S.Nedorezov:Soviet Phys.Solid State 12, 1814 (1971) (some key results of this paper are affected by errors or misprints)
6 A.Baldereschi and N.O.Lipari: Phys.Rev. B8, 2697 (1973)

7 See for example Landolt-Börnstein Numerical Data and Functional Relationships in Science and Technology, Group III, vol.17, O.Madelung ed. (Springer, Berlin 1982)

8 M.Altarelli: "Electronic Structure of Semiconductor Superlattices" in Applications of High Magnetic Fields in Semiconductor Physics, G.Landwehr, ed. (Springer, Berlin 1983) pp.174-185

9 M.Baumgartner, G.Abstreiter and E.Bangert: J.Physics C (Solid State Physics) (in press, 1984). In this paper similar results are derived for hole subbands in p-type inversion layers

10 A.Fasolino and M.Altarelli: in EP2DS V Proceedings, Surface Sci.(in press 1984)

11 L.M.Roth, B.Lax and S.Zwerdling: Phys.Rev.114, 90 (1959)

12 C.R.Pidgeon and R.N.Brown: Phys.Rev.146, 575 (1966)

13 J.C.Maan, Y.Guldner, J.P.Vieren, P.Voisin, M.Voos, L.L.Chang and L.Esaki: Solid State Commun.39, 683 (1981) and private communication

14 P.Voisin, Thesis, Université Paris-Sud 1983 (unpublished)

15 L.L.Chang: J.Phys.Soc.Japan 49, Suppl.A, 997 (1980)

Appendix

The Hamiltonian matrix (12) in terms of the states Ψ_n given by (11) and in the underlying Bloch function basis :

$$|s\uparrow>,\ |J_z=3/2>,\ |J_z=-\tfrac{1}{2}>,\ |s\downarrow>,\ |J_z=\tfrac{1}{2}>,\ |J_z=-3/2>,$$

is given by (the lower half of the matrix follows by Hermitian conjugation):

$$\begin{array}{cccccc}
H_{11} & iP\sqrt{\frac{eB}{c}n} & -iP\sqrt{\frac{eB(n+1)}{3c}} & 0 & -i\sqrt{\frac{2}{3}}Pk_z & 0 \\
 & H_{22} & \frac{eB}{c}\bar{\gamma}\sqrt{3n(n+1)} & 0 & \bar{\gamma}k_z\sqrt{\frac{6eBn}{c}} & 0 \\
 & & H_{33} & i\sqrt{\frac{2}{3}}Pk_z & 0 & -\bar{\gamma}k_z\sqrt{\frac{6eB(n+2)}{c}} \\
 & & & H_{44} & iP\sqrt{\frac{eB(n+1)}{3c}} & -iP\sqrt{\frac{eB(n+2)}{c}} \\
 & & & & H_{55} & \bar{\gamma}\frac{eB}{c}\sqrt{3(n+2)(n+1)} \\
 & & & & & H_{66}
\end{array}$$

where :

$$H_{11} = H_{44} = E_c + \frac{eB}{m^*c}(n+1) + \frac{1}{2m^*}k_z^2$$

$$H_{22} = E_v - (\gamma_1/2 - \bar{\gamma})k_z^2 - \frac{eB}{c}[(\gamma_1 + \bar{\gamma})(n - \tfrac{1}{2}) + 3\kappa/2]$$

$$H_{33} = E_v - (\gamma_1/2 + \bar{\gamma})k_z^2 - \frac{eB}{c}[(\gamma_1 - \bar{\gamma})(n+3/2) - \kappa/2]$$

$$H_{55} = E_v - (\gamma_1/2 + \bar{\gamma})k_z^2 - \frac{eB}{c}[(\gamma_1 - \bar{\gamma})(n + \tfrac{1}{2}) + \tfrac{1}{2}\kappa]$$

$$H_{66} = E_v - (\gamma_1/2 - \bar{\gamma})k_z^2 - \frac{eB}{c}[(\gamma_1 + \bar{\gamma})(n+5/2) - \tfrac{3}{2}\kappa] .$$

Combined Electric and Magnetic Field Effects in Semiconductor Heterostructures

J.C. Maan

Max-Planck-Institut für Festkörperforschung, Hochfeld-Magnetlabor, 166X
F-38042 Grenoble Cedex, France

The subband structure of electrons confined in a parabolic potential well, subjected to magnetic fields at arbitrary angles between the electric and magnetic field, is derived analytically. These theoretical results are used to analyze transport measurements on a thin layer of n-doped GaAs, confined between p layers under these conditions. Furthermore the results of numerical calculations of the Landau level structure in thin (periodicity <120nm) GaAs-GaAlAs doping superlattices are presented for B parallel to the layers. These results show that cyclotron resonance can be observed as long the cyclotron energy is less than the subband width.

I INTRODUCTION

A two-dimensional electronic system, such as can be realized in several semiconductor heterostructures, consists of carriers which are mobile in the plane along the interface and confined in the perpendicular direction. This confinement leads to the formation of discrete subbands. A magnetic field perpendicular to the interface splits each subband into Landau levels and leads therefore to a completely quantized system, which has been extensively studied. This case where the magnetic field is in the same direction as the confining electric field is however a very special one because only then the Hamiltonian can be separated in an electric part leading to subbands and a magnetic part leading to Landau levels. For any other orientation this separation is not possible anymore and the band structure in a magnetic field is more complicated. There exist several experimental results which show the effects of a parallel field component on a two-dimensional (2D) system [1,8]. The problem has been treated theoretically by ANDO[9] for a magnetic field parallel to a SiO_2-Si interface in a MOS transistor with numerical methods. Furthermore, approximate analytic solutions for a triangular potential have been given by BHATTACHARYA [10]. Here analytic solutions will be given for a parabolic potential well which is one of the very few potential shapes where this is possible. Apart from the academic interest in such a potential shape, the solutions turn out to be useful in the analysis of transport measurements of a thin n-doped GaAs layer between two p-type cladding layers.

Next, the results of numerical calculations of the Landau level structure in a Kronig-Penney potential with the magnetic field parallel to the plane of the layers are given. The parameters in the calculation are chosen to correspond to reasonable magnetic fields (<20T), reasonable sample parameters for GaAs-GaAlAs superlattices (periodicity d, 60nm<d<120nm, effective mass $m^* = 0.07m_0$, and a barrier height of 0.3eV). These results show an intriguing band structure with features which seem to be easily experimentally observable, but which have not yet been reported.

II LANDAU LEVELS IN A POTENTIAL WELL IN TILTED MAGNETIC FIELDS.

Now, let us consider the Hamiltonian for carriers in a potential well V(z) and with a magnetic field in the xz plane. The Hamiltonian is given by:

$$H = (\underline{p} + e\underline{A})^2 + V(z). \quad (1)$$

Making use of the translational symmetry in the y direction, the wavefunction can be written as $e^{ik_y y}\Psi(x,z)$. Substituting this wavefunction in eq. 1 and making the usual substitution $x' = x + \hbar k_y / eB_z$, which is valid as long as $B_z \neq 0$ and using $\underline{A} = (0, xB_z - zB_x, 0)$ as a gauge, eq. 1 can be written as:

$$\left[-\frac{h^2}{2m^*}\left(\frac{\partial^2}{\partial x'^2} + \frac{\partial^2}{\partial z^2}\right) + \frac{e^2}{2m^*}(x'B_z - zB_x)^2 + V(z)\right] e^{ik_y y}\Psi(x,z) =$$

$$E\, e^{ik_y y}\Psi(x,z). \quad (2)$$

An interesting aspect of this Hamiltonian is that k_y is not appearing explicitely. Due to the translation symmetry in the x direction the eigenvalues do not depend on x, and therefore they cannot depend on x' either, and consequently all eigenvalues with different k_y values are degenerate with a degeneracy per Landau level of eB_z/h, depending on the perpendicular component of the magnetic field only. Since Shubnikov- de Haas oscillations are determined by the degeneracy per Landau level,these will depend in principle only on the perpendicular component of the magnetic field . The spacing between the Landau levels however will be determined by the total field and therefore a cyclotron resonance experiment will in general show a complicated angular dependence. This effect has experimentally been observed in GaSb/InAs heterojunctions, where Shubnikov-deHaas experiments [11] have clearly show an angular dependence being determined only by the perpendicular component, whereas magnetooptical experiments on the same samples showed a pronounced different behaviour [4,5]. If $B_z = 0$ (magnetic field parallel to the layer) the substitution $x' = x + hk_y / eB_z$ is not anymore possible and instead the eigenvalues will depend explicitely on k_y. In this case no Landau levels are formed but subbands with energies increasing with parallel magnetic field and a field-dependent density of states . This latter fact is observed experimentally since one can change the occupation of subbands by a parallel magnetic field. If for a given sample several subbands are occupied at zero parallel magnetic field less subbands are needed to accompdate the carriers with increasing parallel field. This effect has been observed as a decrease in the resistivity in the GaAs-GaAlAs heterojunction sample with increasing magnetic field [12]. This decrease is caused by the reduction of the intersubband scattering rate and parallel field transport measurements can be used to measure the contribution of the intersubband scattering to the total resistivity.

III LEVEL STRUCTURE IN A PARABOLIC POTENTIAL WELL.

Let V(z) in eq. 1 be Az^2 (parabolic potential) then eq. 2 can be tranformed into:

$$\left[-\frac{\hbar^2}{2m^*}\left(\frac{\partial^2}{\partial\xi^2}+\frac{\partial^2}{\partial\zeta^2}\right)+a\xi^2+b\zeta^2\right]e^{ik_y y}\Xi(\xi)Z(\zeta) = E\, e^{ik_y y}\Xi(\xi)Z(\zeta). \quad (3)$$

a and b are coefficients which depend on A, B_z and B_x in a complex manner. ξ and ζ are related to x' and z through a rotation of the coordinate system through an angle α. Eq. 3 can directly be solved since it can be separated in two harmonic oscillators with eigenvalues $(N_1+1/2)\hbar\omega_1$ and $(N_2+1/2)\hbar\omega_2$ given by:

$$\omega_{1,2} = \left[\frac{1}{2}(\omega_c^2+\omega_0^2) + \frac{1}{2}\left(\omega_c^4+\omega_0^4+2\,\omega_0^2\,(\omega_x^2-\omega_z^2)\right)^{1/2}\right]^{1/2}. \quad (4)$$

Here ω_c is eB_{tot}/m^* $\omega_0=(2A/m^*)^{1/2}$, $\omega_{x,z}$ are the respective components of the cyclotron frequency in the x and z direction.

In Fig.1 the magnetic field dependence of $\hbar\omega_{1,2}$ is shown for different orientations of the magnetic field with respect to the z axis. For B//z the subband separation remains unchanged and every subband splits into Landau levels. For finite angles between B and z the subband separation itself depends on the magnetic field, and in addition splits into Landau levels. It is therefore clear that the separation between successive Landau levels is $\neq\hbar\omega_c$, and depends on the angle. On the other hand, the translational symmetry of k_y, as reflected by the fact that both ξ and ζ through the rotation of the coordinate system contain terms of $x+\hbar k_y/eB_z$, implies that both sets of Landau levels have a degeneracy of $eB_z/\hbar$, as was already mentioned in the previous section.

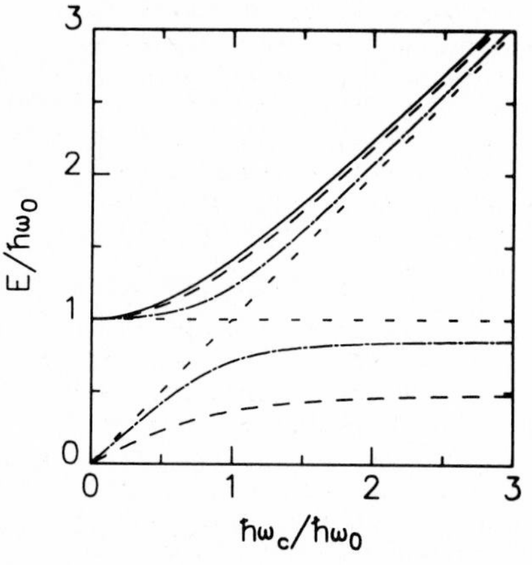

Figure 1

Magnetic field dependence of the eigen-frequencies of the Landau levels in a parabolic potential well (eq. 4) for different angles φ between B and z. φ=0° (short dashed line), φ=30° (dashed dotted line), φ=60° (long dashed line) and φ=90° (drawn line).

Since for a parallel magnetic field $B_z=0$ one cannot perform the translation of x to eliminate k_y from eq. 1 and the eigenvalues depend explicitely on k_y. To solve eq. 1 the wavefunction is chosen as $e^{ik_y y}* e^{ik_x x}\Psi(z)$. After some manipulation eq. 1 can then be transformed into an harmonic oscillator with eigenvalues:

$$E_n = (n+1/2)\,\hbar\,(\omega_c^2+\omega_0^2)^{1/2} + \frac{\hbar^2 k_x^2}{2m^*} + \frac{\hbar^2 k_y^2}{2m^*}\left(\frac{\omega_0^2}{\omega_c^2+\omega_0^2}\right)^{1/2} \quad (5)$$

with n the subband quantum number (0,1,2,3...). Eq.5 implies that the subbands in a parallel magnetic field move to higher energy with increasing magnetic field. The k_y^2 term has a field-dependent coefficient which may be interpreted as a magnetic field dependent effective mass . This anisotropy in the xy plane is induced by the parallel magnetic field which distinguishes the x direction (parallel to the field) from the y direction (perpendicular to the field). Since the effective mass has become anisotropic and field dependent, the density of states is given by:

$$g(E)dE = \frac{m^*}{\pi \hbar^2} \left(\frac{\omega_c^2 + \omega_0^2}{\omega_0^2} \right)^{1/2} dE \qquad (6)$$

and is also field dependent. There is no contradiction between eq. 5 and the fact that at very small deviations from parallel Landau levels are formed with a degeneracy of eB_z/h since one can easily prove that for vanishingly small angles the number of states in a given energy range is equal using either eq.6 or the eigevalues of eq. 4.

IV TRANSPORT MEASUREMENTS ON GaAs pnp STRUCTURES.

A pnp structure consists of alternately grown thin layers of n and p type doping in the same host material. In such a structure the electrons from the donors transfer to the acceptors in the p layers. The resulting space charge leads to a potential well confining the electrons in the center of the n layer. Through selective contacts to the p and the n layers , the potential difference between these layers can be varied externally. This way the carrier density in the n layer can be modulated provided that the n layer thickness is of the order of twice the depletion layer thickness [13]. Since the electron and the holes are spatially separated , the recombination current at low temperatures is negligibly small for reversed bias and for forward bias as long as the applied voltage U_{np} times the elementary charge is small compared to the bandgap. Therefore the change in the carrier density can be considered quasi static. Without external voltage the depletion width d_{depl} in the n layer at the pn junction is approximately given by:

$$d_{depl} = [2\varepsilon E_G N_A / e^2 N_D (N_D + N_A)]^{1/2} \qquad (7)$$

where E_G is the bandgap, ε the dielectric constant, N_D the donor and N_A the acceptor concentration. If d_{depl} is exacly half the doping layer thickness the electrons are confined in a parabolic potential well and equally spaced 2D subbands are formed with a separation of $(e^2 N_D / \varepsilon m^*)^{1/2}$. With a positive (negative) voltage U_{np} carriers are injected (extracted) and therefore impurities are neutralized (ionized) which decreases (increases) the space charge potential leading to a decrease (increase) of the subband separation.

The sample was grown by the technique of Molecular Beam Epitaxy [14] and had the usual Hall bar shape for the transport measurements. Selective n and p contacts were diffused in the layers and were arranged in such a way that the Hall voltage and the resistance of the n channel could be measured as a function of U_{np}. The parameters of the present sample were: thickness of the p-type cladding layers 500 nm and of the central n layer 90 nm with $N_D = N_A = 7\times10^{17}\ cm^{-3}$ and an electron mobility of ($2200 cm^2/Vs$) at 4K [15].

In Fig. 2 the results of the magnetoresistance measurements with

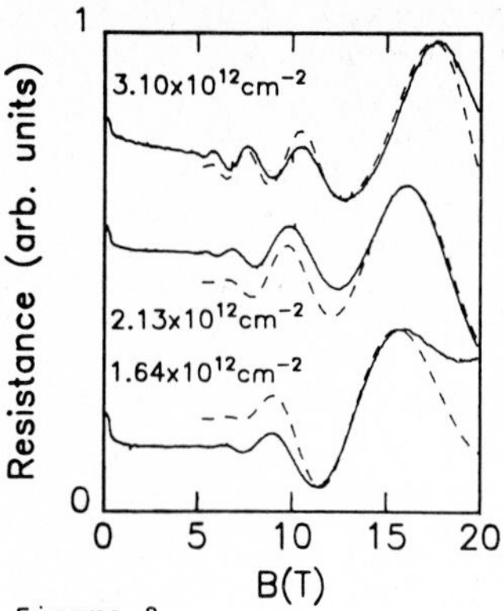

Figure 2

Measured sample resistance as a function of the magnetic field for three different carrier densities (drawn lines) with the magnetic field perpendicular to the layers. The dashed lines represent a theoretical fit as described in the text

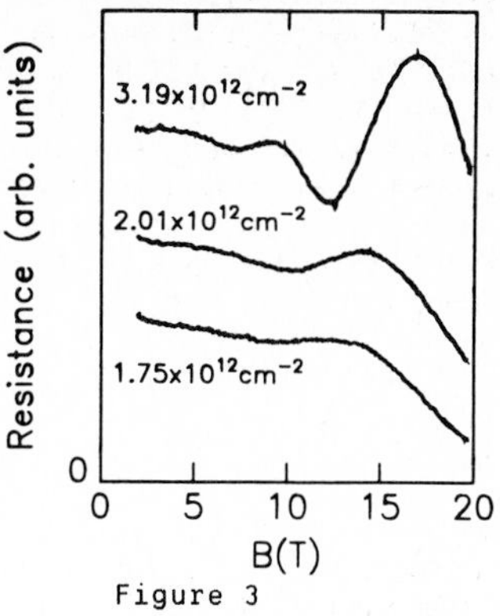

Figure 3

Measured sample resistance as a function of the magnetic field parallel to the layers for three different densities

the magnetic field perpendicular to the layer plane are shown for three different carrier densities . In Fig. 3 the result for B parallel to the layer plane are plotted. The current was always perpendicular to the magnetic field. In both cases well pronounced quantum oscillations are observed but for $B_{//}$ the amplitude is weaker especially at lower fields and the strong maximum at high B values is shifted to lower magnetic fields with respect to the $B_{\perp}$ data. These results clearly show an anisotropy which is incompatible with the three-dimensional GaAs conduction band structure and which demonstrates directly the two-dimensional nature of the conductance in the n channel. Fig. 4 shows the change in the carrier density as determined from the Hall effect as a function of voltage U_{np}.

A model calculation has been used to simulate the $B_{\perp}$ measurements. The two-dimensional conductivity is calculated using the theory of Ando and Uemura[16].This procedure was successfully employed to explain experimental data from the GaAs/$Ga_{1-x}Al_xAs$ heterojunction[17]. This model is extended in a simple manner to allow for the occupation of more subbands. The conductivity tensor element σ_{xx} can be written as:

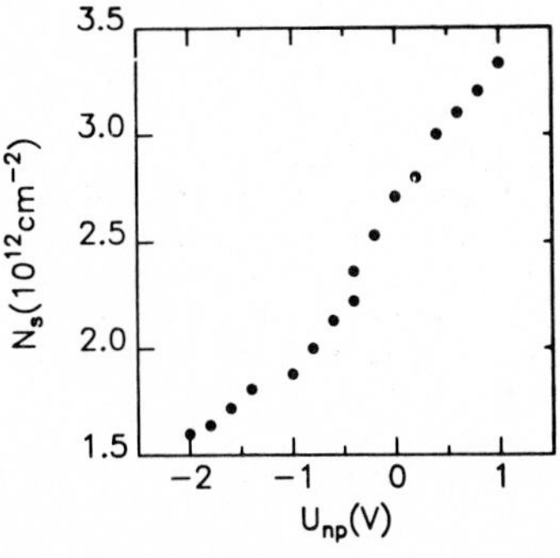

Figure 4

Measured change in the carrier density as a function of the voltage U_{np} between the n and the p layers

$$\sigma_{xx} \propto \int(-\frac{\partial f}{\partial E})\sum_n \sum_N (N+1/2)\exp-\left[\left(\frac{E-E_n-E_N}{\Gamma}\right)\right]^2 dE \; ; \qquad (8)$$

here Γ is a level broadening parameter (a gaussian broadening is assumed) and f the Fermi function. The tensor relation has been used to calculate ϱ_{xx} and σ_{xx} and ϱ_{xy} is taken to be $N_s e/B$, where N_s is the 2D carrier density as determined from the Hall effect. The subband separation $E_{n+1}-E_n=\hbar\omega_F$ is treated as a fit parameter. Representative results for the fit are shown as the dotted lines in Fig. 2. Qualitatively the model reproduces the experimental results remarkably well as witnessed by the good agreement of the position and the amplitude of the observed and calculated extrema. $E_{n+1}-E_n$ as a function of the carrier density obtained in this way is shown in Fig.5 (circles) and is found to decrease with increasing carrier density as expected. It should be noted that the oscillatory pattern does not change drastically as the carrier density increases by a factor two. We explain this by the fact that the observed SdH oscillations result from a superposition of different subbands and that their period in this case is not related to the total 2D carrier density in a simple manner. As a consequence of the simultaneous decrease in the subband separation with increasing carrier density the oscillations move only slightly to higher magnetic fields.

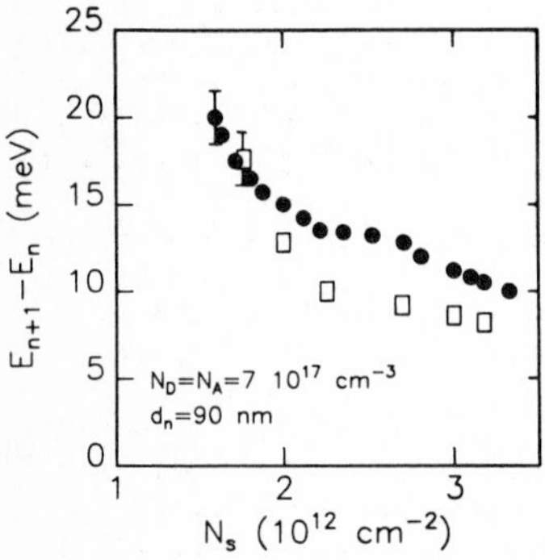

Figure 5

Subband separation as derived from the experiments with B perpendicular (circles) and parallel to the layer plane (squares)

Magnetoresistance oscillations for $B_{//}$ must be explained in another way. As was discussed in section III, the density of states and the subband separation increase with magnetic field and this gives rise to a successive depopulation of the subbands, which is seen as a change in the magnetoresistance due to the associated change in the intersubband scattering rate. It is not a priori clear which feature in the spectra corresponds to the situation where a subband moves through the Fermi energy. It has been found that good agreement between the subband structure obtained for $B_\perp$ and $B_{//}$ can be reached if a maximum in the resistance is assumed to be associated with the case that the Fermi energy is at the edge of a subband. Qualitatively one can argue that the resistance will decrease when a subband is emptied at increasing magnetic field, since the intersubband scattering will be reduced. Furthermore ,at somewhat lower magnetic fields, the density of states in this subband reaches its maximum just before it is emptied, implying a maximum in the intersubband scattering rate as the subband edge crosses the Fermi energy. $E_{n+1}-E_n$ can be calculated from eq.5 and 6 at a magnetic field where a maximum occurs using the measured carrier density. The maximum at the highest magnetic field is assigned to the case where the second subband is emtied. $E_{n+1}-E_n$ as obtained from this analysis is shown by the squares in Fig. 5, and the results agree within experimental error with those obtained for $B_\perp$. In addition, the calculated magnetic field positions for the coincidence of higher subband edges with the Fermi energy using this value for $E_{n+1}-E_n$ agree within measuring accuracy with the resistance maxima observed at lower fields.

V BAND STRUCTURE OF A SUPERLATTICE IN A PARALLEL MAGNETIC FIELD.

A superlattice is a structure consisting of periodically arranged layers of different materials, with layer thicknesses such that the interaction between successive layers plays an essential role. The most simple case is that of GaAs-GaAlAs where the GaAs bandgap is entirely contained within that of GaAlAs, leading to a square wave modulated potential for the valence band and the conduction band. Due to the simple structure of the conduction band of the materials, the conduction band of such a superlattice can adequately be described by the well-known Kronig-Penney model. For parameters as can be experimentally realized in such a system (barrier height 0.3eV, GaAs and GaAlAs layer thickness 3 nm) the interaction between successive wells is such that the subbands shows a dispersion in the growth direction with a width 75 meV in this case. This value is of the order the Landau level separation in GaAs for a realistic magnetic field (30 meV at 20T). It is therefore clear that one cannot use the same approximations as in bulk materials to calculate the level structure in a magnetic field at the band edge, since in that case the bandwidth is much larger than the Landau level spacing. Another way of expressing the same fact is that in the case of a superlattice the lattice parameter in one dimension becomes comparable with the cyclotron orbit.

The dispersion relation in the growth direction has originally been one of the motivations in the investigation and preparation of superlattices [18]. However, very few observations of the finite width of a superlattice have been reported[19], and experiments trying to show the three-dimensional character of a superlattice by measuring the IV characteristic for a current through the layers have been inconclusive [20]. To investigate whether one can observe this three-dimensional aspect of a superlattice with the help of a magnetic field parallel to the layers, calculations of the band structure of a Kronig Penney model in a parallel field have been performed.

It is important to realize that the effect of a parallel magnetic field on a superlattice is somewhat different from that on a single well. In this latter case the effect of the field is that every two-dimensional subband which is characterized by a single subband quantum number is affected by the field and that subbands with higher quantum numbers, due to their larger spatial extension, are more field dependent. Consequently different subbands do not cross each other. The subband in a superlattice is characterized by a subband quantum number and by a wavevector in the superlattice direction which comes from the periodicity of the system. Therefore several magnetic levels will originate from a single subband in this case, and these can interact with each other.

To solve the problem one makes the substitition $z'=z+\hbar k_y/eB_x$ (only a parallel field) in eq.1 which reduces to:

$$\left[-\frac{h^2}{2m^*}\frac{\delta^2}{\delta_z{}'^2}+\frac{e^2B^2z'^2}{2m^*}+V(z)\right]e^{ik_xx}e^{ik_yy}\psi(z) = E\, e^{ik_xx}e^{ik_yy}\psi(z). \quad (9)$$

$V(z-\hbar k_y/eB_x)$ is the Kronig-Penney potential which is shifted by an amount $\hbar k_y/eB_x$ with respect to the center of the harmonic term from the field. Eq.9 can be solved numerically using the finite element method for different values of $\hbar k_y/eB_x$, which represents the position of the CR orbit with respect to the potential. Since the potential is periodic the complete solution is obtained for $0<hk_y/eB_x<d$, where d is the periodicity of a superlattice.

The results of such a calculation are shown in Fig.6 for two different periodicities (6nm and 100 nm) and different magnetic fields and

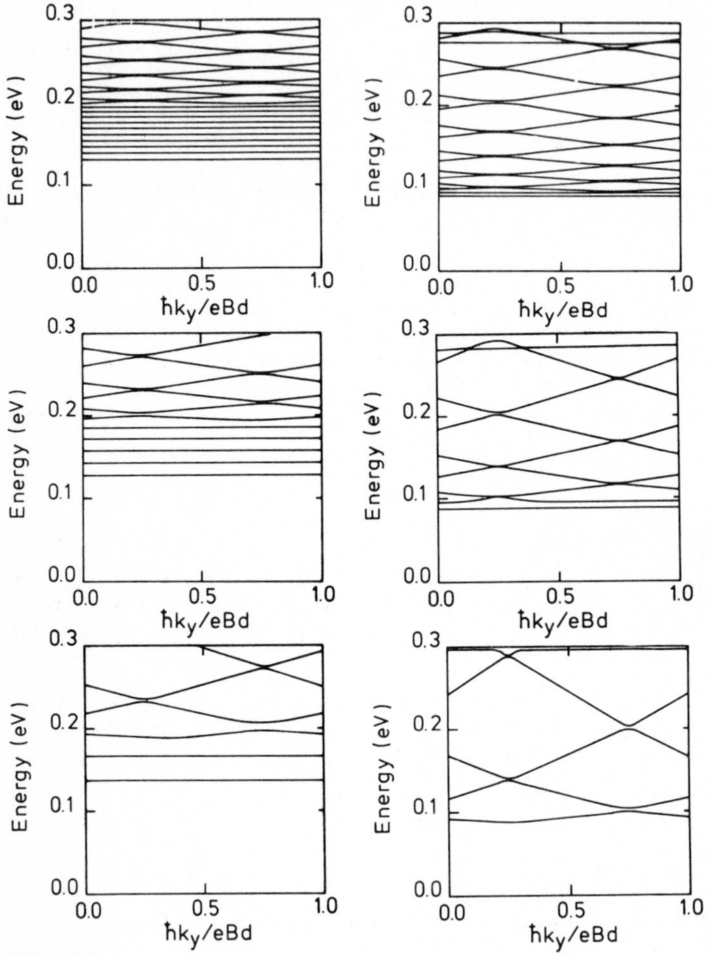

Figure 6

Landau levels in a Kronig-Penney potential with B parallel to the plane of the layers. The barrier height is 0.3 eV in both cases and an effective mass of $0.07m_0$ is used. The left column shows the result for 6nm periodicity with from top to bottom ; B=5,10,20. The right column is the result for d=10nm with from top to bottom B=5,10,20 T. The subband at zero magnetic field is between 120 and 195 meV for 6 nm and between 90 and 98 meV for 10nm.

parameters corresponding to GaAs-GaAlAs with 0.3eV barrier height. One observes a flat Landau level like dispersion in the region of the subbands width (between 120 and 195 meV for 30nm and between 90 and 98meV for 10 nm) and a strong dispersion in the region of the gap in the 10 nm case already at moderate fields. For magnetic field values where flat Landau levels are found , the field dependence of the separation between the levels is close to that of bulk GaAs, but with oscillations around this value.

One can qualitatively understand the results shown in Fig. 6 in the following way . A single well of finite thickness gives rise to a Landau level with an energy which depends quadratic on the distance of the orbit center ($\hbar k_y/eB_x$) to the well. Several wells, regularly arranged, lead to Landau levels of each well crossing those of the other wells in the center of the well and of the barrier (at $\hbar k_y/eB_x d$ 0.25 and 0.75 in Fig. 6). Landau levels of wells which are far apart interact weakly and cross at these points (the higher levels in Fig 6). Landau levels of wells which are nearby (the lowest ones in Fig. 6) do interact strongly and show an anticrossing. The more close the next well is (i.e. the lower the Landau level number, or the smaller the periodicity) the more strong this anticrossing becomes, leading to flat Landau levels in the energy region of the subbands.

It is thought that the behaviour shown in Fig. 6 can easily be observed in a far infrared absorption experiment. One would expect to observe a strong deviation from a linear dependence between resonant field an frequency as soon as the cyclotron energy approaches the subband width. It is felt that such experiments would be interesting in their owm right, but would also be useful to measure an important quantity as the subband width, and thereby determine the barrier height and layer thicknesses upon which the subband width is strongly dependent.

REFERENCES

-1 F.Koch in 'Physics in High Magnetic Fields',Solid State Sciences 24, ed. S.Chikazumi,N.Miura,Springer,Berlin,(1981)

-2 W.Beinvogl,F.Koch,Phys.Rev.Lett 40,1736,(1978)

-3 R.E.Doezema,M.Nealon,S.Whitmore,Phys.Rev.Lett.45,1593,(1980)

-4 J.C.Maan,Ch.Uihlein,L.L.Chang,L.Esaki,Solid State Commun.44,653,(1982)

-5 J.C.Maan in 'Lecture Notes in Physics' 177,ed. G. Landwehr,p 161,Springer Berlin,(1983)

-6 J.H.Craseman,U.Merkt, Solid State Commun.47,917,(1983)

-7 J.H.Craseman,U.Merkt,J.P.Kotthaus, Solid State Commun,to be published

-8 Z.Schlesinger,J.C.M.Hwang,S.J.Allen Jr,Phys.Rev.Lett 50,2089,(1983)

-9 T.Ando, J.Phys.Soc. Japan 39,411,(1975)

-10 S.K.Bhattacharya,Phys.Rev.B 25,3756,(1982)

-11 L.L.Chang,E.E.Mendez,N.J.Kawai,L.Esaki,Surface Sci. 113,306,(1982)

-12 Th.Englert,J.C.Maan, D.C.Tsui,A.C.Gossard,Solid State Commun.45,989, (1983)

-13 K.Ploog,H.Künzel,J.Knecht,A.Fischer,G.H.Döhler,Appl. Phys.Lett.38, 870,(1981)

-14 K.Ploog,A.Fischer,H.Künzel,J.Electrochem.Soc. 128,400,(1981)

-15 J.C.Maan,Th.Englert,Ch.Uihlein,H.Künzel,A.Fischer,K.Ploog,Solid State Commun.47,383,983

-16 T.Ando,Y.Uemura,J.Phys.Soc.Japan 36,959,(1974)

-17 Th.Englert,D.C.Tsui,A.C.Gossard,Ch.Uihlein,Surface Sci. 113,295,(1982)

-18 L.Esaki,R.Tsu, IBM J.Res.Dev.14,61,1970

-19 J.C.Maan,Y.Guldner,J.P.Vieren,P.Voisin,M.Voos,L.L.Chang,L.Esaki,Solid State Commun. 39,683,(1981)

-20 L.Esaki,L.L.Chang,Phys.Rev.Lett.33,495,1974

Strained-Layer Superlattices

Paul Voisin

Groupe de Physique des Solides de l'Ecole Normale Supêrieure
(Laboratoire associé au C.N.R.S.), 24 rue Lhomond
F-75231 Paris Cedex 05, France

We report optical absorption measurements in high quality GaSb-AlSb superlattices. The spectra exhibit free exciton peaks separated by absorption plateaus characteristic of the two-dimensional density of states. The effect of strains induced by the lattice mismatch is manifested by the anomalous energies of the absorption edges and the reversal of the heavy- and light-hole exciton positions. We develop a simple effective mass theory which yields a very good agreement with the experimental data. We finally discuss the importance of strain-induced effects in other superlattice systems.

I Introduction

The development during the last ten years of sophisticated growth techniques such as molecular beam epitaxy (MBE) or metal-organic chemical vapor deposition has allowed the achievement of a variety of superlattice (SL) structures built with two different semiconductors. SL's generally display structural as well as electronic properties indicating a high crystalline integrity, which implies that the small but generally non-zero lattice mismatch of the hosts, $\delta a/a$, is accommodated by elastic deformations rather than by a large density of defects $n_d \sim 2\delta a/a^3$. Recently, successful growth and characterization of SL's built up from materials presenting deliberately large lattice mismatches ($\delta a/a \sim 1\%$) have been reported. Most of the interest has at first been focused [1] on systems such as GaP-$GaAs_xP_{1-x}$ [2] or GaAs-$In_xGa_{1-x}As$ [3,4] in which the small gap material is a ternary alloy. Detailed analysis in these cases is quite difficult, as it requires extended and accurate knowledge of many physical parameters in the alloys, which often are unavailable. Also, in these systems, the bandgaps and lattice constants vary quite rapidly with the alloy composition, which is particularly difficult to measure. Clearly, strained-layer superlattices in which none of the hosts is an alloy are likely to allow an easier and more conclusive study of the strain-induced effects. For that purpose, a very good candidate is the GaSb-AlSb system, which has been recently fabricated by MBE [5] and corresponds to a lattice mismatch equal to 0.65%. Indeed, preliminary optical absorption [6,7] and luminescence [7-9] data have been reported, which [6] established the type I nature of this system and indicated a strain-induced shrinkage of the GaSb band gap. In section II, we report optical absorption measurements performed at low temperature in high quality GaSb-AlSb SL's. The transmission spectra exhibit well-defined exciton peaks separated by absorption plateaus which reflect the two-dimensional nature of the structures investigated. The interpretation of the data from a simple effective mass theory (section III) shows that, as a result of the strains induced by the lattice mismatch, a shrinkage of the GaSb bandgap and a reversal of the heavy- and light-hole subbands occur, an effect hitherto unobserved in SL's. The fit of the results, discussed in section IV, finally allows an accurate determination of band offsets in this system. Section V is devoted to speculative estimates of the importance of the strain-induced effects in other superlattice systems.

II Optical absorption measurements in GaSb-AlSb SL's

The structures that we have investigated were grown on (100) semi-insulating GaAs substrates, with the following growth sequence : a 1000 Å thick GaAs smoothing layer, a 3000 Å thick AlSb layer, n periods of alternating layers of GaSb and AlSb with thicknesses d_1 and d_2 respectively, and a 200 Å thick GaSb protective layer. The layers are of high purity, with an impurity concentration in the 10^{15} cm^{-3} range. The structures were evaluated from X-ray measurements with an automated diffractometer [10] , using the Cu Kα readiation and the (111) planes of Ge for monochromating and collimating the incident beam. Fig.1 shows the standard intensity vs angle plot of the (004) diffraction patterns on a logarithmic scale for the two samples reported here (S1 and S2). It is seen that a large number of high-order satellite peaks are resolved. While the orders cannot be precisely assigned, partly because of contributions from the buffer and the protective layer, their presence and their narrow peak width demonstrate the high structural quality of the samples. In addition, the period can be accurately determined, (632.8 ± 1) Å for S1 and (502.8 ± 1) Å for S2, corresponding to d_1(GaSb) = 181 Å, d_2(AlSb) = 452 Å and d_1 = 84 Å, d_2 = 419 Å, respectively. The number of periods is n = 10 for both samples.

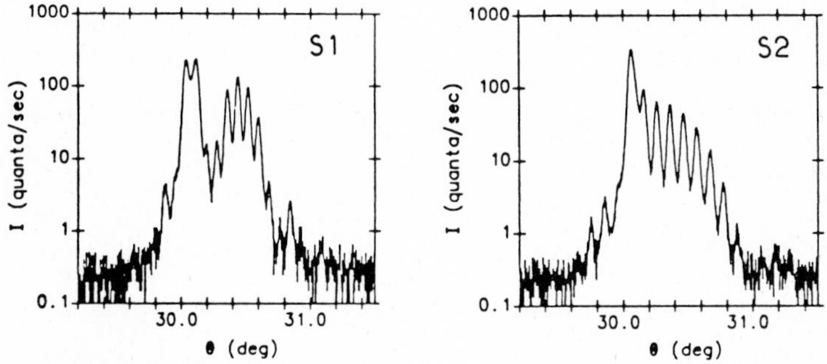

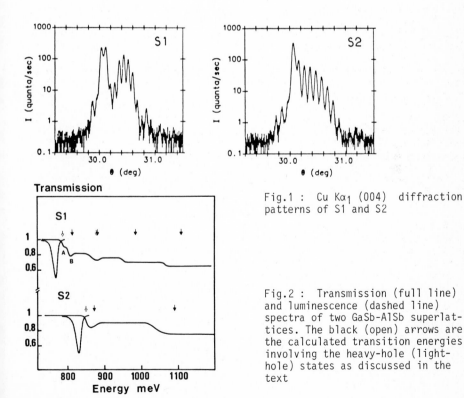

Fig.1 : Cu $K\alpha_1$ (004) diffraction patterns of S1 and S2

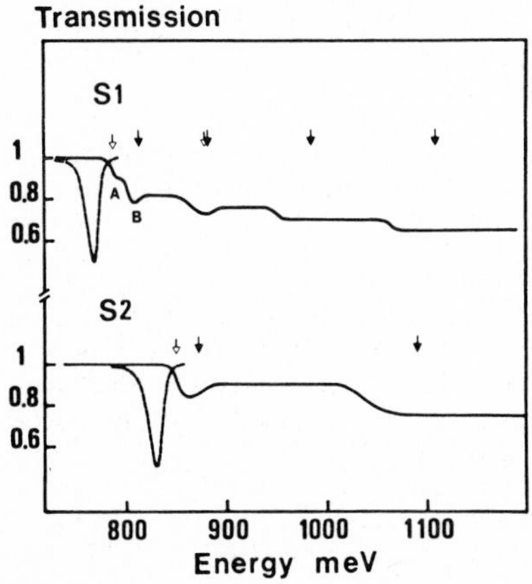

Fig.2 : Transmission (full line) and luminescence (dashed line) spectra of two GaSb-AlSb superlattices. The black (open) arrows are the calculated transition energies involving the heavy-hole (light-hole) states as discussed in the text

Optical transmission spectra were measured using a single beam set-up and compared to the transmission spectrum of a bare semi-insulating GaAs substrate as described in Ref.6. Fig.2 shows the observed spectra which exhibit clearly the step-like behavior characteristic of the two-dimensional density of states, and also exciton absorption peaks at the onset of the steps. The excitonic features are indicative of the high quality of these structures and of the type I nature of this system, i.e., both conduction and valence states are confined in the GaSb quantum wells. The transmission spectrum of S1 presents, at first sight, two intriguing features : (i) the B exciton peak, which is likely to involve the heavy-hole subband HH_1 and the ground conduction subband E_1 because it is the most intense transition, occurs at 810 meV. In a simple quantum well model, the sum $HH_1 + E_1$ of the corresponding confinement energies is easily estimated to be 25 meV, so that this transition is expected to occur at E_g^{GaSb} + 25 meV = 835 meV. The observed energy position is interpreted in terms of a strain-induced GaSb band gap shrinkage. The same effect, in fact, is seen on the transmission spectrum of sample S2 for which $HH_1 + E_1$ = 90 meV. The first exciton peak in this sample is observed at 870 meV, about 30 meV below the strainless predicted value. (ii) A well-resolved structure, A, is observed 20 meV below the B exciton peak, with an absorption coefficient in the same range of magnitude. As discussed below, this structure corresponds to absorption by a free exciton involving the first light-hole level LH_1. As a spectacular consequence of the strain, the ground valence state in this structure is the light-hole level.

III Theory

The interpretation of these data is based on the following theoretical considerations [11]. The lattice constants of GaSb and AlSb at room temperature are a_1 = 6.0959 Å and a_2 = 6.1355 Å respectively. The lattice mismatch, $\delta a/a$ = 0.65%, will retain the same value at low temperature, as easily seen from the comparison of the thermal expansion coefficients [12].

In the superlattice, the GaSb layers stretch in the x, y directions of the layer plane, as a result of a biaxial tensile stress, while the AlSb layers are under equal compressive biaxial stress. This situation can be analysed as well in terms of the sum of hydrostatic dilation plus a uniaxial compression in the (100) z direction (GaSb layers) or a hydrostatic compression plus a uniaxial elongation in the z direction (AlSb layers). As the elastic constants of GaSb and AlSb are essentially the same, the in-layer lattice parameter of the strained layers SL $a_\perp$ is simply given by :

$$a_\perp(d_1 + d_2) = a_1 d_1 + a_2 d_2 \,. \tag{1}$$

The strength of the stress X applied to a GaSb layer is easily expressed in terms of the usual elastic constants K and μ equal [13] to 6.64×10^{10} and 2.16×10^{10} (MKS units) respectively :

$$X = \left(\frac{2}{9K} + \frac{1}{6\mu}\right)^{-1} (a_\perp - a_1)/a_\perp \,. \tag{2}$$

Now, this stress results in the following modification of the GaSb band parameters [14] : (i) A shrinkage of GaSb band gap E_g due to a hydrostatic dilation of strength X :

$$\Delta E_g = - \frac{\delta E_g}{\delta p} X = - 3a(S_{11} + 2S_{12})X \,. \tag{3}$$

ii) An increase of the GaSb band gap and a splitting of the Γ_8 valence band due to the uniaxial compression of strength X in the z direction, resulting in :

$$\Delta(E_c - E_{HH}) = a(S_{11} + 2S_{12})X + b(S_{11} - S_{12})X$$
$$\Delta(E_c - E_{LH}) = a(S_{11} + 2S_{12})X - b(S_{11} - S_{12})X . \qquad (4)$$

In these formulas, a and b are the deformation potentials and S_{11}, S_{12} the elastic compliance constants. For GaSb [15], $\delta E_g/\delta p = 14.27$ eV $N^{-1}m^2$ and $2b(S_{11} - S_{12}) = 8.4$ eV $N^{-1}m^2$. E_C, E_{HH} and E_{LH} refer to the positions of the $|S1/2\rangle$ (conduction) $|P3/2\rangle$ and $|P1/2\rangle$ (valence) band edges, respectively. Thus, under biaxial tensile stress, the GaSb layers are characterized by a fundamental light-hole to conduction band gap :

$$E_c - E_{LH} = E_g^{GaSb} - 2a(S_{11} + 2S_{12})X - b(S_{11} - S_{12})X \qquad (5)$$

and a larger heavy-hole to conduction band gap

$$E_c - E_{HH} = E_g^{GaSb} - 2a(S_{11} + 2S_{12})X + b(S_{11} - S_{12})X . \qquad (6)$$

This is the reason why the ground valence state of the SL may be a light-hole state, despite its larger confinement energy. On the other hand, the AlSb layers present an increased fundamental heavy-hole to conduction band gap and a slightly larger light-hole to conduction band gap. This situation is illustrated in Fig.3a. In the following, we neglect the modifications in the AlSb band structure because in our samples the AlSb layers are much thicker than the GaSb layers and thus they are less strained.

Now, since the spin-orbit splitting $\Delta = 752$ meV is large compared with the strain effects, we can consider that the super-potential arising from the electron affinity and band gap difference between the two host materials

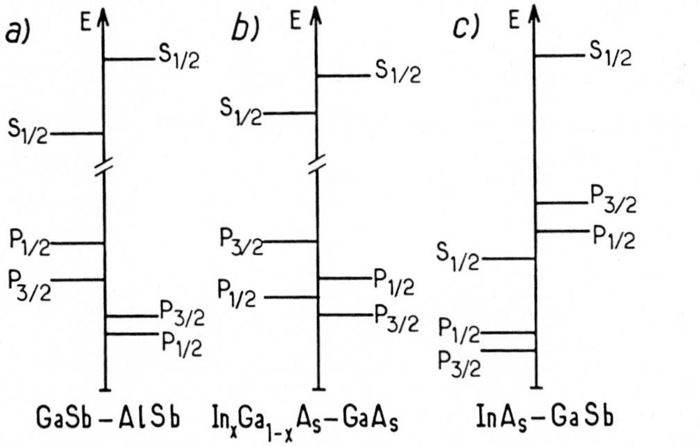

Fig.3 : Schematic band configuration in three "strained-layer superlattices"
(a) GaSb (biaxial tensile) - AlSb (biaxial compression)
(b) $In_xGa_{1-x}As$ (biaxial compression) - GaAs (biaxial tensile)
(c) InAs (biaxial tensile) - GaSb (biaxial compression)

commutes with the strain Hamiltonian. In this case, in the strained layers SL, the $J_z = \pm 3/2$ (heavy-hole) and $J_z = \pm 1/2$ (light-hole) states remain eigenstates of the total Hamiltonian, as long as $\vec{k} = (k_x, k_y) = 0$. Then we calculate the SL eigenenergies E_i, LH_i, HH_i within the envelope wave function formalism [16], using for the light-hole and conduction states the modified value of the (GaSb) band gap and assuming a given value for the light-hole valence band offset ΔE_v^{LH}, and for the heavy-hole states a smaller valence band offset $\Delta E_v^{HH} = \Delta E_v^{LH} - 2b(S_{11} - S_{12})X$. From absorption coefficient strength considerations [6], only the Γ extrema of the host band structure are to be considered. We use Kane's three-band model [17] because the spin-orbit splitting $\Delta = 752$ meV is not large as compared to the band gaps, which are equal to 2200 and 810 meV for AlSb and GaSb, respectively. We take $m^* = 0.041\ m_0$ for the GaSb band-edge conduction mass, and $m_{HH} = 0.33\ m_0$ for the heavy-hole masses in both materials.

IV Discussion

In sample S1, the lattice mismatch obtained from (1) is $(a - a_1)/a = 0.46\%$, so that $E_{LH} - E_{HH} = 30$ meV. In this sample, we observe four transitions involving heavy holes, indicated by black arrows in Fig.1. This implies that there are at least four bound heavy-hole levels in the GaSb quantum well, which leads to $\Delta E_v^{HH} \gtrsim 40$ meV and $\Delta E_v^{LH} \gtrsim 70$ meV. Using these values we get two bound light-hole levels. Assuming the natural parity selection rule [18], the calculated transition energies are $HH_1 \to E_1 = 815$ meV, $HH_2 \to E_2 = 884.5$ meV, $HH_3 \to E_3 = 987$ meV, $HH_4 \to E_4 = 1112$ meV (black arrows in Fig.1), and $LH_1 \to E_1 = 790$ meV, $LH_2 \to E_2 = 881$ meV (open arrows in Fig.1). The calculated transitions towards E_1 are thus in very good agreement with the A and B peaks, and the transitions towards E_2 both contribute to the third experimental peak. These results give support to the predicted reversal of the light- and heavy-hole levels. For $\Delta E_v^{LH} = 90$ meV, for instance, a third bound light-hole level appears. The corresponding transition $LH_3 \to E_3$, expected at 1029 meV, is not observed (Fig.1). More generally, further increase of the valence band offset does not affect the quantitative agreement for the observed transitions, but leads to the prediction of unobserved $LH_i \to E_i$ $(i > 3)$ transitions. Note that the small value of the valence band offset is also supported by the common anion argument. Keeping $\Delta E_v^{HH} = 40$ meV and taking $\Delta E_v^{LH} = 80$ meV to account for the actual value of the lattice mismatch, we find that in sample S2 $LH_1 \to E_1 = 850$ meV and $HH_1 \to E_1 = 873$ meV. Both energies are within the linewidth of the observed exciton peak and, as already pointed out, are significantly smaller than the strainless expected values. Besides, $HH_2 \to E_2 = 1092$ meV is in reasonable agreement with the experiment. Increasing ΔE_v^{LH} up to 100 meV in this case will result in the prediction of an unobserved $LH_2 \to E_2$ transition at 1123 meV.

Finally, we wish to discuss briefly the strength of the absorption coefficient in these structures. As pointed out in Ref.6, the order of magnitude of the observed absorption coefficient on the absorption plateaus is in agreement with the theoretical value. The latter is proportional to an optical matrix element between the atomic part of the wave functions, which is 3 times larger for transitions involving a heavy hole than for those involving a light hole. It is also proportional to the electron-hole reduced mass, which raises the problem of the intricate kinetics of the various hole subbands with respect to the in-layer motion. The $k_\perp$ dispersion relations $HH_i(k_\perp)$ and $LH_i(k_\perp)$ of the valence subbands are not attainable without heavy computation [19]. However, when LH_1 and HH_1 are well energy separated, we may argue that the adequate hole masses are governed by the effect of strain, which is known [20] to exchange the light and heavy masses perpendicular to the strain direction z. Then the reduced mass involving the "light" hole should be about twice that corresponding to the "heavy" hole, and the ratio of the corresponding absorption coefficients should be 2/3. This result is in agreement with the relative strengths of the A band B transitions observed in sample S1, while the opposite assumption predicting a ratio 1/6 does not seem likely.

Fig.1 shows also luminescence spectra (dashed lines) obtained in our SL's. They all lie 30 to 50 meV below the observed exciton absorption transitions. This indicates that these spectra correspond to extrinsic radiative recombination phenomena (most probably band to acceptor recombination) that we have not investigated so far.

V Strains in other superlattice systems

(i) GaAs-$In_xGa_{1-x}As$

A qualitatively new effect is likely to occur in GaAs-$In_xGa_{1-x}As$ ($x \sim 0.1$) SL's [3,4], in which the small gap material ($In_xGa_{1-x}As$) is under biaxial compression, and the large gap material (GaAs) under biaxial tensile stress. This situation leads to the band configurations sketched in Fig.3b. In this system too, Harrison's common anion argument [21] applies, which predicts small valence band offsets, typically in the same range of magnitude (a few tens of meV) as the valence band splittings induced by the strain. Optical absorption measurements [4] have established the type I nature of this system for the heavy-hole to conduction transitions, i.e., the quantum wells for the heavy holes and the conduction electrons both exist in the $In_xGa_{1-x}As$. The valence band splittings $|E_{HH} - E_{LH}|$ increase linearly with the strain, therefore with x, while the centers of gravity $1/2(E_{HH} + E_{LH})$ should remain essentially constant. Thus, for large enough x, the system should be of the type II for the light-hole to conduction transitions, as illustrated in Fig.3b : the quantum wells for the light holes will now exist in the GaAs layers, while those for the heavy holes and the conduction electrons are still in the $In_xGa_{1-x}As$ layers.

(ii) InAs-GaSb

It is well established that the bottom of the InAs conduction band lies in energy at about 150 meV below the top of the GaSb valence band. This results in two important consequences : the InAs conduction band and the GaSb light hole band interact always strongly, giving "large" bandwidths and, at $k_\perp = (k_x, k_y) = 0$, the HH_1 and E_1 subbands cross as the period $d = d_1 + d_2$ is increased. However, from a complete calculation of the SL band structure, Altarelli [19] has concluded that this situation does not correspond to a true semimetallic regime because of the strong repulsion of these bands at finite $k_\perp$. Now, consider the lattice mismatch $\delta a/a = 0.62\%$ with $a_{InAs} < a_{GaSb}$. For a InAs (120 Å) - GaSb (80 Å) SL, we find a strain-induced splitting of the GaSb valence band equal to 26 meV, with the configuration illustrated in Fig.3c. Most probably, the predicted anticrossing behavior at finite $k_\perp$ will remain qualitatively unchanged, but large changes in the numerical results may be expected, which hopefully will lead to a better agreement between theory [22] and experiment [23] .

(iii) HgTe-CdTe

This new system has been recently achieved by MBE and presents several fascinating features [24] . Though only a few experimental data are available as yet, magneto-optical studies in the near infrared have brought indication of a 14 meV shrinkage of the HgTe interaction band gap, which would be in agreement with the estimate of the effect of the 0.31% lattice mismatch.

(iv) GaAs-$Al_xGa_{1-x}As$

This system is by far the most studied, and the state of the art of the growth techniques now allows the realization of very high quality structures presenting luminescence or excitation spectrum linewidths in the range of 1 meV. From the strain point of view, this system presents the same configurations as GaSb-AlSb SL's, but with $\delta a/a = 0.04\%$ only for x = 0,3. Numerical estimate of (4) yields $\Delta(E_C - E_{HH}) = 1.3$ meV and $\Delta(E_C - E_{LH}) = 4.3$ meV, in the thick $Al_xGa_{1-x}As$ barriers limit. Some evidence of such a strain effect in thick sandwich structures has been reported [25], but, surprisingly, no systematic discrepancy between the experimental and strainless predicted positions of the heavy- and light-hole exciton lines has been reported for quantum well structures.

VI Concluding remarks

We have presented a theoretical analysis of the effects of lattice mismatch induced strains on the electronic structure of superlattices. We have shown that strain-induced effects are observed or should be observable in most of the SL systems which have been experimentally achieved. Our analysis is based on two implicit assumptions : (i) the lattice mismatch is fully accommodated by elastic deformations; (ii) the superlattice is in mechanical equilibrium, with no macroscopic strain arising, for example, from lattice mismatch with the substrate. The validity of these assumptions is basically related to the notion of critical layer thicknesses for a given lattice mismatch and to the problem of the relaxation dynamics during the growth. Careful in situ analysis would probably bring information allowing the growth of SL's from hosts, presenting still larger lattice mismatches.

Acknowledgements

I am indebted to my colleagues G. Bastard, C. Delalande, Y. Guldner and M. Voos for their help and encouragement in this work. The GaSb-AlSb superlattices were studied in collaboration with L.L. Chang, A. Segmuller, C.-A. Chang and L. Esaki from IBM Yorktown Heights research center. I also thank J.Y. Marzin for fruitful discussions on the properties of strained layers.

References

1 - G.C. Osbourn, J. Vac. Sci. Technol. B1(2), 379 (1983) and Ref. therein.
2 - P.L. Gourley and R.M. Biefeld, J. Vac. Sci. Technol. B1(2), 383 (1983).
3 - I.J. Fritz, L.R. Dawson and T.E. Zipperian, J. Vac. Sci. Technol. B1(2), 387 (1983).
4 - J.Y. Marzin and E.V.K. Rao, Appl. Phys. Lett. 43, 560 (1983).
5 - C.A. Chang, H. Takaoka, L.L. Chang and L. Esaki, Appl. Phys. Lett. 40, 983 (1982).
6 - P. Voisin, G. Bastard, M. Voos, E.E. Mendez, C.A. Chang, L.L. Chang and L. Esaki, J. Vac. Sci. Technol. B1(2), 409 (1983).
7 - M. Naganuma, Y. Suzuki and H. Okamoto, Inst. Phys. Conf. Ser. N°63, 125 (1981).
8 - E.E. Mendez, C.A. Chang, H. Takaoka, L.L. Chang and L. Esaki, J. Vac. Sci. Technol. B1, 152 (1983).

9 - G. Griffiths, K. Mohammed, S. Subbana, H. Kroemer and J. Merz, Appl. Phys. Lett. 43, 1059 (1983).
10 - A. Segmüller, P. Krisna and L. Esaki, J. Appl. Cryst. 10, 1 (1977).
11 - A detailed account of the theory will be published elsewhere.
12 - S.I. Novikova and N. Kh. Abrikosov, Soviet Physics Solid State 5, 1558 (1964).
13 - From J.R. Drabble, in Semiconductors and Semimetals, edited by R.K. Williardson and A.C. Beer (Academic, New York, 1966) Vol.2, p.75.
14 - H. Pollak and M. Cardona, Phys. Rev. 172, 816 (1968).
15 - C. Benoit à la Guillaume and P. Lavallard, Phys. Rev. B5, 4900 (1972).
16 - G. Bastard, Phys. Rev. B24, 5693 (1981).
17 - G. Bastard, NATO School on MBE and heterostructures, Erice, March 1983. To be published (Martinus Nijhoff Publishers, The Netherlands).
18 - P. Voisin, G. Bastard and M. Voos, Phys. Rev. B 29, 935 (1984).
19 - M. Altarelli, Phys. Rev. B28, 842 (1983).
20 - G.L. Bir and G.E. Pikus, Symmetry and Strain-induced Effects in Semi-Conductors, (J. Wiley and Sons, New York, 1974) p.313.
21 - W.A. Harrison, J. Vac. Sci. Technol. 14, 1016 (1977).
22 - A. Fasolino and M. Altarelli, Proc. Vth Int. Conf. Electronic Properties 2 Dim. Systems. Oxford, 1983 (to be published).
23 - Y. Guldner, J.P. Vieren, P. Voisin, M. Voos, L.L. Chang and L. Esaki, Phys. Rev. Lett. 45, 1719 (1980).
24 - Y. Guldner, this conference.
25 - R. Dingle and W. Wiegmann, J. Appl. Phys. 46, 4312 (1975).

Electronic Properties of HgTe-CdTe Superlattices

Y. Guldner

Groupe de Physique des Solides de l'Ecole Normale Supérieure, 24 rue Lhomond F-75231 Paris Cedex 05, France

We report here magneto-absorption investigations performed in HgTe-CdTe superlattices grown by molecular beam epitaxy. The observed magneto-optical transitions are interpreted in the framework of the envelope function approximation and the band structure of the system is determined, in particular the valence band offset at the HgTe-CdTe interface, and the superlattice energy gap as a function of layer thicknesses.

Introduction

HgTe-CdTe superlattices (SL) exhibit far more diverse characteristics than the conventional GaAs-$Al_xGa_{1-x}As$ SL because of the specific bulk band structures of HgTe and CdTe and of the band line-up of the two host materials. CdTe is a direct gap semiconductor (1.6 eV at low temperature) while HgTe is a zero band gap semiconductor (Fig.1a). This arises from the inverted position of Γ_6 and Γ_8 bands in HgTe. The valence band offset at the HgTe-CdTe interface is very small and the electronic states relevant to the ground conduction and valence SL subbands belong to HgTe and CdTe Γ_8 bands. The two materials closely match in lattice constant (within 0.3%).

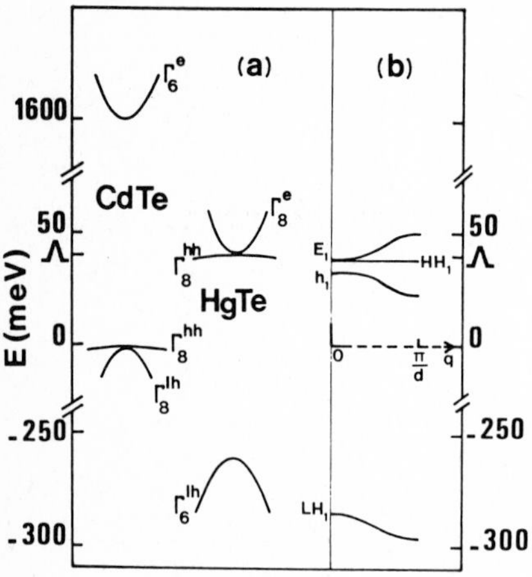

Fig.1 : (a) Band structure of bulk HgTe and CdTe (b) Calculated band structure along the SL wavevector q of the (180 Å) HgTe - (44 Å) CdTe SL at 4 K

We report here magneto-absorption investigations performed on HgTe-CdTe SL grown by molecular beam epitaxy at 200°C [1]. The observed magneto-optical transitions are interpreted in the framework of the envelope function approximation [2] and the band structure of the system is determined, in particular the valence band offset at the HgTe-CdTe interface, and the SL energy gap as a function of layer thicknesses.

Far infrared magneto-optical experiments

The far infrared (FIR) magnetoabsorption experiments were performed at T = 1.6 K, the transmission signal being recorded at fixed photon energies and varying magnetic field in the Faraday configuration. Two kinds of infrared sources were used : a FIR laser and submillimeter carcinotrons. The magnetic field B could be varied continuously between 0 and 10 T. The fig.2 shows typical transmission spectra as a function of B for several FIR wavelengths obtained in a SL sample which consists of one hundred periods of (180 Å) HgTe- (44 Å) CdTe grown on top of a CdTe substrate followed by a CdTe buffer layer [1,3]. The spectra shown on fig.2 were obtained with B perpendicular to the layers (θ = 0). The energy positions of the transmission minima are plotted as a function of B on fig.3a. The observed minima are attributed to interband transitions from Landau levels of the ground heavy-hole subband (HH_1) up to Landau levels of the ground conduction subband (E_1) occuring at the center of the SL Brillouin zone. The extrapolate to an energy $h\nu \sim 0$ at B = 0, but they cannot be due either to electron cyclotron resonance because the SL is p type for $T \lesssim 20$ K, or to hole cyclotron resonance because they would lead to to much too small hole masses. In fact, this indicates that this SL is a quasi-zero-gap semiconductor. We have also studied the θ dependence of the resonance fields of the magneto-optical transitions at fixed wavelengths. These fields follow a$(\cos\theta)^{-1}$ law as evidenced on fig.3b for $\theta = 45°$. Thus a two-dimensional character is induced by the magnetic field.

Fig.2

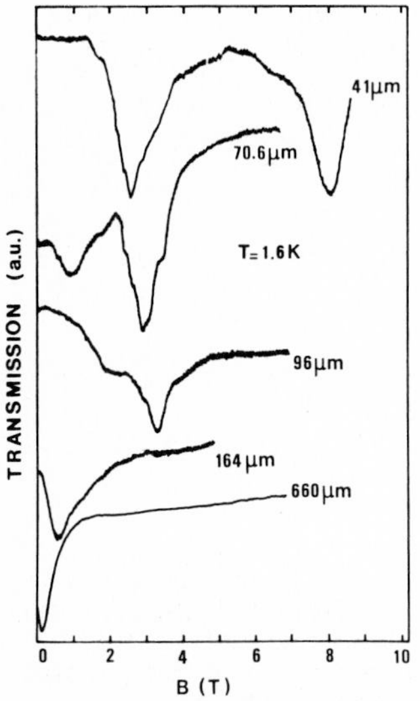

Fig.3

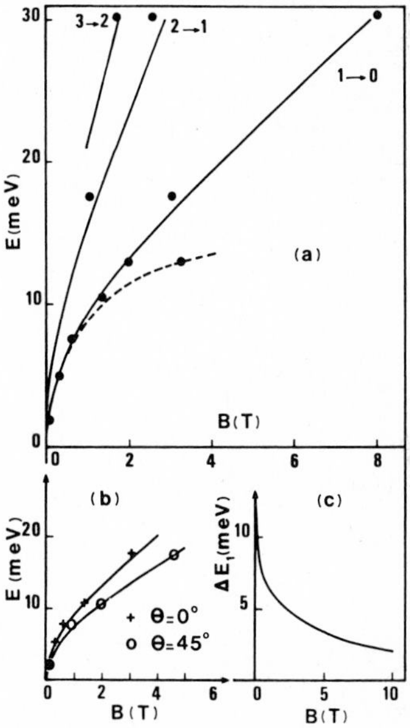

Fig.2 : Typical transmission spectra obtained as a function of the magnetic field B for several infrared wavelengths in the (180 Å) HgTe - (44 Å) $CdTe_{SL}$.

Fig.3 : (a) Energy position of the transmission minima as a function of B observed in the (180 Å) HgTe - (44 Å) CdTe SL. The solid lines are theoretical fits. (b) Transmission minima (0,+) corresponding to the transition 1→0 for two values of θ. The solid line for θ = 0 is the calculated energy and that for θ = 45° corresponds to a perfect two-dimensional behavior (cos θ law). (c) Calculated width ΔE_1 of the E_1 subband vs. B for n = 0 Landau level

SL subband structure calculations :

To calculate the band structure of the SL and the resulting Landau levels under magnetic field, we have used the envelope function approach [2] which provides a simple description of the SL subbands. A SL state is identified by a subband index (E_n, HH_n, h_n...), a SL wavevector q, with $-\pi/d < q < \pi/d$, where d is the SL period, and a two-dimensional wavevector $\vec{k}_\perp$ which is perpendicular to the SL axis. At $\vec{k}_\perp = 0$, there is an exact decoupling between the heavy-hole states ($M_J = \pm 3/2$) and the light-particle states ($M_J = \pm 1/2$). The dispersion relation of the SL light-particle subbands is given by the simple Kronig-Penney expression :

$$\cos qd = \cos k_1 d_1 \cos k_2 d_2 - \frac{1}{2}\left(\xi + \xi^{-1}\right) \sin k_1 d_1 \sin k_2 d_2 \qquad (1)$$

where the subscripts 1,2 refer to HgTe and CdTe respectively, d_1 and d_2 are the layer thicknesses and ξ is given by :

$$\xi = \frac{k_1}{k_2} \frac{E - \varepsilon_2}{E - |\varepsilon_1| - \Lambda} \quad . \qquad (2)$$

In (2), E is the energy measured from the top of the Γ_8 valence band of CdTe, ε_1 and ε_2 are the interaction gap $E_{\Gamma_6} - E_{\Gamma_8}$ in HgTe and CdTe (Table I). Λ is the valence band offset at the HgTe-CdTe interface. The wavectors k_1 and k_2 at a given energy E are obtained from the two-band Kane model [4] in each of the host materials :

$$E(E - \varepsilon_2) = \frac{2}{3} P^2 \hbar^2 k_2^2 \qquad \text{for CdTe}$$

$$(E - \Lambda)(E - \Lambda + |\varepsilon_1|) = \frac{2}{3} P^2 \hbar^2 k_1^2 \quad \text{for HgTe} .$$

P is the Kane matrix element which is assumed to be the same for HgTe and CdTe. For a given energy E, a SL state exists if the right-hand side of (1) lies in the range (- 1, + 1).

Because Λ is positive [3], the CdTe layers are potential barriers for heavy holes. The Kane two-band model does not describe the Γ_8 heavy-hole band of the host materials. The SL heavy-hole subbands are obtained from (1) with $\xi = k_2/k_1$ [2], where k_1 and k_2 are given by :

$$-\frac{\hbar^2 k_1^2}{2m_{hh}} = E - \Lambda$$

$$-\frac{\hbar^2 k_2^2}{2m_{hh}} = E$$

m_{hh} being the heavy-hole mass of the host materials. The model depends on a single parameter, the valence band offset Λ, all the others being well-established bulk parameters. In particular, the values of P and m_{hh} are deduced from the magneto-optical measurements in bulk HgTe [5,6]. As suggested by the common anion argument and the similarity of the host lattice constants, Λ must be very small. The best agreement between experiment and theory is obtained for $\Lambda \sim 40$ meV [3]. Fig.1b presents the calculated band structure of the (180 Å) HgTe - (44 Å) CdTe SL along the SL wavevector q for B = 0. The ground heavy-hole subband HH_1 is almost flat along q at 2 meV below the Γ_8 HgTe band edge. The ground conduction subband E_1 has a mixed electron (HgTe) and light-hole (CdTe) character which explains its energy position and the quasi-zero gap nature of this SL for Λ = 40 meV. In addition another subband h_1 is found to lie in the energy gap $(0,\Lambda)$. Fig.4a gives for HgTe-CdTe SL's the position and width of different subbands as a function of d_2 for $d_1/d_2 = 4$, as obtained from such calculations at T = 4 K. It shows that the cross-over of E_1 and HH_1 at q = 0 occurs for $d_2 \sim 45$ Å. The position of the ground conduction and valence subbands for $d_1 = d_2$ at 4 K is given in fig.4b. The parameter which governs most directly the value of the SL band-gap $E_1 - HH_1$ is the HgTe thickness d_1. The CdTe thickness d_2 governs the width of the subbands which for a given d_1 strongly decreases when d_2 is increased.

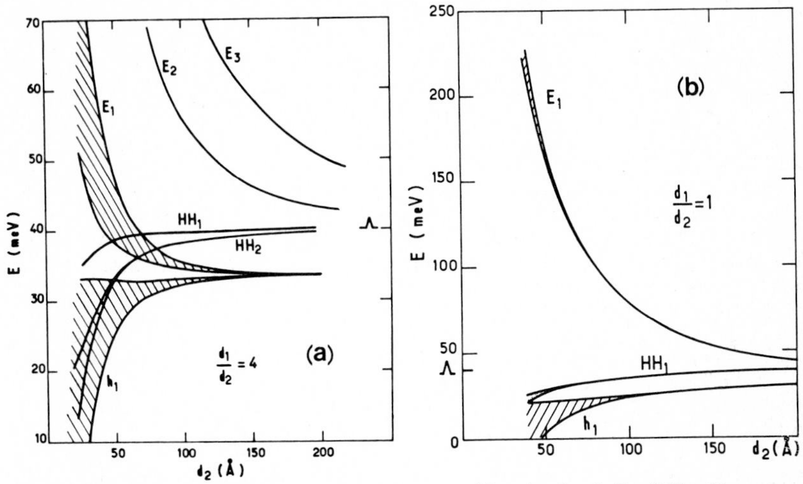

Fig.4 : Energy and width of different subbands in HgTe-CdTe SL as a function of the CdTe thickness d_2.
(a) $d_1/d_2 = 4$ (b) $d_1 = d_2$

Interpretation of magneto-optical data

For finite $\vec{k}_\perp$ or finite B, the situation is much more complicated due to the coupling between heavy-hole and light-particle states. We have not been able to overcome these difficulties and we have used, to calculate the E_1 Landau levels, approximate SL dispersion relations described in Ref. [2] for finite $\vec{k}_\perp^2$. We have neglected spin effects, and we have replaced $\vec{k}_\perp^2$ by $(2n+1)eB/\hbar$, where n = 0, 1... is the Landau level index. For the HH_1 Landau levels, we have considered that they are flat along q, so that their energy is $HH_1(n) = HH_1 - (n+1/2)\hbar eB/m_{hh}$. Theoretical fits to the data observed in the (180 Å) HgTe - (44 Å) CdTe SL are shown in fig.3a. We consider that the Fermi level E_F is close to HH_1 due to the large heavy-hole mass. The curve labelled 1→0 corresponds to transitions from the n = 1 HH_1 Landau level to the n = 0 E_1 Landau level at q = 0. The other curves are analogous transitions with hole and electron Landau indices n equal to 2, 3 and 1, 2, respectively. A good agreement is obtained between experiment and theory for $\Lambda = (40 \pm 10)$ meV. The transition energies are remarkably non-linear as a function of B (in contrast to bulk HgTe), the in-plane apparent electron mass increasing from 0.008 m_0 to 0.03 m_0 between 0.15 and 8 T as a consequence of the $\vec{k}.\vec{p}$ interaction between the E_1, HH_1 and h_1 subbands. Note that the experimental data could be interpreted equally well with the selection rule $\Delta n = +1$ except for the transition 1 → 0. In fact, the transitions n - 1 → n and n + 1 → n would practically coincide since most of the transition energies arise from the conduction levels. Furthermore, the deviation from the theoretical fit of the experimental data (Fig.3a) for the 1→0 transition around 2.5 T is thought to be due to an interband polaron effect, the LO-phonon energy being 16 meV in bulk HgTe. Figure 3c presents the calculated width ΔE_1 of the E_1 subband as a function of B for n = 0. ΔE_1 drops rapidly with B, which is consistent with the two-dimensional behavior observed for B > 1 T (fig.3b) when the Landau levels become nearly flat along q. The variation of ΔE_1 can be understood from the $\vec{k}.\vec{p}$ interaction between E_1, HH_1 and h_1, which are very close at q = 0 and more separated at q = π/d. Applying a magnetic field leads to slower upward shift of E_1 at q = π/d than at q = 0, explaining the decrease of ΔE_1 with increasing B.

The calculated band structure (Fig.1b) is confirmed by the observation of interband transitions from Landau levels of LH_1 (derived from the Γ_6 HgTe states) up to Landau levels of E_1 in the photon energy range 300-400 meV. Figure 5b shows a typical magneto-transmission spectrum observed in this energy range in the Faraday configuration. The position of the transmission minima are presented in fig.5a and the solid lines correspond to the calculated transition slopes from our model. We have taken the same selection rules ($\Delta n = \pm 1$) as those established for $\Gamma_6 \to \Gamma_8$ magneto-optical transitions in bulk HgTe (Faraday configuration) [6]. The observed broad minima (fig.5b) correspond to the two symmetric transitions n → n + 1 and n + 1 → n which are not experimentally resolved. The agreement between theoretical and experimental slopes is rather good. The transitions converge to 344 meV at B = 0 while the corresponding energy $E_1 - LH_1$ at q = 0 is calculated to be 330 meV with our model. The 14 meV difference can be explained by the approximations of the model, in particular the use of the two-band Kane model to describe the band structure of the host materials. It might be also explained by the 0.3% lattice mismatch between HgTe and CdTe which results in an increase of the interaction gap $|\varepsilon_1|$ in HgTe and, therefore, in an increase of $E_1 - LH_1$ [7]. These observations rule out appreciable interdiffusion between HgTe and CdTe because, in the resulting $Hg_xCd_{1-x}Te$ alloy, the corresponding band gap would be smaller than 302.5 meV, its value in bulk HgTe. Important interdiffusion can also be discarded from the FIR results shown in fig.3.

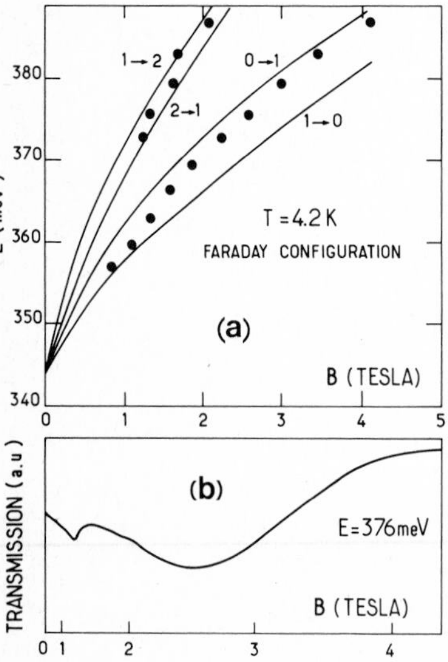

Fig.5 : (a) Energy position of the transmission minima as a function of B corresponding to $LH_1 \to E_1$ transitions in the (180 Å) HgTe - (44 Å) CdTe SL. (b) Typical magneto-transmission spectrum associated to the $LH_1 \to E_1$ transitions

This is evident for non-zero-gap $Hg_xCd_{1-x}Te$ and zero gap $Hg_xCd_{1-x}Te$ would lead to a different nonparabolicity effect [8]. Finally we can deduce that this structure is a true SL and that interdiffusion is less than 40 Å, in agreement with the Auger electron spectroscopy measurements done on this sample [1].

HgTe-CdTe SL : an infrared detector material

It has been suggested recently that HgTe-CdTe SL could be an interesting infrared detector material [9]. Using Λ = 40 meV and the HgTe and CdTe bulk parameters at several temperatures (see Table I), we have calculated the energy gap and the cutoff wavelengths for HgTe-CdTe SL's, with equally thick HgTe and CdTe layers (fig.6). The interesting cutoff wavelengths for infrared detector (8-12 μm) are obtained at 77 K with thicknesses ∿ 50-70 Å which are smaller than the interdiffusion between HgTe and CdTe layers.

Table I : Interaction gap of HgTe and CdTe at several temperatures [10]

T	4 K	77 K	300 K
ε_1(meV)	- 302.5	- 261	- 122
ε_2(meV)	1600	1550	1425

Note that for a given SL period, the energy gap is found to increase when the temperature is increased as observed in the $Hg_{1-x}Cd_xTe$ alloys with similar energy gap [10]. Optical transmission measurements performed at 300 K on a (180 Å) HgTe - (44 Å) CdTe SL and a (60 Å) HgTe - (80 Å) CdTe SL confirm the calculated dependence of the bandgap shown in fig.6.

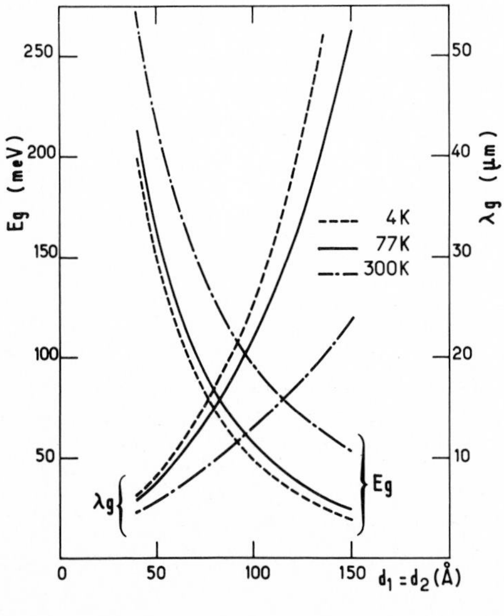

Fig.6 : Energy gap and cutoff wavelength as a function of layer thickness for HgTe-CdTe SL's with equally thick HgTe and CdTe layers

Conclusion

HgTe-CdTe SL's are new and important materials whose properties are only beginning to be investigated. More generally, SL's made from II-VI ternary alloys, such as $Hg_{1-x}Cd$ Te or the semimagnetic semiconductor $Hg_{1-x}Mn_xTe$ and $Cd_{1-x}Mn_xTe$, should present a great fundamental and technical interest.

Acknowledgements

The work presented here is the result of a group effort. It is my pleasure to thank J.P. Faurie from University of Illinois, A. Million from Laboratoire Infrarouge in Grenoble, G. Bastard, J.P. Vieren, M. Voos from Ecole Normale Supérieure.

References

1 - J.P. Faurie, A. Million and J. Piaguet, Appl. Phys. Lett. 41, 713 (1982).
2 - G. Bastard, Phys. Rev. B25, 7584 (1982).
3 - Y. Guldner, G. Bastard, J.P. Vieren, M. Voos, J.P. Faurie and A. Million, Phys. Rev. Lett. 51, 907 (1983).
4 - E.O. Kane, J. Phys. Chem. Solids 1, 249 (1957).
5 - J. Tuchendler, M. Grynberg, Y. Couder, H. Thomé and R. Le Toullec, Phys. Rev. B8, 3884 (1973).
6 - Y. Guldner, C. Rigaux, M. Grynberg and A. Mycielski, Phys. Rev. B8, 3875 (1973).
7 - P. Voisin, this volume.
8 - Y. Guldner, C. Rigaux, A. Mycielski and Y. Couder, Phys. Stat. Sol. b 81, 615 (1977); b 82, 149 (1977); R.S. Kim and S. Narita, Phys. Stat. Sol. b 73, 741 (1976).
9 - D.L. Smith, T.C. McGill and J.N. Schulman, Appl. Phys. Lett. 43, 180 (1983).
10 - M.H. Weiler, in Semiconductors and Semimetals, edited by R.K. Willardson and A.C. Beer (Academic, New York, 1981), Vol.16, p.119.

Strained Layer IV-VI Semiconductor Superlattices

E.J. Fantner and G. Bauer
Institut für Physik, Montanuniversität Leoben, A-8700 Leoben, Austria

The strain in IV-VI semiconductor superlattices (SL) introduces qualitatively new features, originating from the many-valley band structure, not found in III-V SL systems. For the PbTe/PbSnTe system the misfit and substrate induced strain in the constituent layers is determined by X-ray diffractometry. The SL band structure, recently calculated by the envelope function approach is strongly modified by this strain. By varying the thickness ratio of the constituents and thus the strain, the electronic properties can be influenced accordingly.

1. Introduction

Artificial semiconductor superlattices [1-3] composed of a periodic sequence of crystallized layers of IV-VI compounds with alternating composition (e.g. PbTe/$Pb_{1-x}Sn_xTe$) exhibit quite different properties as compared to III-V/III-V, II-VI/II-VI compound superlattices or elemental IV/IV systems. These differences arise from the following properties:

(i) the lead compounds crystallize in the cubic rocksalt structure and not in the zinc blende structure;

(ii) in the electronic band structure the extrema of the conduction and valence band form a minimum direct gap at the L points of the Brillouin zone (BZ). The energy gap is small (Eg (PbTe) = = 190 meV (4.2K)) and apart from influences of higher conduction and valence levels the electron and hole masses are nearly the same. The surfaces of constant energy are ellipsoids of revolution with the <111> direction as main axes.

(iii) At low temperatures the static dielectric constant is extremely high (PbTe:$\varepsilon_s \cong 1300$, $Pb_{0.88}Sn_{0.12}Te$:$\varepsilon_s \cong 2500$) due to a tendency towards a structural phase transition. As a consequence ionized impurity scattering (long-range part) is not effective in limiting the carrier mobility down to the lowest temperatures. The temperature dependence of the mobility is determined by optic and acoustic phonon scattering and at very low temperatures by defect scattering (short-range part). Due to the high ε_s, also the binding energy for hydrogen-like impurities is negligibly small and there are no bound states apart from deep levels.

(iv) A distinctive feature of undoped lead compounds is the presence of at least 10^{16}-10^{17} cm^{-3} carriers in the extrinsic low temperature regime originating from deviations from stoichiometry. E.g. in PbTe tellurium vacancies are responsible for n and lead vacancies for p-type conduction.

High quality single crystalline IV-VI semiconductor films were grown either by molecular beam epitaxy [5] (MBE) or by the hot wall epitaxy method [4] (HWE). Mobilities at liquid helium temperatures up to about $10^6 cm^2/Vs$ were reported for such films.

Two IV-VI superlattice systems were prepared so far: PbTe/$Pb_{1-x}Sn_xTe$ was grown by HWE on BaF_2 or KCl substrates by KINOSHITA et al. [6] and CLEMENS et al. [7]. Carrier concentration n or p of $10^{17}...10^{18} cm^{-3}$ and mobilities up to $2x10^5 cm^2/Vs$ were reported. To achieve a considerable band edge modulation, in these superlattices a Sn content of 0.1-0.2 has been used corresponding to an energy gap of 130-80meV, as compared to 190meV of PbTe. These Sn concentrations change the lattice constant d and a lattice mismatch $\Delta d/d$ equal to $2\text{-}4x10^{-3}$ results. The second superlattice system PbTe/$Pb_{1-x}Ge_xTe$ has been prepared by PARTIN using MBE [8]. The Ge content used was about x=0.03, and a metallurgical characterisation of the superlattice structure has been reported so far.

The PbTe/PbSnTe system has been studied in more detail and magnetotransport and magnetooptical investigations were performed [9,10].

Fig.1 shows the BZ of PbTe or $Pb_{1-x}Sn_xTe$ respectively together with the illustration of the PbTe/$Pb_{1-x}Sn_xTe$ superlattice structure, choosing the [111] direction as growth direction. Thus one valley is oriented with its main axis parallel to the surface normal and the three remaining equivalent valleys are oblique to it.

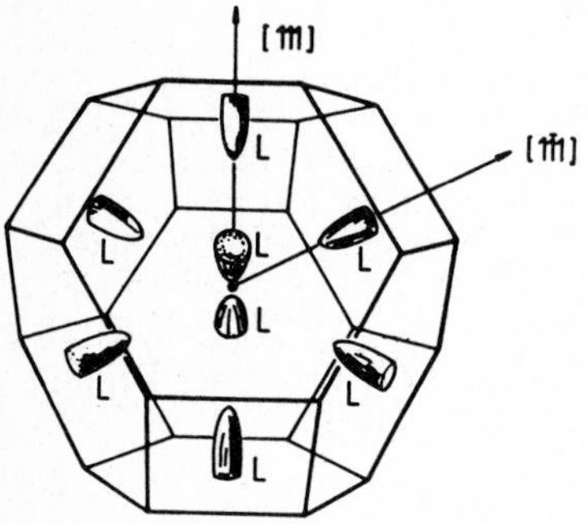

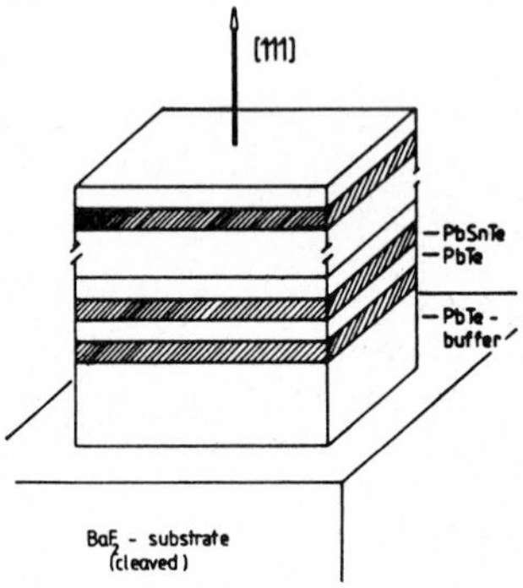

Fig. 1: BZ of IV-VI compounds and schematic layer sequence of SL

2. Crystal Growth

The HWE system as described by LOPEZ-OTERO [11] has been modified for the growth of superlattices by CLEMENS et al. [7] and is shown in Fig.2. It consists of two rotatable substrate furnaces, the wall furnaces, the two source furnaces for PbTe and $Pb_{1-x}Sn_xTe$ and the Te furnaces. A typical temperature profile is shown on the left-hand side of this figure. Cleaved (111) BaF_2 substrates are used. A buffer layer (PbTe,PbSnTe) of a thickness of about 300-400 nm is deposited. The PbTe and PbSnTe layers are deposited by rotating the substrate fur-

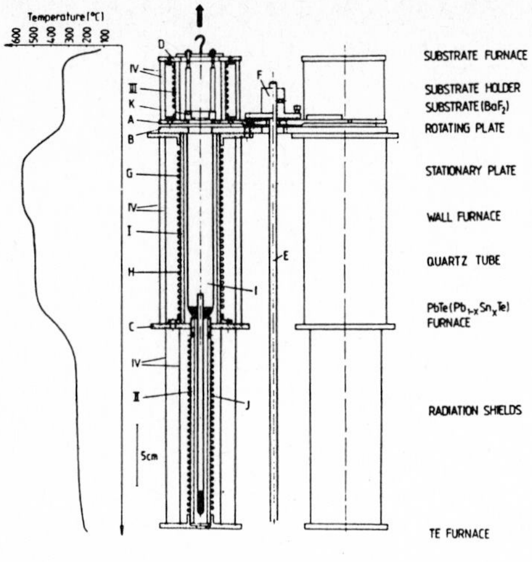

Fig. 2: HWE system for the growth of PbTe/PbSnTe superlattices

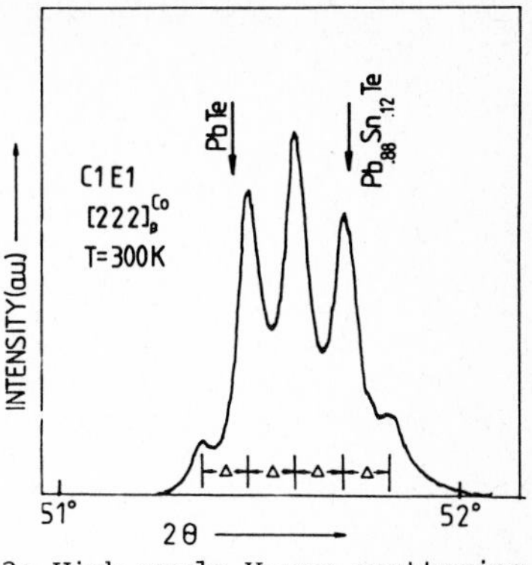

Fig. 3: High angle X-ray scattering for superlattice period determination

naces on a rotating plate over the corresponding source furnaces. Using this apparatus up to 100 periods (200 layers) were deposited with superlattice periods D ranging from 20nm to 300 nm. With this system two superlattice samples are grown simultaneously with complementary sequence of the constituents. Typical growth rates are 1-5 µm/h and thus comparable to the corresponding ones in MBE.

3. Characterization

The composition of the superlattices is checked by lattice constant measurements of the PbTe and $Pb_{1-x}Sn_xTe$ layers using X-ray diffraction. The line widths of the diffraction peaks of about 0.05° prove the high crystalline quality of the epitaxial films, comparable to that of the substrate.

The periodicity of the superlattices is confirmed by high angle X-ray diffraction [12] for superlattice periods D<200 nm. The superlattice structure affects the phase of the X-rays diffracted from the PbTe and PbSnTe layers. The superposition of these waves leads to an interference pattern which reflects the spatial periodic modulation in the superlattice. A typical example is shown in Fig.3. From the apparent periodicity, D is given by:

$$D = \frac{\lambda}{2} \frac{1}{\sin\delta_{i+1} - \sin\delta_i} \qquad (1)$$

where λ denotes the X-ray wave length and δ_i the angular position of the i-th fringe.

This procedure for evaluating the superlattice periodicity was repeated for Bragg reflections up to the 8th order for various wavelengths. Fig.4 summarizes results on two superlattices with periods of 33 nm (P4F1: 100 layers) and of 117 nm (C1F1: 40 layers). The well-resolved structure proves the excellent periodicity of the superlattices and allows an accurate control of the growth rate if combined with a Sn-profile analysis. This Sn profile is obtained by Ar^+-ion sputtering and Auger electron spectroscopy. Fig.5 shows as an example the dependence of the

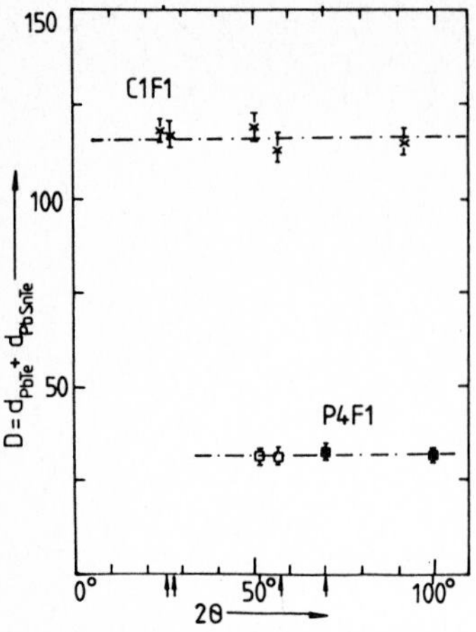

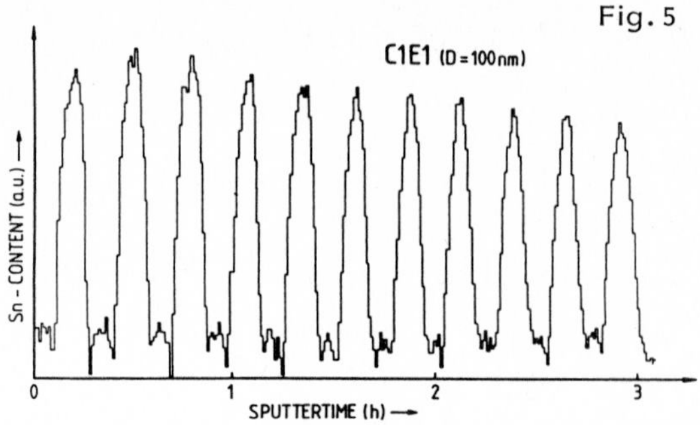

Fig. 4: Comparison of SL-period determination for different reflections

Fig. 5: Determination of SL period by Ar-ion sputtering and AES (Sn-437eV-line)

Sn 437eV peak intensity as a function of sputtering time. The sputtering depth is more than 1 µm and the total sputtering time about 3h. The Sn profile as shown in Fig.5 thus not only reflects effects of interdiffusion but is also smeared out by sputtering effects (increasing surface roughness). Diffusion experiments indicate that interdiffusion effects are negligible in $PbTe/Pb_{0.88}Sn_{0.22}Te$ superlattice growth for substrate temperature $T_s \leq 300\ ^oC$.

4. Elastic Strain

The lattice mismatch $\frac{\Delta d}{d}$ between PbTe and $Pb_{1-x}Sn_xTe$ for $0.1<x<0.2$ is $2\text{-}4\times10^{-3}$. This mismatch can cause misfit strain and above a critical layer thickness misfit dislocations [13]. Since the lattice constant of PbTe is larger than that of $Pb_{1-x}Sn_xTe$, the misfit strain ought to be compressive for PbTe and tensile for $Pb_{1-x}Sn_xTe$ in the film plane as illustrated in the insert of Fig.6. For lattice planes tilted relative to the interface direction, the angle of inclination Ψ is increased for a compressive and decreased for a tensile strain.

For the sample configuration given in Fig.1, $\Delta\psi$ is given by [14]

$$\Delta\psi = \frac{1}{2}(\varepsilon_{xx}-\varepsilon_{zz})\sin(2\psi) \qquad (2)$$

where ε_{xx} and ε_{zz} are components of the strain tensor defined in sec.5. $\Delta\psi$ is measured by X-ray diffraction in an Ω goniometer. Data for three different samples with periods D = 60, 200 and 600 nm and $d_{PbTe} = d_{PbSnTe}$ are shown in Fig.6.

The components ε_{xx} and ε_{zz} are related to each other via the elastic constants C_{11}, C_{12} and C_{44} of PbTe and $Pb_{1-x}Sn_xTe$, respectively. For T = 300K, using the values of Ref.[15], $\varepsilon_{xx}=$

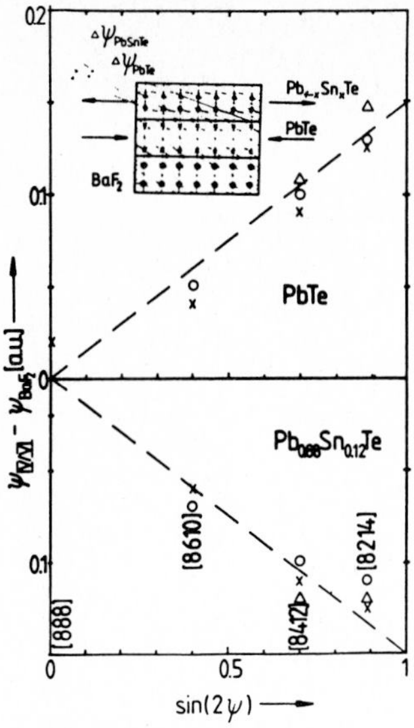

Fig.6: Inclination Ψ for three samples: D=60nm (o), 200nm (×), and 600nm (Δ)

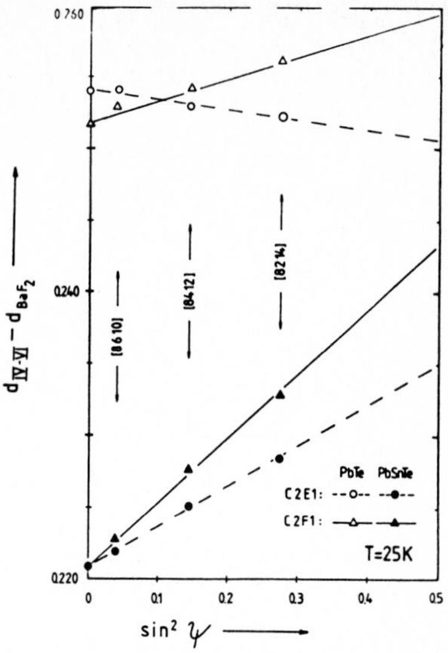

Fig.7: Determination of strain in two SL's: C2E1 (d_A=d_B=100nm), C2F1 (d_A=175nm =3d_B)

$= -1.06\ \varepsilon_{zz}$. For T→0, $\varepsilon_{xx} = -0.93\ \varepsilon_{zz}$ according to Ref. [16]. The deduced strain shows that the total misfit of 2.5×10^{-3} is shared between PbTe and $Pb_{1-x}Sn_xTe$ [17].

At room temperature and above the thermal expansion coefficients of PbTe, PbSnTe and the substrate BaF_2 are nearly the same: $\alpha \cong 2\times10^{-5}K^{-1}$. At lower temperatures, the thermal expansion coefficient of BaF_2 is considerably smaller and thus the relatively thick substrate (0.5...1mm) induces an additional tensile strain (for T=50K: $1.1\ \times10^{-3}$) [18]. The resulting change of the lattice constant of the PbTe and PbSnTe layers is shown in Fig.7 for two superlattice samples. A tensile strain is manifested by a maximum lattice constant in the interface plane and by a minimum value in the perpendicular direction (insert Fig.6). The relation between $(\Delta d/d)_\psi$ and ψ is given by

$$\left(\frac{\Delta d}{d}\right)_\psi = (\varepsilon_{xx}-\varepsilon_{zz})\ \sin^2\psi+\varepsilon_{zz} \qquad (3)$$

if the strain is isotropic in the film plane.

For the sample C2E1 (d_{PbTe}=d_{PbSnTe}), the misfit strain is found to be shared between the constituents. Consequently the lattice constant of the upper layers of the buffer layer has to match - by strain or dislocations- to the average lattice constant of the SL constituents. At low temperatures the substrate-induced strain reduces the compressive part in the PbTe and enhances the tensile one in the PbSnTe layers in agreement with calculated data (broken and solid lines). In sample C2F1, (d_{PbTe}/d_{PbSnTe}=3) the tensile strain in PbSnTe is three times as large as the compressive misfit strain. The substrate-induced strain converts the strain in the PbTe layers into a tensile one.

The data show that the X-ray diffraction methods used are particularly suitable for precise determination of the strain in the constituent layers of a superlattice, separately.

5. Electronic Structure

For a calculation of the superlattice band structure the positions of the band edge of PbTe and $Pb_{0.88}Sn_{0.12}Te$ relative to each other have to be known. These were obtained by KOCEVAR [19] by an empirical LCAO calculation, the results of which are shown in Fig.8. For x=0.12 (Eg=120meV) the conduction band discontinuity is 20meV and the valence band discontinuity 50meV. Charge transfer among the constituents will result in an electrostatic potential. Due to the high static dielectric constant of the materials involved, the depletion length is of the order of 150...300 nm for the samples of interest (N_A, $N_D \cong 2\times10^{17}cm^{-3}$).

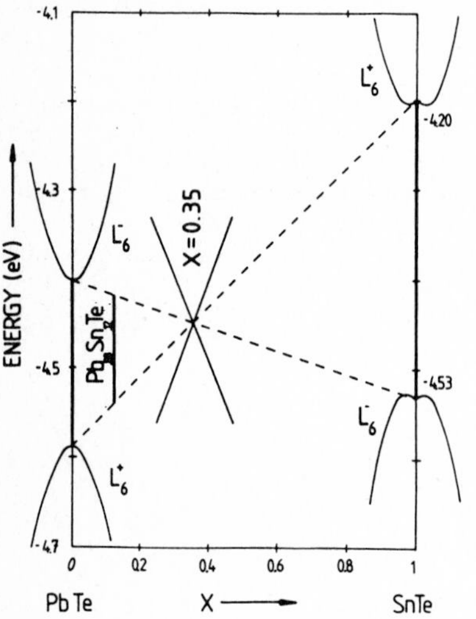

Fig. 8: Absolute conduction and valence band extrema in the PbTe-SnTe system as obtained from results of an LCAO calculation by KOCEVAR [19]

As a consequence also the electrostatic potential is extremely small of the order of 1meV. Thus the approximation of constant band edges within each layer can be used.

The energy dispersion relation for $PbTe/Pb_{1-x}Sn_xTe$ superlattices was recently calculated by KRIECHBAUM [20] within the framework of the envelope function approach (EFA) [21,22]. Propagating or evanescent envelope functions are matched at the boundaries of consecutive layers and the proper $\vec{k}.\vec{p}$ Hamiltonian for cubic IV-VI compounds at the L point of the BZ is used. This Hamiltonian is given by

$$\begin{pmatrix} E_c+A_c k_\perp^2+B_c k_3^2 & P_\perp(\sigma_1 k_1+\sigma_2 k_2)+P_{||}\sigma_3 k_3 \\ h.c. & E_v+A_v k_\perp^2+B_v k_3^2 \end{pmatrix} . \qquad (4)$$

E_c und E_v are the band edge energies which change from layer to layer. $P_\perp$ and $P_{||}$ denote the two-band interband momentum matrix elements, the indices 1,2,3 refer to the two perpendicular axes and the [111] axis. σ_i denote the Pauli spin matrices and A_c, B_c, A_v, B_v higher band contributions. According to the usual procedure of EFA, the continuity condition for the Schrödinger function requires the continuity of all four envelope functions f_i, which are solutions of the eigenvalue problem Hf=Ef with the Hamiltonian Eq.4: $f_i^A = f_i^B$ (i: 1...4) at all interfaces, A,B referring to the two constituents. The superlattice periodicity $D = d_{PbTe} + d_{PbSnTe}$ imposes the Bloch condition

$$f_i(z+D) = e^{iKD} f_i(z) \tag{5}$$

with the superlattice Bloch vector K.

The energy dispersions $E(k_x k_y, K)$ were calculated by KRIECHBAUM and are given in Figs.9 and 10 for superlattices with different periodicities and layer thicknesses d_A and d_B. For small layer thicknesses the superlattice energy bands are not independent of K. For larger values of d_A, d_B, E does not exhibit any dispersion in the direction of K. The surfaces of constant energy become then cylinders with their main axis parallel to growth direction (for [111] one circular and three approximately elliptical cylinders). This behaviour indicates a transition to the multi-quantum well regime. For small periods or energies higher than the depth of the wells not only E(K) shows pronounced dispersion but also $E(k_x, k_y)$ is no longer monotonically increasing with energy and different electric subbands mix for nonvanishing momentum.

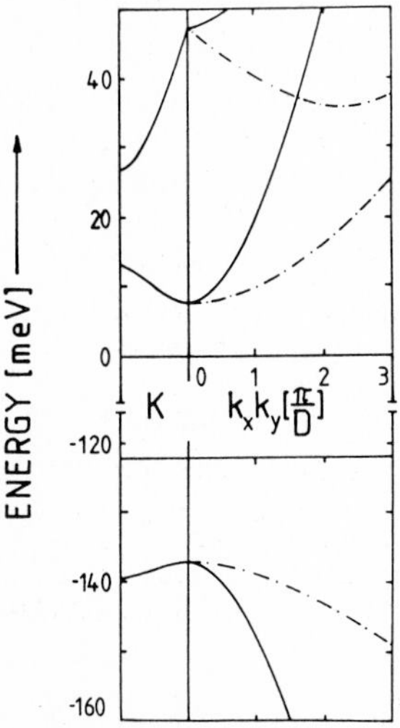

Fig. 9

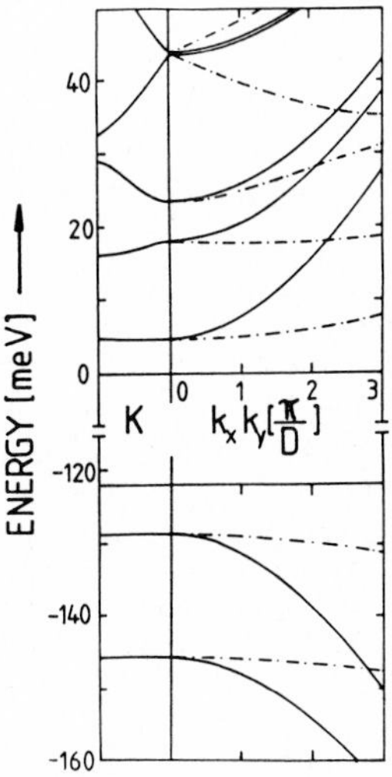

Fig. 10

Fig. 9: SL band structure of conduction band for D=40nm, oblique valleys, K || [111], k_x, k_y in (111) plane; ▲: [11$\bar{2}$], ●: [1$\bar{1}$0]

Fig. 10: SL band structure of conduction band for D=87nm, oblique valleys. Energy zero: PbSnTe CB edge

The height of the wells due to the band edge discontinuities at the boundaries are, however, not only given by the results shown in Fig.8, but are shifted in the epitaxial films due to the presence of strain.

6. Strain-Induced Shifts of the Energy Levels

Using the coordinate system x|| [1$\bar{1}$0],y|| [11$\bar{2}$], z|| [111], the strain tensor is given by:

$$\varepsilon_{ij}=\begin{pmatrix} \varepsilon_{xx} & 0 & 0 \\ 0 & \varepsilon_{xx} & 0 \\ 0 & 0 & \varepsilon_{zz} \end{pmatrix} \tag{6}$$

$\varepsilon_{yy}=\varepsilon_{xx}$ since in the plane of the film the strain is isotropic. The strain values (ε_{zz}) measured at low temperatures by X-ray diffractometry were used to calculate the energetic shifts of the conduction and valence band edges of the PbTe and PbSnTe layers for the different samples. These shifts are determined by the acoustic deformation potentials of the conduction (c) and valence (v) states of the L points of BZ. For the strain as given in Eg.6, as a consequence the four equivalent L states become non- equivalent and the [111] and the three oblique [11$\bar{1}$], [1$\bar{1}$1] and [$\bar{1}$11] valleys are shifted with respect to each other. The energy shift of the kth valley can be written

$$\delta E^{(k)} = \sum_{ij} D_{ij}^{(k)} \cdot \varepsilon_{ij} \tag{7}$$

where ε_{ij} is the strain tensor and $D_{ij}^{(k)}$ the deformation potential tensor which can be written in the cubic axis system as

$$D_{ij}^{(k)} = D_d \cdot \delta_{ij} + D_u \cdot u_i \cdot u_j \, . \tag{8}$$

D_d and D_u represent the dilatational and uniaxial acoustic deformation potential constants in the notation by HERRING and VOGT and the u's are the direction cosines of the angles between the major axis of the kth valley and the cubic ([100], [010] and [001] axes. Transforming the strain tensor (Eq.6) in the cubic axes system yields:

$$\begin{pmatrix} \varepsilon_d & \varepsilon_s & \varepsilon_s \\ \varepsilon_s & \varepsilon_d & \varepsilon_s \\ \varepsilon_s & \varepsilon_s & \varepsilon_d \end{pmatrix} \tag{9}$$

with

$$\varepsilon_d=\tfrac{2}{3}\varepsilon_{xx}+\tfrac{1}{3}\varepsilon_{zz}, \quad \varepsilon_s=\tfrac{1}{3}(\varepsilon_{zz}-\varepsilon_{xx}) \, .$$

E.g. the components of D_{ij} for the [111] valley are then given by: $D_{11}=D_{22}=D_{33}=D_d+\frac{1}{3}D_u$; $D_{12}=D_{13}=D_{23}=\frac{1}{3}D_u$; for the [1$\bar{1}$1] valley: $D_{11}=D_{22}=D_{33}=D_d+\frac{1}{3}D_u$; $D_{12}=\frac{1}{3}D_u$, $D_{23}=D_{13}=-\frac{1}{3}D_u$. Thus the resulting shifts are given by:

$$\delta\ E^{c,v}[111] = (3D_d{}^{c,v}+D_u)\cdot\varepsilon_d+2D_u\cdot\varepsilon_s \tag{10}$$

and for the oblique valleys:

$$\delta\ E^{c,v}\ \langle\bar{1}11\rangle = (3D_d^{c,v} + D_u)\cdot\varepsilon_d-\frac{2}{3}\ D_u\cdot\varepsilon_s\ . \tag{11}$$

Values for $D_d^{c,v}$ and $D_u^{c,v}$ were calculated by FERREIRA: D_u^c = 8.3eV, D_d^c = - 4.4eV, D_u^v = 10.5eV und D_d^v = - 8.9eV.

However these deformation potential values are too high. A comparison with measurements of the energy gaps by interband magnetooptics at different L points in strained single crystalline PbTe films on BaF_2 yields values which are by a factor of four smaller, keeping the ratio unaltered. Since the acoustic deformation potentials $D_d{}^{c,v}$, $D_u{}^{c,v}$ are not known for the PbTe-SnTe alloy system, we chose for x=0.12 the same values as for x=0 (PbTe). In Fig.11 the band discontinuities for the conduction and valence bands are indicated using Eqs.10,11 for the calculations of the strain-induced shifts $\delta\ E^{c,v}$ and values for ε_d and ε_s as derived from the X-ray strain measurements.

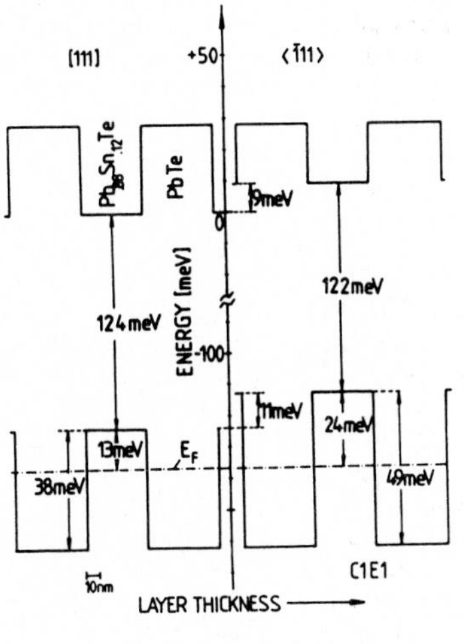

Fig.11: Conduction and valence band discontinuities in SL band structure including the effects of elastic strain: [111] valley and oblique valleys are shifted with respect to each other and band discontinuities are changed as compared to unstrained PbTe/PbSnTe SL

7. Magnetooptical Intraband Transitions

Far infrared laser transmission experiments were performed in Faraday geometry($\vec{B}\,||\,\vec{k}\,||$ [111]) as well as for $\vec{B}$ in the plane of the layers [10]. Typical results for an n-type SL are shown in Fig.12 in comparison to a bulk PbSnTe sample. For 70.6µm, the laser frequency is above the LO-phonon frequency and transmission minima appear, whereas for longer wavelengths dielectric anomalies are associated with the resonances. For the two subband systems ([111],<1̄11>) the cyclotron masses for $\vec{B}\,||$ [111] are given by m_t and $m_t/3\cdot\sqrt{(1+8m_l/m_t)}$, respectively, for the quasi two-dimensional carriers in ellipsoidal approximation (m_t, m_l: transverse, longitudinal mass). For the interpretation of magnetooptical transitions in the PbTe/PbSnTe SL's KRIECHBAUM has extended the EFA in order to introduce the quantizing magnetic field B for the band

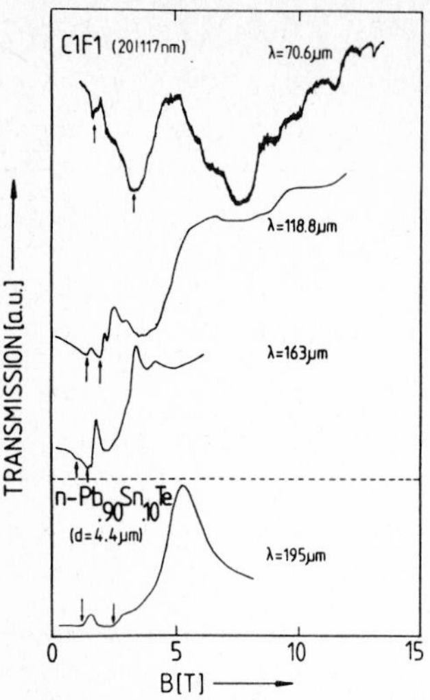

Fig.12:FIR cyclotron resonance for $\vec{B} \| \vec{k} \| [111]$ of a SL sample for various laser wavelengths in comparison to a PbSnTe film

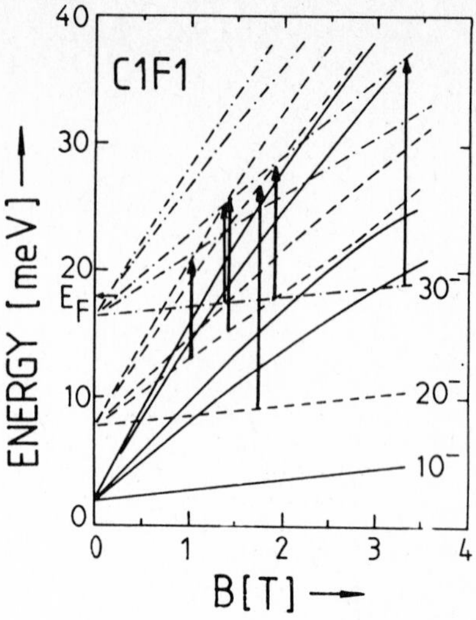

Fig.13:Landau level structure of [111] valley in the conduction band of C1F1 SL (D=117nm).Arrows: experimental data

structure of Fig.11. For the [111] valley this calculation yields the results shown in Fig.13 where the various electric subbands are denoted by the first number(2nd number: Landau quantum number,± sign indicates spin).The agreement with observed resonances is satisfactor. Shubnikov-de Haas oscillations of the carriers confined to the PbSnTe layers, which are very sensitive to the relative position of the valleys, are in agreement with the reported SL band structure.

Apart from the X-ray method reported, other methods [24], and ion chanelling methods [25] were used to investigate strain in III-V SL systems. The knowledge of the elastic strain in IV-VI SL's and its variability by the layer thickness ratio allows an intentional modification of the SL band structure.

Acknowledgments: We thank M.Kriechbaum,B.Ortner, M.von Ortenberg, M.Maier, K.E.Ambrosch, and P.Pichler for helpful discussions and H.Clemens for growing the samples. Work supported by Fonds zur Förderung der wissenschaftlichen Forschung, Project No.4474,5321.

References

1. L. Esaki and R. Tsu: IBM J.Res.Dev. 14, 61 (1970)
2. L. Esaki: in "Molecular Beam Epitaxy and Heterostructures" Proc.Int. NATO School Erice 1983 (North Holland,Amsterdam 1984)
3. K. Ploog and G. H. Döhler: Advances in Physics 32, 286 (1983)
4. A. Lopez-Otero: Thin Solid Films 49, 3 (1978)
5. H. Holloway: in "Physics of Thin Films" ed. G. Haas and M.H. Francombe (Academic N.Y. 1980) Vol.11, p.106
6. H. Kinoshita and H. Fujiyasu: J.Appl.Phys. 51, 5845 (1980); 52, 2869 (1981)

7. H. Clemens, E.J. Fantner and G. Bauer: Rev.Sci.Instr. 54, 685 (1983)
8. D. L. Partin: J.Vac.Sci. Technology 21, 1 (1982)
9. H. Kinoshita, S. Takaoka, K. Murase and H. Fujiyasu: Proc.2nd Symp. MBE Tokyo: MBE-CST-2 (1982) p.61; H. Fujiyasu et al: Surface Science, in print
10. K. E. Ambrosch, H. Clemens, E.J. Fantner, G. Bauer, M. Kriechbaum, P. Kocevar: Surface Science, in print
11. A. Lopez-Otero and D. L. Haas: Thin Solid Films 23, 1 (1974)
12. A. Segmüller, P. Krishna and L. Esaki: J.Appl.Cryst. 10, 1 (1977)
13. C. A. B. Ball and J. H. van der Merwe: in "Dislocations in Solids", ed. F.R.N. Nabarro (North Holland Publ.Comp. Amsterdam 1983) Vol.6, p. 121
14. B. Ortner: in "Eigenspannungen" ed. E. Macherauch and V. Hauk (Deutsche Gesellschaft für Metallkunde, Oberursel, 1983) Vol.II, p.49
15. A. A. Chudinov: Sov.Phys.Solid State 4, 553 (1962)
16. B. Houston, R. E. Strakna and H. S. Belson: J.Appl.Phys. 39 3913 (1968)
17. E. J. Fantner, H. Clemens and G. Bauer: Adv. in X-Ray Analysis Vol.27, (1984) in print
18. R. F. Bis: J.Appl.Phys. 47, 736 (1976)
19. P. Kocevar: to be published
20. M. Kriechbaum: to be published
21. G. Bastard: Phys.Rev. B25, 7594 (1982)
22. M. Altarelli: Lecture Notes in Physics 177, 174 (1983)
23. L. G. Ferreira: Phys.Rev. 137, A1601 (1965)
24. W. J. Bartels and W. Nijman: J.Cryst.Growth 44, 518 (1978)
25. S. T. Picraux, L. R. Dawson, G.C. Osbourn, R.M. Biefeld, W. K. Chu: Appl.Phys.Lett. 43, 1020 (1983)

Part IV

Doping Superlattices

Preparation and Characterization of GaAs Doping Superlattices*

Klaus Ploog

Max-Planck-Institut für Festkörperforschung
D-7000 Stuttgart 80, Fed. Rep. of Germany

The space-charge induced periodic modulation of the energy bands in GaAs doping superlattices grown by molecular beam epitaxy leads to a confinement of electrons and holes in alternate layers (indirect gap in real space). Due to the effective spatial separation the recombination lifetimes of excess carriers are enhanced and large deviations of electron and hole concentrations from thermal equilibrium become quasi-stable. As a result, doping superlattices exhibit unique tunable electronic properties. This fundamental difference from familiar semiconductors is exemplified by results on the tunability of bipolar conductivity and of the two-dimensional subband structure by carrier injection via selective electrodes and by photoexcitation.

1. Introduction

Artificial semiconductor quantum well structures and superlattices provide potential wells of tailored depth and width within which carriers are effectively confined [1, 2]. The free carriers in these systems become essentially two-dimensional (2D) in nature, and a characteristic series of discrete energy levels or subbands is formed if the width of confinement is of the order of 20 nm. This paper reviews some of the unusual electronic properties of a new superlattice formed by periodic alternate doping with n- and p-type impurities in an otherwise homogeneous semiconductor material. In n-p doping superlattices in GaAs the desired one-dimensional (1D) periodic band-edge discontinuities are achieved purely by space-charge effects. The resulting space-charge induced potential leads to a confinement of electrons and holes in alternate layers, thus giving rise to a very effective spatial separation of those carriers. As a conseqeuence of this indirect energy gap in real space, the recombination lifetimes of excess carriers in doping superlattices are strongly enhanced, and large deviations of electron and hole concentrations from thermal equilibrium become quasi-stable. The charge of mobile excess carriers, however, partly compensates the bare impurity space-charge potential. The free-carrier concentration as well as the effective energy gap are thus no longer constant material parameters in a given GaAs doping superlattice specimen, but they are tunable by external carrier injection or extractions. In addition, the mobile excess carriers are screening any statistical impurity potential fluctuation so effectively that 2D subbands are formed in these potential wells even though they are induced by pure space-charge effects.

---

* This work was sponsored by the Bundesministerium für Forschung und Technologie of the Federal Republic of Germany.

2. Origin of Superlattice Potential and Correlation with Design Parameters

The periodic layer sequence of a GaAs doping superlattice and the modulation of the energy bands in real space is schematically shown in Fig. 1. For clarity, we restrict our considerations to the prototype structure consisting of homogeneously n- and p-doped regions of equal layer thicknesses $d_n = d_p$ and with equal doping concentrations $n_D = n_A$. The electrons from the donor impurities transfer into the p regions, thus leading to periodic space-charge induced parabolic potential wells for electrons in the n layers and corresponding potential wells for holes in the p layers. In the ground state all impurities are ionized as shown for the macroscopically compensated situation of Fig. 1b. In this case the amplitude V_0 of the periodic parabolic space charge potential is simply given by

$$V_0 = (4\pi e^2/\kappa_0) n_D d_n^2/8 \qquad (1)$$

where κ_0 is the static dielectric constant (= 12.5 for GaAs). V_0 is thus proportional to the doping concentration and to the square of the constituent layer thickness. The effective energy gap E_g^{eff} defined as the energy spacing between the lowest conduction subband $\varepsilon_{c,0}$ and the uppermost valence subband $\varepsilon_{v,0}$ becomes

$$E_g^{eff} = E_g^0 - 2V_0 + \varepsilon_{c,0} + |\varepsilon_{v,0}| . \qquad (2).$$

To a good approximation we can neglect any potential fluctuations arising from the actual randomly located nonuniform impurity distribution. If we then take the solutions of the harmonic oscillator within the parabolic part of the respective potential well to calculate the subband energies,

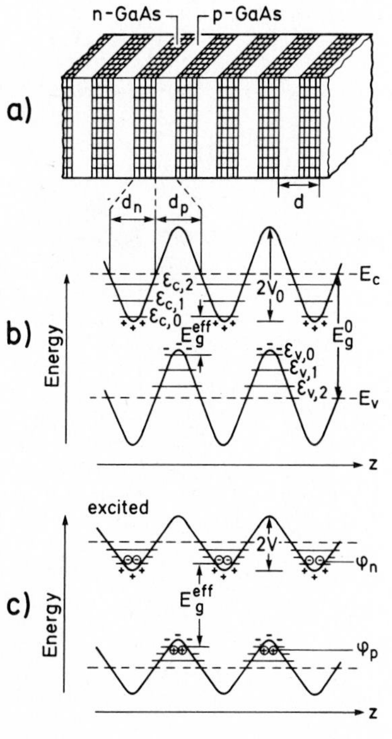

Fig. 1 (a) Schematic illustration of GaAs doping superlattice with constituent layer thicknesses d_n and d_p and doping concentrations n_D and n_A, respectively

(b) Schematic real-space energy band diagram in the ground state. Plus signs indicate ionized donor levels in the n layers near the conduction band edge and minus signs the negatively charged acceptor levels in the p layers above the valence band edge

(c) Modified energy band diagram in the excited state with excess electrons in the n layers and holes in the p layers and separate quasi-Fermi levels for electrons and for holes, respectively we obtain

$$\varepsilon_{c,\mu} = \hbar(4\pi e^2 n_D/\kappa_0 m_c)^{1/2} \ (\mu + 1/2) \quad (3)$$

for electrons and corresponding expressions for the light and heavy holes (m^* = effective mass). For illustration we present some numerical values for a GaAs doping superlattice. Using the design parameters $n_D = n_A = 2 \times 10^{18}$ cm^{-3} and $d_n = d_p = 25$ nm, the amplitude of the space-charge potential becomes $2V_0 = 0.45$ eV, yielding an effective energy gap of $E_g^{eff} = 1.05$ eV in the ground state. The subband spacings in this configuration are 58 meV for electrons, 53 meV for light holes, and 22 meV for heavy holes.

Equations (1)-(3) provide the basis to preselect the electronic properties of GaAs doping superlattices by appropriate choice of the design parameters during epitaxial growth. With increasing doping concentration or layer thickness the potential amplitude V_0 becomes larger, thus yielding a reduced effective gap E_g^{eff}. The strong impact of the design parameters on E_g^{eff} and upon the electron and heavy-hole subband energies is illustrated for two specific examples in Fig. 2 [4]. In Fig. 2a we consider the effect of increasing layer thickness ($d_n = d_p$) for a fixed doping concentration of $n_D = n_A = 2 \times 10^{18}$ cm^{-3} (this concentration gives rise to relatively deep quantum wells at $d_n = d_p > 20$ nm). While the relative positions of the subband levels within each set show only little variation (negligible energy dispersion with k), the effective energy gap as well as the subband energy clearly exhibit the square dependence according to (1) - (3). The effective gap becomes zero, i.e. $2V_0 \simeq E_g^0$, at layer thicknesses of $d_n = d_p \sim 48$ nm. Beyond this layer thickness E_g^{eff} remains zero, however, even in the ground state there are already free electrons in the n layers and free holes in the p layers. Figure 2b shows the dependence of E_g^{eff} and of the subband energies on the doping concentration $n_D = n_A$ for fixed layer thicknesses of $d_n = d_p = 40$ nm. According to (1) - (3) we observe a linear dependence of the energy levels

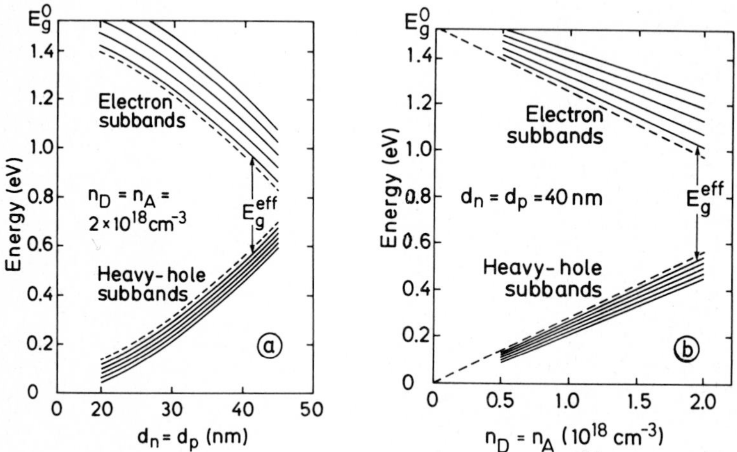

Fig. 2 Relationship between effective energy gap E_g^{eff} and subband energies as a function of design parameters in a GaAs doping superlattice; (a) dependence on layer thickness $d_n = d_p$ at a given doping concentration of $n_D = n_A = 2 \times 10^{18}$ cm^{-3}; (b) dependence on doping concentration $n_D = n_A$ at constant layer thicknesses of $d_n = d_p = 40$ nm. The dashed lines represent the original conduction band minima and valence band maxima in the respective layers

on $n_D = n_A$, and the effective gap is significantly reduced as the doping concentration is increased. E_g^{eff} vanishes at a value of $n_D = n_A \simeq 3 \times 10^{18}$ cm^{-3}. In the thickness relationship the relative positions of the subband energies within each set change considerably due to the energy dispersion with k. The subband spacing increases with the depth of the potential well.

3. Material Preparation

The technique of molecular beam epitaxy (MBE) provides highly improved spatial control of dopant incorporation in GaAs in the direction of layer growth on a microscopic scale [5]. The application of Si as donor and Be as acceptor impurities yields a uniform doping within each individual layer. The intentional impurity concentration in the Si- and Be-doped regions of the growing GaAs films is regulated by the intensity of the respective dopant molecular beam via the effusion cell temperature. Abrupt n-p and p-n junctions are obtained by actuating the mechanical shutters in front of the effusion cells. The shutters are connected to servo-motor driven UHV rotary feedthroughs which are operated by a desk-top calculator via a newly designed IEC Bus programmable shutter controller [6]. At growth temperatures of $530 < T_s < 600\ ^{o}C$ the transition regions ("interfaces") between the n- and p-doped GaAs layers remain extremely smooth down to atomic steps, and a controlled thickness with constant periodicity can be achieved also for a large number of alternating layers in sequence.

These improvements in the growth technique enabled us to prepare GaAs doping superlattices covering a wide range of design parameters in a reproducible manner [7]. In the previous section we have shown that the design parameters determine the 1D periodic space charge potential superimposed on the crystal potential. Therefore, different types of doping superlattices may be produced intentionally by preselecting those parameters. In the case of moderate doping concentration and a small superlattice period, as shown in Fig. 1, twice the potential amplitude $2V_0$ is smaller than the gap of the homogeneous host material. With equal doping densities in the constituent layers, i.e. $n_D d_n = n_A d_p$, all electrons from the n layers recombine with all holes in the p layers. This structure represents the special situation of a compensated intrinsic superlattice with no free carriers in the ground state and having the Fermi level ϕ in the middle of the effective gap. In the more general case, however, the doping densities differ at least slightly in the two layer types, i.e. $n_D d_n \neq n_A d_p$. If $n_D d_n > n_A d_p$, the n layers contain excess electrons of density $n_0^{(2)} = n_D d_n - n_A d_p$ even in the ground state while the p layers are totally depleted. In this n-type superlattice the Fermi level ϕ is shifted to the conduction band minima. Correspondingly, if $n_A d_p > n_D d_n$, the superlattice is p type with the Fermi level close to the valence band maxima. If the values of $n_D d_n$ and $n_A d_p$ are large, $2V_0$ reaches its maximum value of slightly more than the energy gap E_g^0 of the host material and the effective gap E_g^{eff} totally disappears. In this case there are free carriers in the respective layers even in the ground state, and the superlattice resembles a semimetal. This semimetallic structure may be n type or p type depending on the choice of the doping densities in the respective layers.

Figure 3 shows different types of GaAs doping superlattices, which all fulfil $n_D = n_A$ and $d_n = d_p$, and the dependence on the design parameters. They can all be produced with realistic values for the doping concentrations and the layer thicknesses. Curve a corresponds to a compensated intrinsic superlattice where the potential amplitude $2V_0$ is equal to half the energy gap of bulk GaAs. Curve b shows the transition from a compensated to a semimetallic superlattice with $2V_0 \simeq E_g^0$. In the case of curve c there are 1×10^{12} cm^{-2} free carriers per layer in the semimetallic superlattice in the ground state.

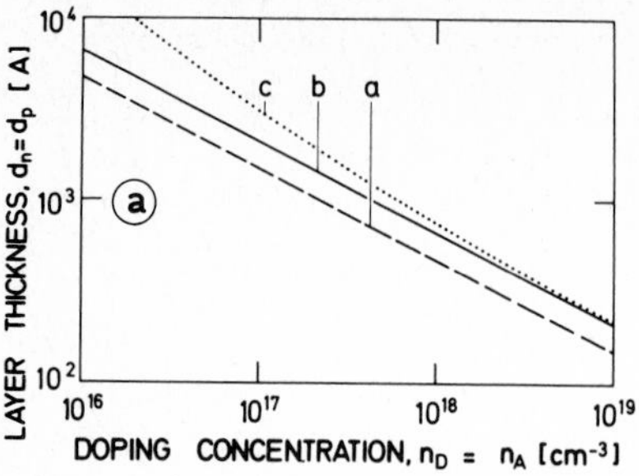

Fig. 3 Dependence of the electronic properties of GaAs-doping superlattices on the design parameters; (a) $2V_o \simeq \frac{1}{2} E_g^0$; (b) $2V_o \simeq E_g^0$, (c) semimetallic with 10^{12} cm^{-2} free carriers per layer

The relatively small amounts of impurities required for doping of GaAs (typically 10^{17} to 10^{19} cm^{-3}) induce only a minor distortion of the lattice of the host material. Doping superlattices thus do not contain any of the typical material interfaces present in compositional superlattices. Instead, they exhibit tunable electronic properties due to the indirect gap in real space. We demonstrate the peculiarities of GaAs doping superlattices in the next section by presenting results on the tunability of bipolar conductivity and on the tunability of the 2D subband structure by carrier injection via selective electrodes and by photoexcitation.

4. Tunable Electronic Properties

The unusual tunability of the electronic properties of GaAs doping superlattices arises from the very effective spatial separation of electrons in the n layers and holes in the p layers by half a superlattice period. The effective energy gap of the material is thus an indirect gap in real space, and there is only an exponentially small overlap between electron and hole states. As a consequence, the recombination lifetimes of excess carriers may become very long (the actual value depends on the design parameters) [8]. Owing to these strongly enhanced lifetimes we are able to populate the subbands with a large number of free carriers, and even large deviations of electron and hole concentrations from thermal equilibrium may become quasi-stable. This non-equilibrium excited state is characterized by quasi-Fermi levels ϕ_n for electrons and ϕ_p for holes (see Fig. 1c).

The most crucial effect of the enhanced recombination lifetimes is the strong modification of the bare space charge potential of the ionized impurities. Due to a certain compensation of the positive and negative space charges by the mobile excess carriers in the respective layers, the original potential amplitude V_o is reduced, thus yielding a considerable widening of the effective energy gap. This flattened potential modulation of the energy bands in the excited state of a doping superlattice is depicted in Fig. 1c. The important result is that we can modulate the effective gap of a given doping superlattice specimen by variation of the number of electrons and holes in the respective layers, i.e. the tunability of the gap is given by

$$\Delta E_g^{eff} \simeq (e^2 d_n/\kappa_o)\, \Delta n \qquad (4)$$

with $\Delta n = \Delta p$ because of macroscopic neutrality. In a similar manner we can modulate the bipolar conductivity $\sigma_{nn}^{(2)}$ and $\sigma_{pp}^{(2)}$ parallel to the respective layers by carrier injection or extraction according to the relation

$$\sigma_{nn}^{(2)} = e\mu_n n^{(2)} \qquad (5)$$

for the electrons and a similar one for the holes ($\mu_{n,p}$ = carrier mobility).

Finally, the mobile excess carriers also modify the shape of the original space charge potential and thereby the energetic spacing of the subbands. With increasing population of the subbands the potential amplitude flattens and the subband spacing decreases. This means nothing else but the tunability of the 2D subband structure in GaAs doping superlattices by variation of the excess carrier concentration.

We will discuss two possibilities to generate excess (free) electrons and holes in GaAs doping superlattices. The first method involves the injection (or extraction) of free carriers by applying an external bias U_{np} between the constituent n and p layers via selective electrodes (see inset of Fig. 4). Electrons and holes are injected or extracted until the quasi-Fermi level difference corresponds to the applied external potential difference, i.e. $eU_{np} = \phi_n - \phi_p = \Delta\phi_{np}$. The major advantage of this method is the possibility to deplete a semimetallic doping superlattice even below the ground state and to achieve a quasi-stable non-equilibrium situation with $\Delta\phi_{np} < 0$, i.e. a negative effective gap. The second method to inject free carriers is given by the electron-hole pair generation during absorption of photons. The efficiency for this process is high ($\mu \simeq 1$). The relaxation times of the photoexcited carriers for thermalization to the subband system are much shorter (in the order of ps) than the electron-hole recombination lifetimes. Therefore, this relaxation process leads to a spatial separation between electrons collected in the n layers and holes collected in the p layers by the internal space charge field. The advantage of this method is that the reduced effective gap in GaAs doping superlattices allows for absorption of light with energy even below the gap of bulk GaAs.

4.1 Tunability of Bipolar Conductivity

The tunability of bipolar conductivity in GaAs doping superlattices was first demonstrated by measuring the simultaneous variation of the n- and p-layer conductances $G^{(2)}_{nn}(U_{np}) = \Delta I_{nn}/\Delta U_{nn}(U_{np})$ and $G^{(2)}_{pp}(U_{np}) = \Delta I_{pp}/\Delta U_{pp}(U_{np})$ as a function of the external bias U_{np} [9]. The results obtained from a representative sample with the design parameters $d_n = d_p = 70$ nm and $n_D = n_A = 1 \times 10^{18}$ cm^{-3} are displayed in Fig. 4. The inset of this figure shows the schematic arrangement of the selective n± and p± electrodes to the respective layers which are formed by alloying small Sn and Sn/Zn balls, respectively, from the surface of the superlattice device (rectangular piece of ∿ 0.4 cm^2 area cleaved from the as-grown wafer). The electron and hole conductances parallel to the respective layer can be modulated simultaneously over a wide range by only slight changes of the controlling bias U_{np}. Below threshold U_{np}, the constituent n layers are totally depleted, whereas in the p layers a minor conductance persists due to $n_A d_p > n_D d_n$ in this sample. The electron conductance can be tuned between zero and 3.4×10^{-4} Ohm^{-1} per layer and the hole conductance between 1.8×10^{-6} Ohm^{-1} and 5.3×10^{-5} Ohm^{-5} per layer. This example clearly demonstrates the operation of a GaAs doping superlattice as a bipolar multijunction FET which is actually a bulk device in contrast to the conventional JFET.

If we neglect quantum size effects in this structure we can analyze the experimental conductance data of Fig. 4 by means of the following equations for the electron concentration $n^{(2)}$ and the hole concentration $p^{(2)}$

$$n^{(2)} = n_D \frac{d}{2} \left[1 - \left(\frac{V_{bi} - eU_{np}}{V_{bi} - eU^{th}_{np}} \right)^{1/2} \right] \qquad (6)$$

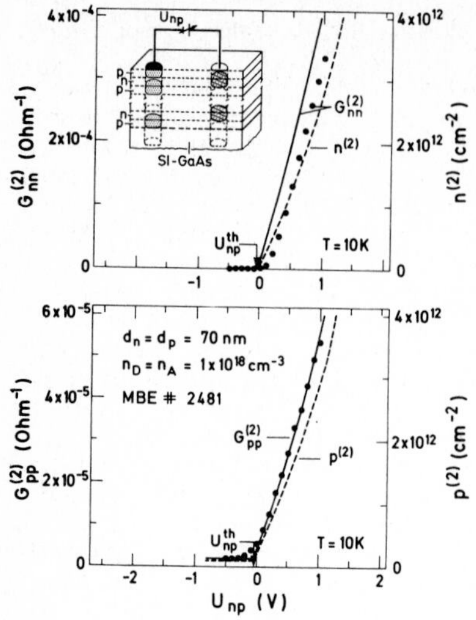

Fig. 4 Observed simultaneous variation of the conductance $G_{nn}^{(2)}$ and $G_{pp}^{(2)}$ parallel to the constituent layers of a GaAs doping superlattice as a function of applied bias U_{np}. The full lines indicate the calculated conductance behaviour as derived from the free-carrier concentration (dotted lines) and assuming a constant mobility. The inset illustrates the arrangement of the selective $n^{\pm}$ and $p^{\pm}$ electrodes to the respective layers

$$p^{(2)} = n_A \frac{d}{2} \left[1 - \frac{n_D}{n_A} \left(\frac{V_{bi} - eU_{np}}{V_{bi} - eU_{np}^{th}} \right)^{1/2} \right] \tag{7}$$

which are valid for $n_A > n_D$, $d_n = d_p = d/2$ and $eU_{np} \geq eU_{np}^{th}$. Here

$$eU_{np}^{th} = V_{bi} - \frac{\pi e^2}{2\kappa} n_D \frac{d}{4} \left(1 + \frac{n_D}{n_A} \right) \tag{8}$$

is the threshold voltage at which the electron concentration in the n layers becomes just zero if $n_A d_p > n_D d_n$. The built-in voltage

$$V_{bi} = E_g - (E_c - \phi)_n - (\phi - E_v)_p \tag{9}$$

is defined in the usual way as the energy gap reduced by the distance of the Fermi level from the respective band edges in the n and p layers. The theoretical calculations for the dependence of $G_{nn}^{(2)}$ and $G_{pp}^{(2)}$ upon bias U_{np} are based on the proportionality of $\sigma_{nn}^{(2)}$ (and $\sigma_{pp}^{(2)}$) from (5) to the respective quasi-2D carrier concentrations $n^{(2)}$ and $p^{(2)}$, assuming concentration-independent mobilities. The best fit to the experimental conductance data of our sample,which uses n_D, n_A and the corresponding carrier mobilities as variable parameters, is indicated by the full lines in Fig. 4. The doping concentrations are used to fit the threshold voltage U_{np}^{th}, the residual hole concentration below threshold, and the magnitude of conductance increase with bias U_{np}, while the mobilities describe only the increase of conductance with U_{np}. The dotted lines in Fig. 4 indicate the calculated variation of quasi-2D electron and hole concentration with bias.

For application of GaAs doping superlattices in JFET devices, a steep slope of $G_{nn}^{(2)}(U_{np})$ and $G_{pp}^{(2)}(U_{np})$ is highly desirable. A steeper slope can be achieved by increasing the doping concentration of the constituent layers, while an increase of the layer thicknesses shifts the whole conductance characteristics to higher bias values. Consequently, the slope of the conductance curve and also the threshold voltage can be tailored by the appropriate choice of the design parameters during MBE growth of the doping superlattice configuration [9]. In this way, depletion-mode as well as enhancement-mode multijunction FET type devices can be designed arbitrarily. They can be operated with a wide range of predetermined threshold voltages. Another important aspect for device application is the room-temperature behaviour of the conductance characteristics. Our detailed investigations revealed [9] that even at 300 K the original concept of tunability of bipolar conductivity is valid and that the n-layer conductance can be totally suppressed by carrier extraction at $U_{np} < U_{np}^{th}$. The only modification is given by the threshold voltage which shifts to lower U_{np} values with enhanced temperature due to the temperature dependence of the energy gap of GaAs.

While deviation from the ground state induced by external bias U_{np} include both carrier injection and extraction, the photoexcitation of doping superlattices allows only carrier injection, i.e. only $\Delta n^{(2)}, \Delta p^{(2)} > 0$ is possible. The advantage of photoconductivity measurements is the excellent homogeneity of excitation and thereby of $\Delta\phi_{np}$ which can be achieved over the hole superlattice specimen even at high excitation levels. At the beginning of photoexcitation by illumination with monochromatic light, the carrier densities in the respective layers increase drastically with time because of the low recombination rates. When the photogenerated carriers are more and more compensating the bare space charge potential, the recombination rates are continuously enhanced. Finally, a steady state of the whole system is reached when the recombination losses balance the free-carrier generation during photoexcitation. The density of photoexcited carriers is then given by the difference between the rate of carrier generation by the absorption of $(I_\omega/\hbar\omega)\alpha^{NIPI}(\omega)$ d photons per layer at a light intensity I_ω and the recombination rate, i.e.

$$\Delta n^{(2)} = (I_\omega/\hbar\omega)\alpha^{NIPI}(\omega)\, d - R(\Delta\phi_{np}) \qquad (10)$$

where $\alpha^{NIPI}(\omega)$ is the absorption coefficient.

We have used Hall effect measurements to characterize the photoexcited steady state of GaAs doping superlattices [10]. The results obtained from a representative n-type sample with $d_n = d_p = 190$ nm, $n_D = 3 \times 10^{17}$ cm^{-3} and $n_A = 2 \times 10^{17}$ cm^{-3} are shown in Fig. 5. These four-point measurements via selective electrodes do not only allow us to determine the conductivity; they also yield the two factors that govern this quantity, i.e. both the free-carrier concentration and the mobility can be deduced independently. In Fig. 5 we have plotted the photon-energy dependence of the measured parameters in the constituent layers only. As required from macroscopic neutrality, an increase of the hole concentration in the p layers by the same amount was observed (the dark value of $\sigma_{pp}^{(2)} = 9.3 \times 10^{-6}$ Ohm^{-1} increased to 7.8×10^{-5} Ohm^{-1} by photoexcitation). The important result of Fig. 5a is that the observed optically enhanced conductivity as a function of $\hbar\omega$ directly reflects the dependence of the absorption coefficient $\alpha^{NIPI}(\omega)$ on the photon energy. At photon energies close to the effective gap E_g^{eff}, optical absorption in GaAs doping superlattices is very low due to the reduced overlap between valence and conduction band wavefunction involved in this process. However, $\alpha^{NIPI}(\omega)$ increases exponentially with enhanced photon energy up to E_g^0 [11]. The absorption coefficient for photon energies $\hbar\omega < E_g^0$ is thus no longer a constant quantity in GaAs doping superlattices, and it varies over a wide range if the effective gap (or the quasi-Fermi level difference $\Delta\phi_{np}$) is modulated by changing the excess carrier concentration.

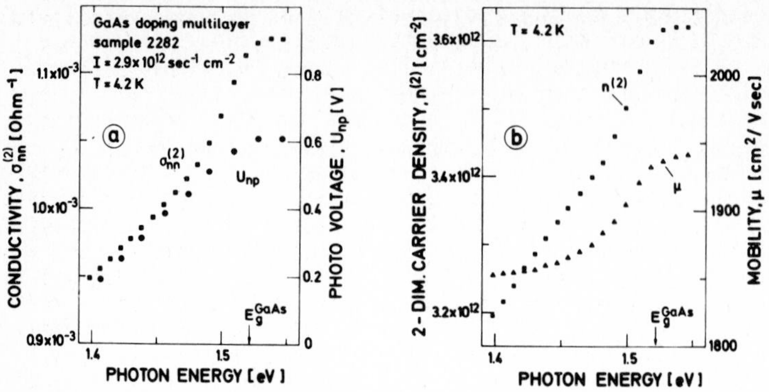

Fig. 5 Photon-energy dependence of the steady-state electronic properties of a strongly populated n-type GaAs doping superlattice with d_n = d_p = 190 nm, $n_D = 3 \times 10^{17}\ cm^{-3}$ and $n_A = 2 \times 10^{17}\ cm^{-3}$; (a) variation of n-type conductivity parallel to the layers and of photovoltage; (b) variation of free-electron concentration and of electron mobility

A second method to detect optically induced change of the carrier concentration in doping superlattices is the measurement of the photovoltaic response. The splitting of the quasi-Fermi level for electrons ϕ_n and for holes ϕ_p during photoexcitation can be measured directly as a voltage U_{np} via selective electrodes to the n and p layers. We have inserted the results of such experiments into Fig. 5a. A maximum value of $U_{np} \simeq 0.6$ V was achieved for $\hbar\omega$ close to E_g^0. Besides its photon-energy dependence, the photovoltage strongly relies on the photon flux. At high excitation levels, a value of eU_{np} in the order of E_g^0 can be achieved.

Finally, inspection of the results depicted in Fig. 5b reveals that the observed increase of the photoconductivity in the studied sample is partly due to an increase of mobility, but mainly due to an increase of the free-electron concentration $n^{(2)}$. The maximum increase of $n^{(2)}$ obtained at $\hbar\omega$ = E_g^0 is $\Delta n^{(2)}$ = 21%. This value also depends strongly on the applied photon flux. The carrier mobility μ_n, on the other hand, increases with optical excitation only by about 8% at $\hbar\omega = E_g^0$. Therefore, as a result, the increase in carrier concentration and mobility both contribute to the measured conductivity enhancement [10].

These results on transport parallel to the n and p layers exemplify the tunability of the electronic properties of GaAs doping superlattices for the quasi-continuum limit. In the next section we describe experiments to demonstrate the occurrence of quantum-size effects in this superlattice.

4.2 Tunability of 2D Subband Structure

The first observation of a 2D subband structure in GaAs doping superlattices and a study of its population and tunability was reported for the case of resonant inelastic light scattering (Raman) experiments [11]. The striking result of this study was that the statistical impurity potential fluctuations in this purely space-charge induced superlattice are so effectively screened by the mobile excess electrons that they do not prevent the formation of subbands. The Raman investigations further demonstrated that with increasing photoexcitation of doping superlattices the subbands merge and a continuous transition from 2D to a 3D system occurs [12]. Details of the inelastic light scattering experiments are reported in a separate paper of these Proceedings [13] and will not be repeated here.

Recently, detailed magnetotransport measurements confirmed the existence of a quasi-2D electronic system in GaAs doping superlattices [14]. In particular, the analysis of the observed quantum oscillations allowed a determination of the subband spacings in a region of low excess-carrier densities, a region that is difficult to access by optical methods. For these experiments the superlattice configuration with $d_n = d_p$ and $n_D = n_A$ was designed to have a finite excess carrier density even in the ground state at $U_{np} = 0$ (semimetallic). The chosen doping level of $\sim 7 \times 10^{17}$ cm^{-3} was a compromise between the requirement of a sufficiently high electron mobility, which implies a relatively low doping concentration, and a maximum subband separation, which requires a high doping concentration according to (3). The samples had the usual Hall bar shape with eight contacts as indicated in the inset of Fig. 6. Six selective n electrodes and two selective p electrodes are arranged such that the conductivity and the Hall voltage of the n layers can be measured as a function of bias U_{np}. With the magnetic induction B normal to the superlattice layers the system gets completely quantized, and each subband is split into discrete Landau levels. If we measure the resistance of the n layers at 4.2 K as a function of $B_\perp$ we observe clear quantum oscillations (Shubnikov-de Haas effect), as shown in Fig. 6. The strong maximum at high B values is shifted to higher magnetic induction with increasing U_{np}. In addition, we observed a distinct anisotropy for the orientation of the magnetic induction with respect to the superlattice plane (not shown in the figure), which exemplifies the 2D nature of the conductance in the n layers. In the perpendicular configuration the oscillations are caused by changes of the intra Landau level electron scattering rate when the Landau levels cross the Fermi energy at increasing magnetic induction, and thus reflect the density of states. Inspection of Fig. 6 reveals that the shift of the Shubnikov-de Haas oscillations with U_{np} is smaller than expected from the actual increase of 2D carrier densities with U_{np} as deduced from additional Hall effect measurements on this sample. We interpret this result as an indication that at

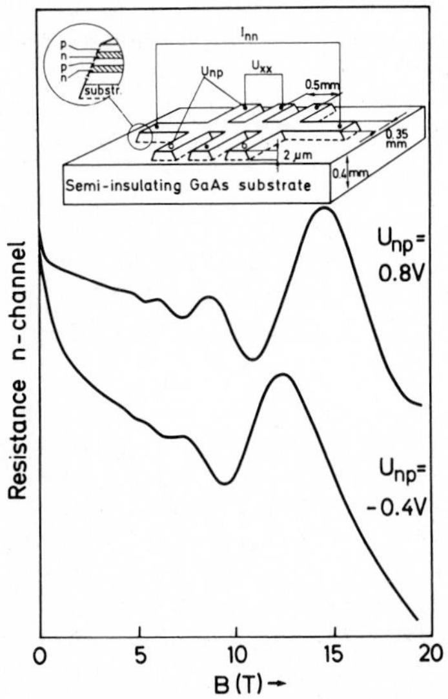

Fig. 6 Observed magnetoresistance of the n layers of a GaAs doping superlattice ($d_n = d_p = 90$ nm, $n_D = n_A = 7 \times 10^{17}$cm^{-3}) as a function of magnetic induction perpendicular to the layers for two different biases U_{np}. The inset shows the Hall bar shape of the sample and the arrangement of the selective electrodes (schematic)

low U_{np} primarily the lowest subband is occupied while at increased U_{np} more subbands are populated and thus the positions of the maxima of the magnetoresistance cannot be correlated with the carrier density in a simple manner [14].

MAAN et al. [14] developed a model which also allows for the occupation of more subbands to analyze the magnetoresistance data in great detail. In the calculation of the 2D magnetoconductivity the subbands are assumed to be equally spaced in a parabolic potential well (a reasonable assumption for low and moderate carrier densities). The calculated position and amplitude of the magnetoresistance extrema are in good agreement with the experimental results of Fig. 6. In addition, the carrier densities and the subband spacings could be deduced from a fit of the theoretical curves to the experiment for different values of U_{np}. The results clearly demonstrate the expected decreasing subband spacing with increasing carrier density, i.e. the tunability of the subband structure in GaAs doping superlattices by carrier injection via selective electrodes.

Further details of magnetotransport experiments on GaAs doping superlattices, in particular with the magnetic induction parallel to the layers, are presented in a separate paper of these Proceedings [15].

5. Concluding Remarks

In this paper we described the peculiar electronic properties of GaAs doping superlattices which arise from the efficient spatial separation of electrons and holes by the periodic space charge potential (indirect gap in real space). We presented a few selected results on transport measurements to show that even large nonequilibrium carrier concentrations are quasi-stable under weak excitation conditions and to demonstrate the validity of the concept of tunable bipolar conductivity and of tunable 2D subband spacing in this new class of semiconductor materials.

The original concept of doping superlattices has now been extended. We prepared a new artificial semiconductor superlattice with tunable electronic properties and with a significant mobility enhancement of both 2D electrons and 2D holes by MBE [16]. The structure consists of a periodic sequence of n-Al_xGa_{1-x}As/i-GaAs/n-Al_xGa_{1-x}As/p-Al_xGa_{1-x}As/i-GaAs/p-Al_xGa_{1-x}As stacks with undoped Al_xGa_{1-x}As spacers between the intentionally doped Al_xGa_{1-x}As and the nominally undoped i-GaAs layers. In this new <u>heterojunction doping superlattice</u> we have for the first time achieved <u>a spatial separation of electrons</u> and holes by half a superlattice period as well as a spatial separation of both types of free carriers from their parent ionized impurities. These unique properties were experimentally demonstrated by the strongly increased tunability of bipolar conductivity with bias. In addition, the observed temperature dependence of Hall mobilities provided direct evidence for a strong mobility enhancement of both electrons and holes in the spatially separated 2D accumulation channels formed in the lower band gap material.

References

[1] L. Esaki, R. Tsu: IBM J. Res. Develop. 14, 61 (1970)
[2] R. Dingle: Festkörperprobleme, Vol. XV, Ed. H.J. Queisser (Vieweg, Braunschweig, 1975) p. 21
[3] K. Ploog, G.H. Döhler: Adv. Phys. 32, 285 (1983)
[4] S. Brand, R.A. Abram: J. Phys. C16, 6111 (1983)

[5] K. Ploog: Ann. Rev. Mater. Sci. 11, 171 (1982)
[6] A. Fischer, K. Graf, M. Hafendörfer, H. Künzel, K. Ploog: Techn.Mess. tm 49, 403 and 467 (1982)
[7] K. Ploog, A. Fischer, H. Künzel: J. Electrochem. Soc. 128, 400 (1981)
[8] K. Ploog, H. Künzel: Microelectron. J. 13, No. 3, p. 5 (1982)
[9] H. Künzel, K. Ploog: J. Vac. Sci. Technol. B2, 72 (1984)
[10] H. Künzel, G.H. Döhler, K. Ploog: Appl. Phys. A 27, 1 (1982)
[11] G.H. Döhler, H. Künzel, D. Olego, K. Ploog, P. Ruden, H.J. Stolz, G. Abstreiter: Phys. Rev. Lett. 47, 864 (1981)
[12] C. Zeller, B. Vinter, G. Abstreiter, K. Ploog: Phys. Rev. B26, 2124 (1982)
[13] G. Abstreiter: These Proceedings
[14] J.C. Maan, T. Englert, C. Uihlein, H. Künzel, K. Ploog, A. Fischer: J. Vac. Sci. Technol. B1, 289 (1983)
[15] J.C. Maan: These Proceedings
[16] H. Künzel, A. Fischer, J. Knecht, K. Ploog: Appl. Phys. A30, 73 (1983)

Luminescence and Inelastic Light Scattering in GaAs Doping Superlattices

Gerhard Abstreiter

Physik-Department, Technische Universität München
D-8046 Garching, Fed. Rep. of Germany

Periodic doping multilayer structures of GaAs exhibit new and exciting semiconductor properties like tunable effective band gap, long lifetime of photoexcited carriers, and formation of electric subbands due to space charge induced potential wells. Photo- and electroluminescence experiments demonstrate the wide tunability of the effective gap. Resonant inelastic light scattering is used to obtain subband splittings and Coulomb screening effects. The resonance behavior leads to information on the scattering processes. Contributions of different subband transitions can be extracted from the dependence of the excitations on laser energy.

1. Introduction

During the past ten years extensive theoretical and experimental studies of artificial semiconductor superlattices have been performed. Such superlattices are composed of a periodic sequence of ultrathin semiconducting layers with alternating composition. Various exciting optical and electronic properties of these new, "man-made" semiconductors have been suggested by Esaki and Tsu /1/. Most of the early work in this field was devoted to the study of superlattices which consist of two different types of semiconductors with different fundamental band gaps (e.g., GaAs-(Al_xGa_{1-x})As /2/, InAs-GaSb /3/). However, already in 1972 Döhler /4/ reported on theoretical studies of another type of artificial semiconductor superlattice which also had been suggested already in Ref. /1/. It is composed of ultrathin n- and p-doped layers of an otherwise homogeneous semiconductor. The development of abrupt doping techniques in combination with molecular beam epitaxy made feasible the growth of periodic GaAs doping multilayer structures of the required quality a few years ago /5/. In these types of superlattices, also called nipi crystals, the electrons of the donor impurities in the n-type layers are transferred in space to the acceptor sites in the p-type layers. Due to the ionized impurities a space charge potential perpendicular to the layers is formed. Since the first successful growth of such crystals a large variety of different experimental and theoretical studies of the exciting electric and optical properties has been performed (see for example Ref. /6/). The present paper discusses new experiments related to the optical properties of doping superlattices. Both photo- and electroluminescence, far in the infrared region, as well as resonant inelastic light scattering experiments are used to get information on the tunability of the gap, the influence of impurities, the two-dimensional nature of the excited carrier system, and the resonance behavior.

In chapter 2 some basic concepts of nipi crystals are reviewed. Chapter 3 summarizes the present status of experimental knowledge on optical recombination of excited carriers in doping superlattices and electronic excitations of electrons which occupy subbands in the conduction band.

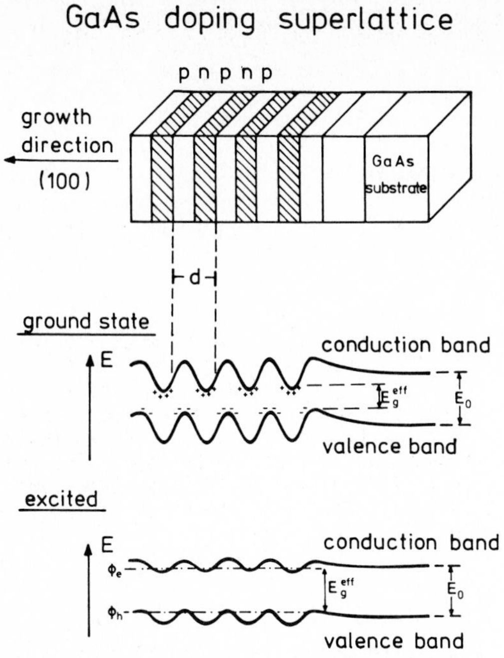

Fig. 1:

Schematic illustration of a periodic doping multilayer structure. Also shown is the periodic modulation of the conduction and valence band edges in the groundstate and under excitation

2. General properties of doping superlattices

The periodic doping in nipi crystals gives rise to purely space charge induced potential wells. The spatial variation of the conduction and valence band edges of an np superlattice is shown schematically in Fig. 1. The structure is simply an alternation of p-n and n-p junctions. It is characterized by the concentration of donors N_D and acceptors N_A and by the thickness of the n-type layers d_n and the p-type layers d_p. If $d_n N_D = d_p N_A$, the nipi crystal is called compensated. The potential variation in the ground state of such a crystal is shown in Fig. 1. The electrons from the donors are attracted by the acceptors in the p-type layers, resulting in a periodic rise and fall of the conduction and valence band. The total depth of this variation depends on the distribution of donors and acceptors and on d_n and d_p. The potential is obtained from Poisson's equation:

$$\frac{d^2V}{dz^2} = \frac{e^2}{\varepsilon} (N_D(z) - N_A(z)) . \tag{1}$$

The total depth is then given by

$$V_{max} = \frac{e^2}{8\varepsilon} (N_D d_n^2 + N_A d_p^2) . \tag{2}$$

The potential fluctuations due to the statistical distribution of the doping impurities in the individual layers have been neglected. These periodic structures exhibit the special feature of a semiconductor with an "indirect gap in real space", because the lowest conduction band states are shifted by half a period from the highest states in the valence band. Excited electrons are therefore separated in space and can have extremely long recombination lifetimes. This leads to large deviations of electron and hole concentrations from equilibrium. The effective band gap is increased for the excited situation. This is shown schematically in the lower part of Fig. 1.

In order to calculate the periodic potential for the excited case, the free carriers have to be taken into account. This has been done by Zeller et al. /7/ and by Ruden and Döhler /8/ who have solved the Schrödinger and the Poisson equation self-consistently. The Schrödinger equation

$$\left(-\frac{\hbar^2}{2m^*}\frac{d^2}{dz^2} + V(z) - \varepsilon_n\right)\Psi_n(z) = 0 \qquad (3)$$

$$n = 0, 1, 2, \ldots$$

yields the subband energies ε_0, ε_1, ε_2, ... of the one-dimensional potential well. For low excitation intensities the superlattice potential is deep and the electrons are strongly bound in each layer. In this case the superlattice potential can be treated as a series of independent wells in which the carriers behave two-dimensionally. The tunability of the effective band gap with excitation results in unique optical properties of such crystals. The excitation can be performed either optically or electrically via selective contacts. In the following sections we discuss luminescence and inelastic light scattering experiments which demonstrate unambiguously the large tunability of the effective band gap and the two-dimensional character of the photoexcited carriers for low excitation intensities. The electronic transitions involved in both types of experiments are shown schematically in Fig. 2. The quasi Fermi level for the holes is assumed to lie in a narrow acceptor impurity band, which seems reasonable because of the large effective mass and the acceptor binding energy of about 27 meV. It will be shown later, however, that also deeper impurities play a role in the infrared luminescence of the samples studied. The important parameters of the np superlattices discussed in the present work are given in table 1. The samples have been grown using molecular beam epitaxy.

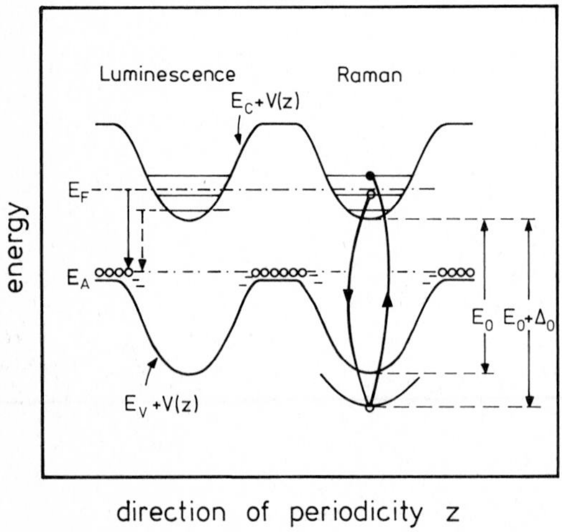

Fig. 2: Real space energy diagram of an excited nipi crystal. The Raman and luminescence processes are shown schematically

Table 1:

Sample number	N_D (Si) $[10^{18} cm^{-3}]$	N_A (Be) $[10^{18} cm^{-3}]$	d_n [Å]	d_p [Å]	V_{max} [eV]
2228) 2432)	1	1	400	400	0.55
2431	2	2	400	400	1.10
2434	4	4	400	400	(2.21)

3. Results and discussion

3.1. Photo- and electroluminescence

The tunability of the effective band gap of periodic GaAs doping multilayers has been demonstrated experimentally using photoluminescence experiments /7,9,10/. The position of the asymmetric luminescence line was found to shift strongly to the infrared with decreasing excitation intensity. This behavior has been observed for various samples with different doping concentrations and/or different layer thicknesses. The luminescence process has been interpreted as the recombination of electrons in the conduction subbands with holes in the acceptor impurity band. These transitions can be considered as indirect in real space. Due to the increased overlap of the wavefunctions of electrons in higher subbands with the acceptor wavefunction,the spectra are dominated by recombination energies of the highest occupied subband. The sharp energy cutoff therefore reflects the effective energy gap which depends strongly on the excitation intensity or on the concentration of photoexcited carriers. The position of the luminescence consequently is directly related to the carrier concentration in the sample. This has been used for the interpretation of resonant inelastic light scattering experiments /7,9/.

Künzel et al. /11/ performed the first electroluminescence studies of np superlattices. They also observed a tunability of the peak similar as reported for the photoluminescence. All these "early" experiments, however, were limited to the energy range $\hbar\omega \gtrsim 1.2$ eV. Therefore the maximum detectable shift of the nipi luminescence was approximately 300 meV.

Recently Kirchstetter /12/ has performed both electro- and photoluminescence measurements in the energy range $0.7\ \text{eV} \lesssim \hbar\omega \lesssim 1.4$ eV. The dominant features of these studies are discussed in the following. For samples with large V_{max} it is possible to shift the nipi luminescence with decreasing excitation intensity through the whole energy region available (see Figs. 3 and 4). Thus the tunability of the effective band gap could be demonstrated down to energies as small as half the direct band gap of GaAs. In addition, impurity assisted luminescence peaks have been identified below the energy of the "fundamental" nipi luminescence. They show the same tunable behavior (Fig. 4). It is believed that these peaks originate from recombinations of electrons in the electric subbands with deep impurity levels in the p-type layers. In the electroluminescence the intensity of these impurity-induced "indirect" transitions is larger than in the photoluminescence. This is probably due to the different excitation mechanisms. At $\hbar\omega \simeq 0.8$ eV a broad emission line is observed which shows only little dependence on excitation intensity. Therefore it is related to direct transitions in real space, however, also via deep levels. The strength of all these impurity-induced

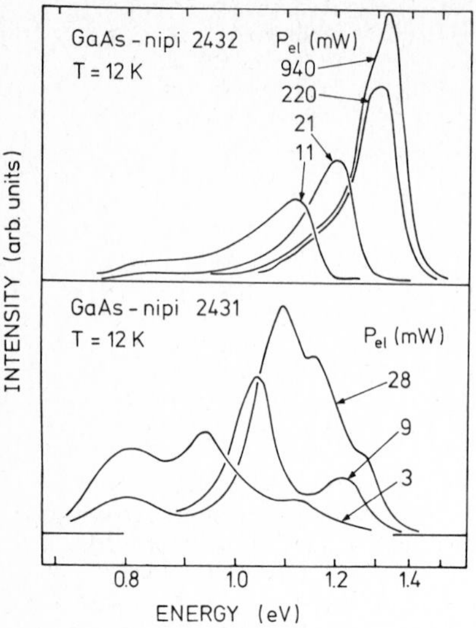

Fig. 3:

Electroluminescence of two nipi samples for different electrical input powers (from Ref. /12/)

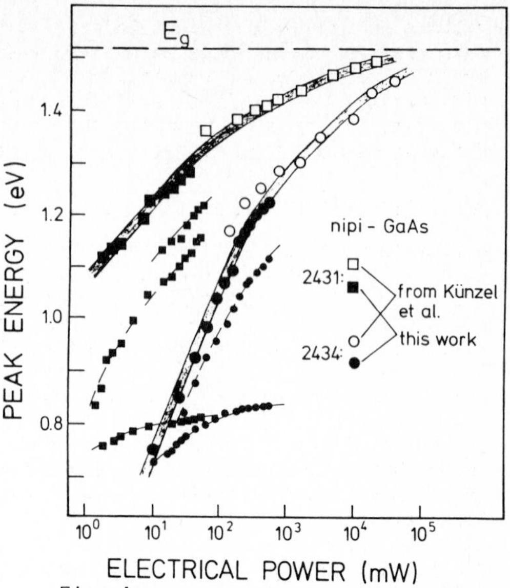

Fig. 4:

Peak energies of the electroluminescence versus electrical input power. The small-size points and squares are related to transitions involving deep levels (from Ref. /12/)

transitions is increasing with increasing doping concentration. An exact identification of their nature was not possible so far. They might be related to As and Ga vacancies and to transition metal impurities, e.g., Cr originating from the Cr doped substrate. Some of these impurity or defect levels have also been observed in tunneling experiments /13/ and in direct photoluminescence studies /14/. Spectra which exhibit various features just discussed are shown in Fig. 3. In Fig. 4 the energy of electroluminescence peaks is plotted versus electrical power. Also shown are the results of Künzel et al. /11/ which have been obtained at higher energies using the same samples. There is a remarkably good overlap, considering the given uncertainty of the excitation intensity in the two experiments. As expected from the transition matrix elements and from the sample parameters, the nipi luminescence of the higher doped sample (number 2434) shifts much faster to smaller energies than that of lower doped samples (e.g., number 2431). Energies for impurity-induced transitions are also plotted in Fig. 4.

Another important feature of these studies is the clear evidence of the contribution of different electric subbands in the recombination processes. It can be seen directly by analyzing the lineshape of the spectra for P_{el} = 220 mW and P_{el} = 940 mW of the nipi crystal 2432 (Fig. 3). For the larger excitation intensity an additional peak appears at the high energy side. Model calculations can reproduce this lineshape and show that in this case it is caused by the occupation of the third subband /12/. The total carrier concentration per layer is increasing from approximately 1.95 x 10^{12} cm^{-2} for P_{el} = 220 mW to 2.1 x 10^{12} cm^{-2} for 940 mW. The occupation of the third subband is changing from 2 % to 6 %. The subband splitting ε_{12} at this energy is about 25 meV. Despite the low occupation. however, the luminescence spectrum is already dominated by electrons from the third subband, a fact which is due to the strongly increased overlap of the wavefunctions which leads to an increase of the transition matrix element by about one to two orders of magnitude. Plots of the luminescence intensity versus energy exhibit also sharp bends whenever a new subband is occupied. This can be taken as further evidence for the contribution of different subbands in the luminescence spectra.

3.2. Resonant inelastic light scattering

Direct spectroscopical information on subband splittings in GaAs doping superlattices has been obtained using resonant inelastic light scattering techniques /7,9/. This method was applied first for the investigation of electronic excitations in GaAs-(Al_xGa_{1-x})As heterostructures /15/. It yields separate spectra of single-particle and collective excitations. Therefore it can be used for the determination of electron energy levels and Coulomb screening effects. For details of these features see Ref. /16/.

In GaAs nipi crystals it has been shown that spin-flip single-particle intersubband excitations are directly related to the bare subband splitting. At low excitation intensities several distinct peaks have been observed. These have been identified as $\Delta = 1$, $\Delta = 2$, $\Delta = 3$ intersubband transitions of photoexcited electrons in the nipi crystal. $\Delta = 1$ stands for all possible transitions between first-nearest subbands, $\Delta = 2$ between second, and $\Delta = 3$ between third nearest. The peaks shift with excitation intensity and are in excellent agreement with self-consistent calculations of the subband splittings /7,9/. In Fig. 5 both spin-flip single-particle excitations and collective excitations, which are coupled to the LO phonons due to their polarization field, are shown as obtained from the nipi crystal 2228. The carrier concentration determined experimentally is 2.8×10^{12} cm^{-2}. The theoretical values for various transition energies are marked by arrows. For the collective excitations only the strongest lines are indicated. The dashed

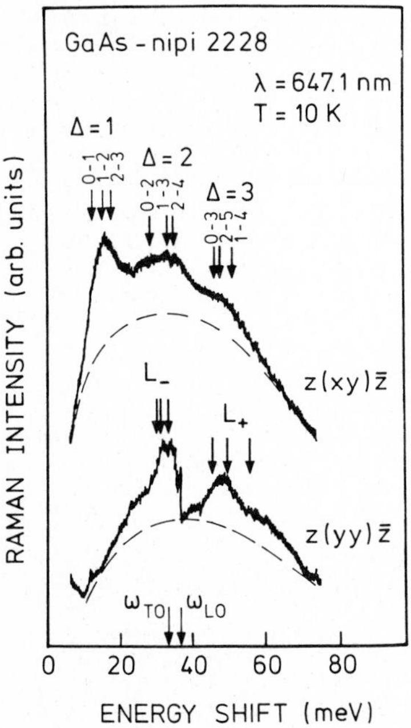

Fig. 5:

Single-particle ($z(xy)\bar{z}$) and collective ($z(yy)\bar{z}$) light scattering spectra at an excitation density of $\simeq$ 100 mW/cm^2. The dashed lines indicate the $E_0 + \Delta_0$ hot luminescence background. The arrows mark calculated transition energies

line is the estimated luminescence background. The large intensities of the higher order transitions are related to the relatively large $k_\perp$ involved in the excitation process. At high excitation intensity a change to three-dimensional behavior has been observed /17/.

All these experiments require excitation under resonance conditions, otherwise the scattering intensity is too weak. In GaAs this resonance condition can be fulfilled best with laser excitation lines close to the $E_0 + \Delta_0$ energy gap. The exact resonance energy depends on the subband occupation and the empty subband which is relevant for the scattering process. This is shown schematically for the 0 - 1 and 0 - 2 excitation in Fig. 6. The resonance behavior has been investigated recently in detail using a tunable dye laser as excitation source /18/. In Fig. 7a the intensity of the $\Delta = 1$, $\Delta = 2$, and $\Delta = 3$ transitions is plotted versus laser energy. The resonance profiles show a sharp onset at the high energy side and a long tail to the low energy side. The arrows mark the calculated resonance maxima where also the curvature of the valence band is included (Fig. 6). For the higher transitions the resonance peak is shifted to higher energies, a fact which is born out by the experiment.

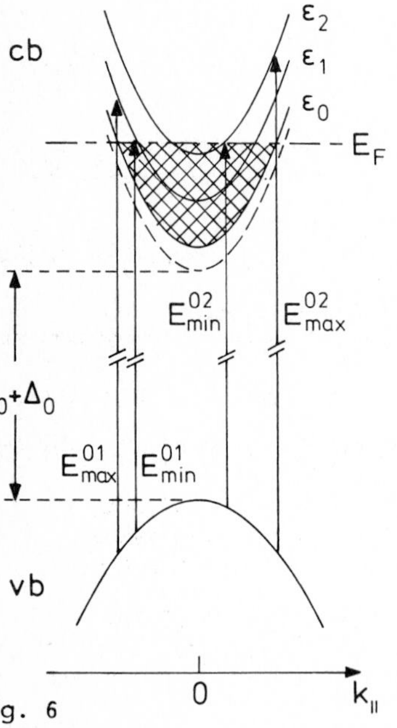

Fig. 6

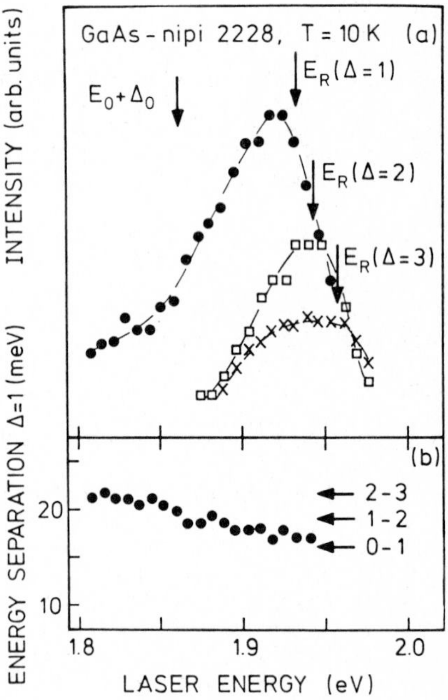

Fig. 7

Fig. 6: Schematic diagram showing relevant transitions for resonant inelastic light scattering in GaAs multilayer structures. The conduction band is split into subbands ε_0, ε_1 ... Parabolic dispersion in $k_\parallel$ is assumed (from Ref. /18/)

Fig. 7: a) Intensity of the $\Delta = 1$, $\Delta = 2$, and $\Delta = 3$ intersubband transitions versus laser energy. The arrows mark the calculated resonance positions. b) Measured $\Delta = 1$ subband splitting versus laser energy. The arrows show the expected subband separations ε_{01}, ε_{12}, and ε_{23} (from Ref. /18/)

In Fig. 7b the measured value for the Δ = 1 subband excitation is plotted versus laser energy. With increasing energy a slight decrease of the subband splitting is observed. This has been explained /18/ by taking into account that different subbands contribute to the Δ = 1 line (see also Fig. 5). The arrows mark the calculated subband separations ε_{01}, ε_{12}, and ε_{23}. The resonance condition of these excitations varies slightly. While at small energies the 2 – 3 transition is dominating, the 0 – 1 transition should contribute more strongly at high energies. This causes the shift of the Δ = 1 peak with increasing laser energy. The experimental observation is in excellent agreement with the self-consistent calculations. It measures directly the deviation of V(z) from a series of harmonic oscillator potential wells which would lead to equidistant energy separations.

Acknowledgements

It is a pleasure to thank K. Ploog and his group at the Max-Planck-Institut für Festkörperforschung in Stuttgart for the excellent cooperation and for providing the samples. The theoretical and experimental work in Munich was a collaborative effort of our group. I want to mention especially the work of Ch. Zeller, B. Vinter, H. Kirchstetter, and W. Einberger. Their contributions are also listed in the references.

References:

/1/ L. Esaki and R. Tsu, IBM J. Res. Develop. 14, 61 (1970)
/2/ R. Dingle, in "Festkörperprobleme", ed. by H.J. Queisser (Vieweg, Braunschweig, 1975), Vol. XV, p. 21
/3/ L.L. Chang and L. Esaki, Surf. Sci. 98, 70 (1980)
/4/ G.H. Döhler, Phys. Status Solidi B 52, 79 (1972); 52, 533 (1972)
/5/ K. Ploog, A. Fischer, and H. Künzel, J. Electrochem. Soc. 128, 400 (1981)
/6/ K. Ploog and G.H. Döhler, Advan. Phys. 32, 285 (1983)
/7/ Ch. Zeller, B. Vinter, G. Abstreiter, and K. Ploog, Phys. Rev. B 26, 2124 (1982)
/8/ P. Ruden and G.H. Döhler, Phys. Rev. B 27, 3538 (1983)
/9/ G.H. Döhler, H. Künzel, D. Olego, K. Ploog, P. Ruden, H.J. Stolz, and G. Abstreiter, Phys. Rev. Lett. 47, 864 (1981)
/10/ H. Jung, G.H. Döhler, H. Künzel, K. Ploog, P. Ruden, and H.J. Stolz, Solid State Commun. 43, 291 (1982)
/11/ H. Künzel, G.H. Döhler, P. Ruden, and K. Ploog, Appl. Phys. Lett. 41, 852 (1982)
/12/ H. Kirchstetter, Diplomarbeit (TU Munich, 1984), unpublished; H. Kirchstetter, G. Abstreiter, and K. Ploog, to be published
/13/ Ch. Zeller, G. Abstreiter, and K. Ploog, Surf. Sci. (in press)
/14/ H. Jung (private communication)
/15/ G. Abstreiter and K. Ploog, Phys. Rev. Lett. 42, 1308 (1979)
/16/ G. Abstreiter, M. Cardona, and A. Pinczuk,in "Light Scattering in Solids IV", eds. M. Cardona and G. Güntherodt, Topics in Applied Physics 54 (Springer, Heidelberg, 1984)
/17/ Ch. Zeller, B. Vinter, G. Abstreiter, and K. Ploog, Physica 117B & 118B (North-Holland Publishing Company, 1983), p. 729
/18/ W. Einberger, Diplomarbeit (TU Munich, 1983), unpublished; W. Einberger, Ch. Zeller, and G. Abstreiter, to be published

Part V

Quantum Hall Effect

The Quantum Hall Effect

K. v.Klitzing and G. Ebert

Physik-Department, Technische Universität München
D-8046 Garching, Fed. Rep. of Germany

1. Introduction

Usually the Hall effect is used to measure the concentration of free electrons in semiconductors. The classical expression for the Hall resistivity ρ_{xy} of a two-dimensional electron system is:

$$\rho_{xy} = \frac{B}{N_s e} \; . \tag{1}$$

The direction of the magnetic field B is perpendicular to the x-y plane of the electron gas with the two-dimensional carrier density N_s. The notation two-dimensional electron gas (2DEG) means that the energy spectrum of the electron is exclusively determined by the motion within the x-y plane. Such a 2DEG or at least an energy spectrum with well-separated energy levels for the electron motion perpendicular to the x-y plane is essential for the observation of the quantum Hall effect. Silicon MOSFETs and GaAs/$Al_xGa_{1-x}As$ heterostructures are typical devices used for studies of the electronic properties of a 2DEG /1/.

The expected linear relation between ρ_{xy} and B is not observed at low temperatures and strong magnetic fields. Figure 1 shows a typical result for $R_H = \rho_{xy}$ (which is the ratio between the Hall voltage and the current through

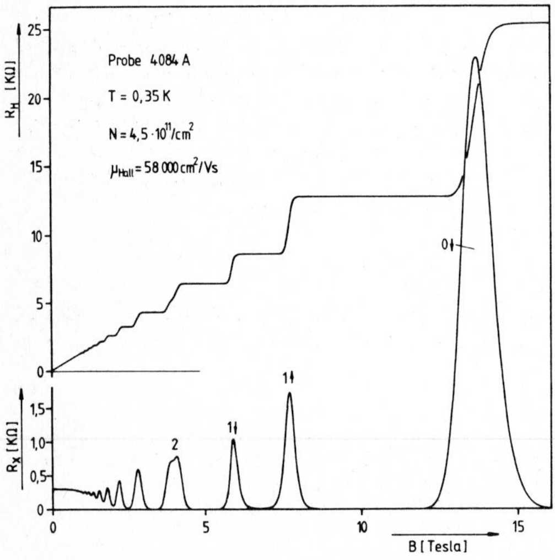

Fig. 1: Experimental result for the diagonal resistance R_x and the Hall resistance R_H of a GaAs-$Al_xGa_{1-x}As$ heterostructure

the sample) and the diagonal resistivity $\rho_{xx} \sim R_x$ as a function of the magnetic field for a GaAs-$Al_{0.3}Ga_{0.7}As$ heterojunction. The plateaus in ρ_{xy} are typical for the quantum Hall effect (QHE) with the peculiarity that the Hall resistivity ρ_{xy} of these plateaus is independent of experimental details and given by the expression

$$R_H = \rho_{xy} = \frac{h}{ie^2} \approx \frac{25812{,}8\ \Omega}{i} \qquad \begin{array}{l} i = 1, 2, 3, \ldots \\ h = \text{Planck constant}. \end{array} \qquad (2)$$

Equation (2) seems to be correct within the experimental uncertainty of less than 10^{-6} but deviations from the quantized value can easily be observed under nonideal experimental conditions. This paper discusses mainly phenomena related to QHE experiments obtained under conditions which are usually not considered in theoretical discussions of the QHE, like finite temperature, inhomogeneous carrier density, etc.

Chapter 2 reviews some theoretical aspects of the QHE, Chapter 3 summarizes the status of high precision measurements of the quantized Hall resistance and discusses the influence of the finite temperature on the measurements, and in Chapter 4 the potential drops close to the current contacts in the QHE regime and some peculiarities related to a network of different quantized Hall resistances will be discussed.

2. Theory

The QHE is always connected with a vanishing resistivity ρ_{xx} corresponding to a Hall angle of 90^0. This means that a dissipationless current is flowing through the sample, which can be explained only if the Fermi energy is located within an energy gap or at least within a mobility gap. Therefore the following discussion of the QHE starts with an analysis of the question under which conditions gaps in the energy spectrum of electrons in a two-dimensional system are expected and which value for the Hall resistivity ρ_{xy} results, if the Fermi energy is located within such an energy gap. The most relevant gaps are those between Landau levels.

2.1. Landau levels of a two-dimensional electron gas

All microscopic theories of the QHE start with the idealized model of free electrons moving in a plane perpendicular to a magnetic field. The electron-electron interaction seems to be important for an explanation of the fractional Hall quantization and will be discussed in a separate paper by R.B. Laughlin /2/. If the magnetic field is in z direction, the eigenfunctions and eigenvalues are usually calculated for electrons moving within an area L_xL_y with periodic boundary conditions. Including the magnetic field by a Peierls substitution, the well-known Landau level spectrum

$$E_n = (n + \frac{1}{2}) \frac{\hbar eB}{m} \qquad n = 0, 1, 2, \ldots \qquad (3)$$

is obtained. The energy is independent of the center coordinate of the cyclotron motion which leads to an N-fold degeneracy of each Landau level

$$N = e \cdot B \cdot L_x \cdot L_y / h . \qquad (4)$$

This degeneracy is equal to Φ/Φ_0, where $\Phi = BL_xL_y$ is the flux through the sample and $\Phi_0 = h/e$ the flux quantum. Energy gaps at the Fermi energy of a two-dimensional electron system appear, if each flux quantum is occupied by an integer number i of electrons.

$$N_{tot} = i \frac{\Phi}{\Phi_0} \qquad N_s = i \frac{eB}{h} . \tag{5}$$

N_{tot} is the total number of electrons within the area $L_x L_y$ and $N_s = N_{tot}/L_x L_y$ is the carrier density. The corresponding Hall resistivity ρ_{xy} under the condition that the Fermi energy lies in a gap can be deduced from (1) and (5) and leads to the quantized Hall resistance $\rho_{xy} = h/ie^2$. The validity of (1) for the calculation of the Hall resistivity may be questionable, but for the special condition that the Fermi energy lies in a real gap it has been shown /3/ that

$$\frac{1}{\rho_{xy}} = e \left. \frac{\partial N_s}{\partial B} \right|_{E=E_F} \tag{6}$$

which leads with (5) to the correct value for the quantized Hall resistance.

The periodic lattice potential is usually included within the effective mass approximation which changes the wavefunction and the energy spectrum but not the fundamental relation between B and n for a filled Landau level (Eq. (5)).
The influence of impurity potentials, which may lead to localized states within the x-y plane, has been discussed by numerous authors /4-7/. The conclusions are that the value of the quantized Hall resistance is not modified by localized states, since (in a simplified picture) potential fluctuations lead to local areas in the x-y plane with an equipotential line as a boundary which acts like a real hole in the sample with no influence on the Hall voltage-Hall current relation /8,9/.

2.2. Bloch electrons in a magnetic field

Not only the Landau quantization but also the periodic lattice potential can open energy gaps at the Fermi energy, if the total number N_{tot} of electrons within the area $F = L_x L_y$ is equal to an integer number j of unit cells

$$N_{tot} = j \frac{F}{\Omega} . \tag{7}$$

Ω is the area of the unit cell.
Combination of (5) and (7) gives a more general expression for the relation between the filling factor $\nu' = N_{tot}\Omega/F$ of a tight-binding energy band and the number of flux quanta per unit cell $\alpha = eB\Omega/h$ for which energy gaps are expected /10/:

$$N_s\Omega = j + i \frac{eB}{h} \Omega \qquad \nu' = j + i\alpha . \tag{8}$$

Each energy gap is characterized by the two quantum numbers i and j, but only the integer i determines the value of the Hall resistivity $\rho_{xy} = h/ie^2$ (Eq. 6). Independent of the origin of the energy gap, the Hall resistivity adopts always a quantized value $\rho_{xy} = h/ie^2$, but the integer i may jump between quite different values, if the carrier density is varied at a fixed magnetic field. This behavior has been discussed in detail by Thouless et al. /11/, and for a magnetic field corresponding to $\alpha = 7/11$ one gets i = -3, +5, +2, -1, -4, ... for increasing filling factor ν'.

In real experiments, α is extremely small ($\alpha < 10^{-3}$) and the filling factor is close to zero (electrons) or one (holes). The energy gaps are usually classified by the Landau quantization with a fixed band index j = 0 (electrons) or j = 1 (holes).

3. Experiments

3.1. High precision measurements of the quantized Hall resistance

The quantized Hall resistance has been observed for a large number of two-dimensional systems, but only silicon MOSFETs and GaAs-$Al_xGa_{1-x}As$ heterostructures have been used for high precision measurements. Yoshihiro et al. /12/ measured the ratio of quantized Hall resistance values at different integers i and found

$$2 \times R_H(i = 8)/R_H(i = 4) = 0.99999993\ (18)$$
$$3 \times R_H(i = 12)/R_H(i = 4) = 0.99999994\ (28).$$

A comparison of the quantized Hall resistance between different GaAs-$Al_xGa_{1-x}As$ devices from different sources showed that identical results are obtained within the experimental uncertainty of 4.6×10^{-8} /13/. Similar results are found from a comparison between GaAs-$Al_xGa_{1-x}As$ heterostructures and silicon MOSFETs /14,15/. Absolute values for the quantized Hall resistance are relatively inaccurate, since the value of the reference resistor in SI ohms is known only within some parts of 10^{-7}. Published values for the quantized Hall resistance are

$$R_H(i = 2) = 12906.4030\ (35) \quad /13/$$
$$R_H(i = 4) = 6453.2004\ (11) \quad /16/$$
$$R_H(i = 8) = 3226.6001\ (7) \quad /12/.$$

These data agree with the value for h/e^2 determined from the fine-structure constant α /17/. All experiments indicate that the quantized Hall resistance depends exclusively on the fundamental constant h/e^2. This means that the Hall resistance R_H can be used as a material- and time-independent resistance standard,and a recommended value for R_H could lead to international uniformity in all resistance measurements, similar to the application of the Josephson effect with respect to voltage measurements. High precision measurements of the fine-structure constant based on QHE experiments are in progress - the main problem is the calibration of the resistance values in SI ohms.

3.2. Corrections to the quantized Hall resistance

All theories which predict that the equation for the quantized Hall resistance $\rho_{xy} = h/e^2 i$ is strictly correct are based on the assumption $\rho_{xx} = 0$ and periodic boundary conditions. An influence of the finite size of the sample on the value of the quantized Hall resistance has not been observed experimentally. Such corrections should be a function of the ratio ℓ/L, where $\ell = \sqrt{\hbar/eB}$ is the magnetic length and L a characteristic sample dimension. From the experiments one can conclude that the correction $\Delta\rho_{xy}/\rho_{xy}$ must be smaller than ℓ/L, but terms like $\Delta\rho_{xy}/\rho_{xy} \sim \exp(-\ell/L)$ or $\Delta\rho_{xy}/\rho_{xy} \sim (\ell/L)^n$ with $n \geq 2$ may be possible. Whereas such corrections seem to be unimportant at the present level of experimental accuracy for devices with typical dimensions of some millimeters, a finite resistivity $\rho_{xx} \neq 0$ may lead to quite different corrections in ρ_{xy} measurements /18/.

a) The Hall angle $\tan\theta = \rho_{xy}/\rho_{xx}$ becomes smaller than 90°, if $\rho_{xx} \neq 0$, which leads to well-known corrections in Hall effect measurements on devices with finite aspect ratios /19,20/.
b) A finite resistivity leads always to a finite slope of the Hall plateaus. If the resistivity ρ_{xx} in the Shubnikov-de Haas minima is thermally activated (activation across the energy gap), the following relation is observed /21/:

$$\frac{d\rho_{xy}}{dN_s} \sim \frac{d\rho_{xy}}{dB} = \alpha\rho_{xx}^{min} \quad . \tag{9}$$

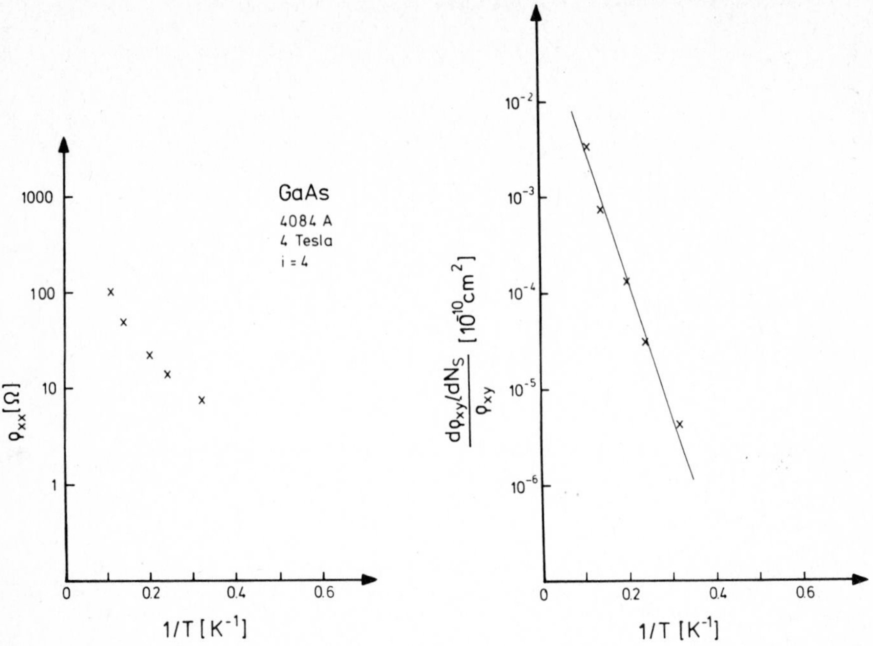

Fig. 2: Temperature dependence of both the resistivity ρ_{xx} at a filling factor $i = 4$ and the slope of the corresponding Hall plateau for a GaAs-$Al_xGa_{1-x}As$ heterostructure. The resistivity becomes weakly temperature dependent at low temperatures due to variable range hopping, whereas $d\rho_{xy}/dB$ remains thermally activated corresponding to the cyclotron energy

The proportional constant α is only weakly dependent on temperature and magnetic field. The slope $d\rho_{xy}/dB$ remains thermally activated /22/ even if the temperature dependence of the minimal resistivity becomes much weaker at low temperatures due to variable range hopping (VRH). Experimental data for a GaAs-$Al_xGa_{1-x}As$ heterostructure are shown in Fig. 2. This result agrees with theoretical calculations which demonstrate that the contribution of VRH processes in ρ_{xy} measurements is negligibly small /23/.

A finite slope of the Hall plateaus provokes the question at which magnetic field position B_i of the $\rho_{xy}(B)$ curve the quantized value is expected. Experimentally the magnetic field position where the resistivity ρ_{xx} has a minimum is used in high precision measurements of the quantized Hall resistance. However, this position changes along the sample, if inhomogeneities are important, so that uncertainties in the B_i position and therefore uncertainties in the ρ_{xy} value up to $\Delta\rho_{xy} = 0.1\ \rho_{xx}$ are realistic.

c) Another correction which leads to a shift of the Hall plateaus to lower ρ_{xy} values arises from an electrical bypass. If the current is not flowing exclusively in the plane of the 2DEG but also, for example, in the highly doped $Ga_xAl_{1-x}As$ layer of a heterostructure (if the thickness of this layer is larger than the depletion length or if the device is illuminated with infrared radiation), corrections to the quantized Hall resistance are expected. Figure 3 shows a typical result. The resistivity minima do not decrease monotonically with increasing magnetic field but

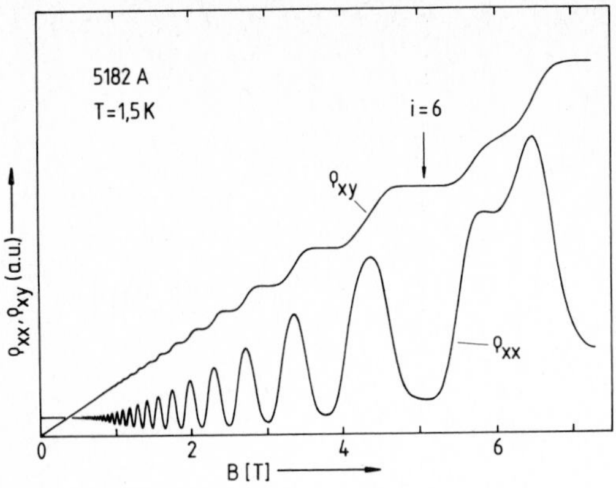

Fig. 3:

Anomalous increase of the minimal resistivity ρ_{xx} at high magnetic fields due to an electrical bypass. The quantized Hall resistance for i = 6 is 1 ‰ smaller than the theoretical value

show a dependence approximately described by $\rho_{xx} \sim B^2$. Simultaneously, the resistivity in the Hall plateaus is reduced ($\Delta\rho_{xy}/\rho_{xy}$ = 1 ‰ for the i = 6-plateau in Fig. 3).

This result can be explained, if one assumes that the Hall voltage is short-circuited by an electrical bypass. The interpretation is based on the following information about the voltage distribution in a network of Hall resistors.

4. Equipotential lines in a network of Hall resistors

A current I flowing under QHE conditions from source to drain (source potential $V_S = 0$) leads to a potential distribution at the device boundary as shown in Fig. 4a. Singularities in the potential appear at the right-hand or left-hand corner (depending on the magnetic field direction) of the current contacts. Any two-terminal resistance with contacts at a common boundary of the sample (e.g., the source-drain resistance of a Hall device, but not the source-drain resistance of a Corbino device) is identical with the Hall resistance /24/.

If a bypass resistance R_x between the potential probes 3-5 and 4-6 in Fig. 4a exists, a Hall current

$$I_H = \frac{U_H}{R_H + R_x} \approx \frac{R_H}{R_x} \cdot I \qquad (R_H << R_x) \tag{10}$$

will flow which produces a voltage drop $U_{34} = R_H \cdot I_H \sim R_H^2$. This leads to the misinterpretation that the resistivity in the plateau region $\rho_{xx} \sim U_{34}$ increases proportional to the square of the magnetic field (see Fig. 3). Simultaneously, the Hall voltage is reduced. A complete short-circuit of the Hall voltage ($U_{35} = U_{46} = 0$) corresponding to $R_x = 0$ leads to a source-drain voltage $V_{SD} = 3\,U_H$ and $U_{34} = U_{56} = U_H$.

A more complicated situation is present if different parts of the sample have different filling factors i and j. Such a sample has been realized with a GaAs-$Al_xGa_{1-x}As$ heterostructure covered partly by a Schottky gate which allows a reduction of the carrier density. The experimental result for i < j is shown in Fig. 4b for the two magnetic field directions B^+ and

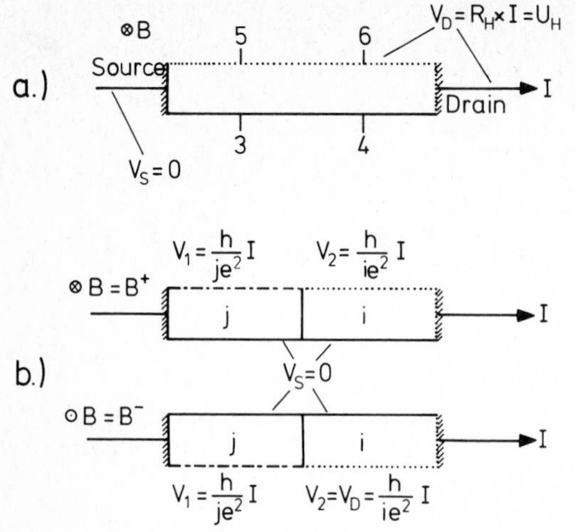

Fig. 4: Potential distribution at the device boundary under QHE conditions
a) standard device with Hall geometry
b) combination of quantized Hall resistances with different filling factors i and j at different directions of the magnetic field (i < j)

B^- perpendicular to the 2DEG. The source-drain resistance is exclusively determined by the lower filling factor i, and the measured voltage drop across the boundary between the different regions with different filling factors depends on the magnetic field direction.

The experimental result can be deduced directly from the potential distribution shown in Fig. 4a, if one assumes that the boundary between two regions of the sample with different filling factors can be replaced by an infinite number of electrically separated metallic connections. Such a model leads to the correct potential distribution and to the information that the total source-drain current is flowing across the j-i boundary at the edge of the device where the potential drop V_2-V_1 occurs.

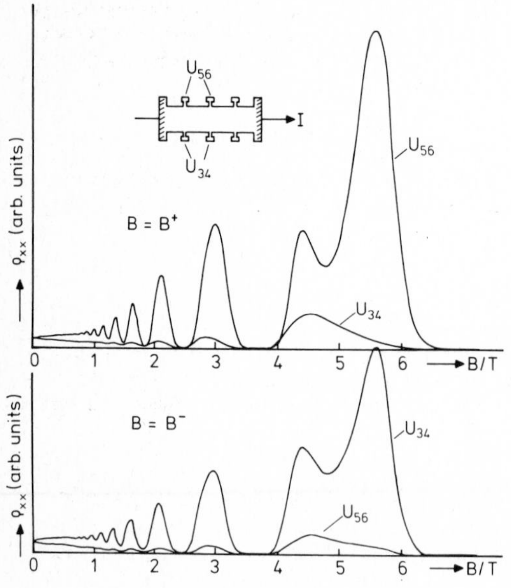

Fig. 5:

Experimental result of resistivity measurement on a device with a carrier density gradient in current direction of about 10 %. The magnetic field direction influences strongly the amplitude of the Shubnikov-de Haas oscillations

Experimental data similar to the result shown in Fig. 4b are found also for a device with a carrier density gradient in the current direction. The experimental data shown in Fig. 5 originate from a device with a carrier density varying by about 10 % ($U_{35} < U_{46}$). The amplitude and the shape of the Shubnikov-de Haas oscillations depend strongly on the magnetic field direction. Especially the shape of spin-split levels is influenced by a fluctuation in the carrier density. An interpretation of measured quantities in terms of the components ρ_{xx} and ρ_{xy} of the resistivity tensor may be incorrect, if the distribution of the carrier or potential fluctuations within the sample is not known. Therefore it will be difficult to calculate the coefficients α_i in the expression /18/

$$\Delta\rho_{xy}/\rho_{xy} = \alpha_0 + \alpha_1(\rho_{xx}/\rho_{xy}) + \alpha_2(\rho_{xx}/\rho_{xy})^2 + \dots \quad (11)$$

The experimental data obtained from a large number of different devices demonstrate that the coefficient α_0 seems to be negligibly small and the value α_1 is negative with a value smaller than 0.5. Other authors found $\alpha_1 \approx -0.1$ /12/ and $|\alpha_1| = 0.2 - 0.4$ /18/.

Finally we would like to point out that macroscopic inhomogeneities or domains within the sample may lead to complicated structures in ρ_{xy} measurements. We believe that for high quality devices long-range fluctuations lead to a spatially varying filling factor ν (ν = number of electrons per flux quantum). A mean filling factor $\nu = 0.8$ can be interpreted for example as a filling factor of $\nu = 2$ within 10 % of the area of the sample, $\nu = 1$ within 60 % of the sample, and no carrier within the remaining part of the device. Since the spin-splitting for electrons in GaAs is extremely small /25/, relatively small potential fluctuations may lead to regions with a filling factor $\nu = 2$ even if the mean value for ν is smaller than $\nu = 1$.

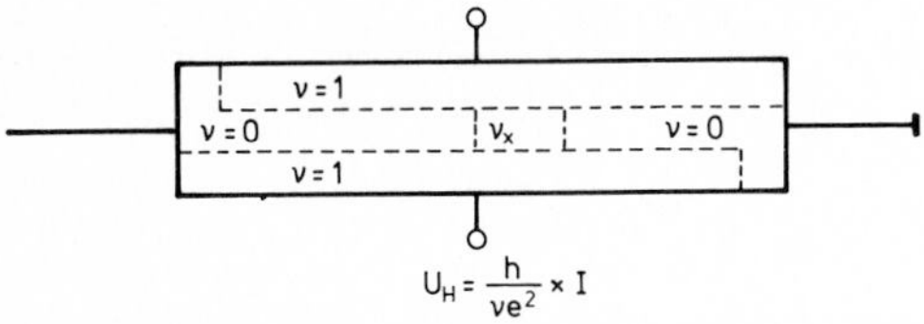

Fig. 6: Combination of different parts of a device with different filling factor which leads to "fractional" filling factors for the Hall resistivity

Figure 6 shows a simple model for an inhomogeneous carrier distribution which produces plateaus in the Hall resistance at fractional filling factors ν:

a) $\nu = 1/3$ for $\nu_x = 1$
b) $\nu = 2/5$ for $\nu_x = 2$
c) $\nu = 2/3$ for $\nu_x = 2$.

In case c) a boundary between the regions with the filling factors $\nu = 1$ and $\nu = \nu_x = 2$ is assumed which is similar to the situation in Fig. 4b, whereas a) and b) are calculated on the basis of a metallic interface between the regions with different filling factors. We do not believe that the fractional Hall resistance /26/ originates from a carrier distribution as shown in Fig. 6, because for example a Hall plateau at $\nu = 1/2$ can easily be constructed (which has not been observed experimentally). We would, however, like to point out that the data of transport measurements, especially at very low temperatures, may be influenced by some kind of domain formation.

A theory of the quantized Hall resistance which describes the real experimental situation (T ≠ 0 K, boundary conditions corresponding to a Hall geometry, inhomogeneities) is not available. Even for an ideal system with periodic boundary conditions in one direction (for which the correct value for the quantized Hall resistance is expected /27/), a quantitative expression for the Hall current distribution across the width of the device is not known /28/.

Up to now it is not clear whether edge currents are significant for an interpretation of the quantum Hall effect, and it seems that this quantum phenomenon is not only important for practical applications in metrology, but is also very valuable for the development and critical test of transport theories in strong magnetic fields.

Acknowledgement

High quality devices are necessary for QHE experiments, and we would like to thank G. Weimann, K. Ploog, and G. Dorda for the production of such samples.
This work has been supported by the Deutsche Forschungsgemeinschaft via SFB 128.

References

/1/ For a review see: Proceedings of the Int. Conf. on Electronic Properties of Two-Dimensional Systems, Surf.Sci. 58 (1976), 73 (1978), 98 (1980), 113 (1982).
/2/ R.B. Laughlin, Phys. Rev. Lett. 50, 1395 (1983), and article in this book.
/3/ P. Streda, J.Phys.C: Solid State Physics 15, L 717 (1982).
/4/ T. Ando, Y. Matsumoto, and Y. Uemura, J. Phys. Soc. Jpn. 39, 279 (1975)
/5/ R.E. Prange, Phys. Rev. B 23, 4802 (1981)
/6/ H. Aoki and T. Ando, Solid State Commun. 38, 1079 (1981)
/7/ W. Brenig, Z. Phys. B - Condensed Matter 50, 305 (1983)
/8/ D.C. Tsui and S.J. Allen, Phys. Rev. B 24, 4082 (1981)
/9/ S. Luryi and R.F. Kazarinov, Phys. Rev. B 27, 1386 (1983)
/10/ G.H. Wannier, Phys. Stat. Sol. b 88, 757 (1978)
/11/ D.J. Thouless, M. Kohmoto, M.P. Nightingale, and M. Nijs, Phys. Rev. Lett. 49, 405 (1982)
/12/ K. Yoshihiro, J. Kinoshita, K. Inagaki, C. Yamanouchi, J. Moriyama, and S. Kawaji, J. Phys. Soc. Jpn. 51, 5 (1982)
/13/ L. Bliek, E. Braun, F. Melchert, P. Warnecke, W. Schlapp, G. Weimann, K. Ploog, and G. Ebert, Phys. Bl. 39, 157 (1983)
/14/ R.J. Wagner, M.E. Cage, R.F. Dziuba, B.F. Field, A.C. Gossard, and D.C. Tsui, Bull. American Phys. Soc. 28, 365 (1983)
/15/ L. Bliek, E. Braun, F. Melchert, P. Warnecke, W. Schlapp, G. Weimann, K. Ploog, G. Ebert, and G. Dorda, to be published
/16/ D.C. Tsui, A.C. Gossard, B.F. Field, M.E. Cage, and R.F. Dziuba, Phys. Rev. Lett. 48, 3 (1982)
/17/ K. v.Klitzing, in Atomic Physics 8, ed. I. Lindgren, A. Rosen, and S. Svanberg, Plenum Publishing Corp., 1983, p. 43
/18/ V.M. Pudalov and S.G. Semenchinsky, JETP Letters 38, 173 (1983)
/19/ K. v.Klitzing in Festkörperprobleme XXI (Advances in Solid State Physics, p. 1, J. Treusch (ed.), Vieweg, Braunschweig 1981
/20/ R.W. Rendell and S.M. Girvin, Phys. Rev. B 23, 6610 (1981)
/21/ K. v.Klitzing, B. Tausendfreund, H. Obloh, and T. Herzog, in Lecture Notes in Physics 177, p. 1, G. Landwehr (ed.), Springer-Verlag 1983
/22/ B. Tausendfreund and K. v.Klitzing, Surf.Sci. (1984) (Proceedings 2DEG Conf., Oxford 1983)

/23/ K.I. Wysokinski and W. Brenig, Z. Phys. B - Condensed Matter 54, 11 (1983)
/24/ F.F. Fang and P.J. Stiles, Phys. Rev. B 27, 6487 (1983)
/25/ D. Stein, K. v.Klitzing, and G. Weimann, Phys. Rev. Letters 51, 130 (1983)
/26/ H.L. Störmer, A. Chang, D.C. Tsui, J.C.M. Hwang, A.C. Gossard, and W. Wiegmann, Phys. Rev. Letters 50, 1953 (1983)
/27/ R.B. Laughlin, Surf. Sci. 113, 22 (1983)
/28/ A.H. MacDonald, T.M. Rice, and W.F. Brinkman, Phys. Rev. B 28, 3648 (1983)

Two-Dimensional Electron Transport Under Hydrostatic Pressure

J.L. Robert, J.M. Mercy, C. Bousquet, A. Raymond

GES-USTL, Place E. Bataillon, F-34060 Montpellier Cedex, France

J.C. Portal, G. Gregoris, and J. Beerens

LPS INSA, F-31077 Toulouse, France, and
SNCI-CNRS, F-38042 Grenoble Cedex, France

ABSTRACT : We present an original method to investigate the properties of two-dimensional electronic systems. This method uses the variations of different physical quantities related to the host materials of the heterostructure under hydrostatic pressure in order to obtain a modification of the "situation" of the 2D gas.

1. INTRODUCTION

An obvious advantage presented by the use of hydrostatic pressure, associated with high magnetic fields, for the study of semiconductors lies in the fact that the "intrinsic" properties of the crystals are not changed. A very convenient way to study their properties is then provided. Such a method is well adapted to three-dimensional systems and can be a useful tool to investigate the properties of 2D structures.
In the first part of this lecture, we will give a short review of the high pressure effects in bulk semiconductors, then we will concentrate on 2D systems. In this last part, we will consider for the first time the case of $Ga_{0.7}\,Al_{0.3}$ As/GaAs heterojunctions on which classical transport experiments under hydrostatic pressure have been done : the main feature is the linear decrease of the 2D electron density as the pressure increases [1] . We will interpret this behaviour by invoking the pressure dependence of the depth of the Si impurity level in GaAlAs. We will look at the consequences of this effect on the energy band diagram of the structure and we will discuss some new possibilities in studying the magneto-transport effects like Shubnikov-de Haas or quantum Hall effect.
We will also be interested in the magnetophonon resonance observed under hydrostatic pressure on AlInAs/GaInAs heterostructures. We will show that this kind of experiment offers an interesting possibility to test the theoretical models giving the effective mass in 2D systems.

2. EFFECTS OF HYDROSTATIC PRESSURE ON BULK SEMICONDUCTORS

Hydrostatic pressure applied to a crystal leads to a change in the lattice constant. The relative variation is rather small (10^{-3}/Kbar) but the effect on the electronic properties can be important. Pressure induces a modification of the band structure ; in particular, in the III-V an increase of the gap, and therefore the effective mass, with pressure is observed. This effect is particularly strong in the narrow gaps : for example, in InSb the Γ point effective mass is changed by a factor of 3 when the pressure varies from 0 to 31 Kbar.

A second effect, which is directly connected with the increase of the effective mass, is the corresponding reduction of the wave function overlaps when a magnetic field B is superimposed, and for a given value of B, this leads to an increase of the binding energy and a decrease of the free carrier concentration. This method has been used to study the metal–non-metal transition, on the same sample [2]; usually, such observations are made on several samples, with different compensation or doping.

The hydrostatic pressure is also useful in studying the localized impurity levels and the impurity states coupled to the lattice [4] . Localized impurity levels are associated with satellite conduction bands (L or X), i.e. these levels will follow the movement of the X or L minimum under pressure. The depth of the impurity level will then vary at a rate of the order of 10 meV/Kbar (relative to the Γ minimum). The effect of pressure can therefore be fairly large on the carrier concentration. In contrast with this, the shallow impurity states (associated with Γ) will have a relatively small pressure coefficient (< 0.1 meV/Kbar even for a small gap like InSb).

When deep levels are accompanied by a large distortion of the surrounding lattice, the impurity states are described by the configuration coordinates, diagram Fig. 1 , which change when pressure is applied.

InSb is probably the most illustrative example of this kind of study. We will briefly look at the different points discussed above on this material :

a) The magnetic field induced metal-non-metal transition under hydrostatic pressure.

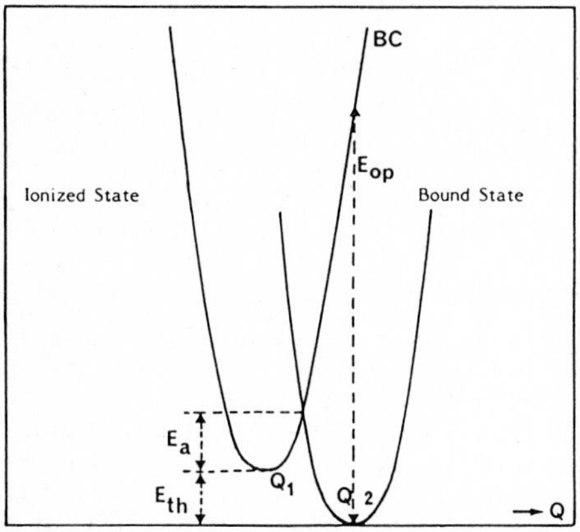

Fig. 1. Configuration coordinate diagram for a large lattice relaxation impurity level

The metallic behaviour of InSb has been attributed to various causes :

- the impurity states merge with the conduction band even in the lightly doped material [5]
- the metallic behaviour is governed by an impurity band slightly separated from the conduction band (ε_2 process) [6] .

It was difficult to choose between these two alternatives before a study of the metal-non-metal transition under pressure and high magnetic fields was made [2].

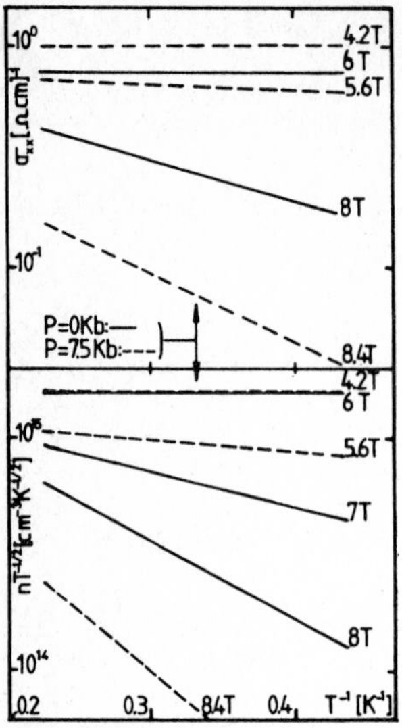

Fig.2. Temperature dependence of σ_{xx} and $nT^{-1/2}$ for various magnetic fields

This study showed that the transition occurs for the same value of the critical density under hydrostatic pressure but that the conductivity increases as pressures goes up, Fig.2 . According to Mott's theory, if the transition involves an ε_2 process, the minimum metallic conductivity should be pressure independent. So, these experiments clearly demonstrated that we are dealing with a situation which cannot be described by an ε_2 process.

b) Deep impurity states in InSb under hydrostatic pressure.

As we said, strong variations of the band structure are obtained in a narrow gap semiconductor. Due to the relative shift of the various minima of the conduction band, it has been possible to observe three impurity states in the present samples of n-InSb under pressure [4] :

- a shallow, Γ -like level. The pressure coefficient of this level is small (the previous experiments on the metal-non-metal transition are related to this level).

- two deep impurity levels. At zero pressure these two levels, respectively associated with L and X conduction band minima,are located at 85 meV and 140 meV (above the Γ minimum) and their pressure coefficients are 10,5 meV/Kbar and 20 meV/Kbar.

The analysis of magneto-transport measurements under pressure demonstrates the existence of the L-like impurity state [7] . A sharp increase in the freezing out of the electrons is observed as soon as the pressure is enough to bring the L-like state below the hydrogenic (Γ-like) level, Fig.3 .

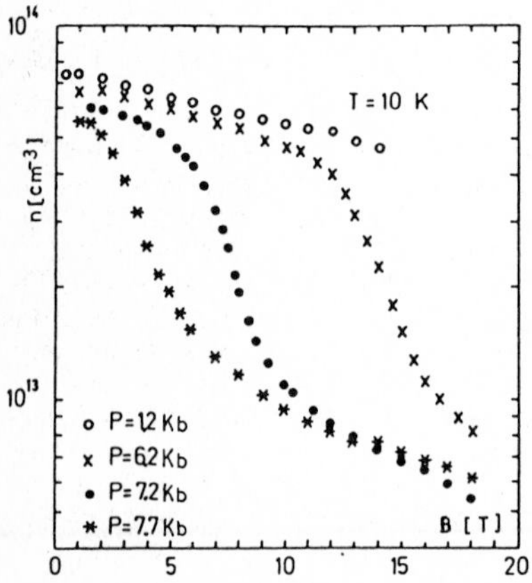

Fig.3. Electron concentration versus B

The second level exhibits a metastable character, and therefore is strongly coupled to the lattice. If a hydrostatic pressure of at least 13 Kbar is applied at room temperature, the electrons located in the conduction band are trapped on this level. By cooling the sample below 100 K, we prevent any reexcitation of the carriers, even when the pressure is removed ; this indicates clearly that a large potential barrier exists separating localized states from band states (see Fig.1). If the temperature is raised near 100-110 K a slow relaxation is observed,providing a way to change the free carrier concentration over a certain range [8] , Fig.4 .
Another illustrating example is the case of bulk GaAlAs largely investigated by Saxena [9].

3. $Ga_{0.7}Al_{0.3}As/GaAs$ HETEROJUNCTIONS UNDER HYDROSTATIC PRESSURE

MBE and MOCVD techniques allow the possibility to obtain a semiconductor heterojunction in the Anderson sense - i.e. an a abrupt semiconductor-semiconductor interface of a high degree of monocrystallinity.

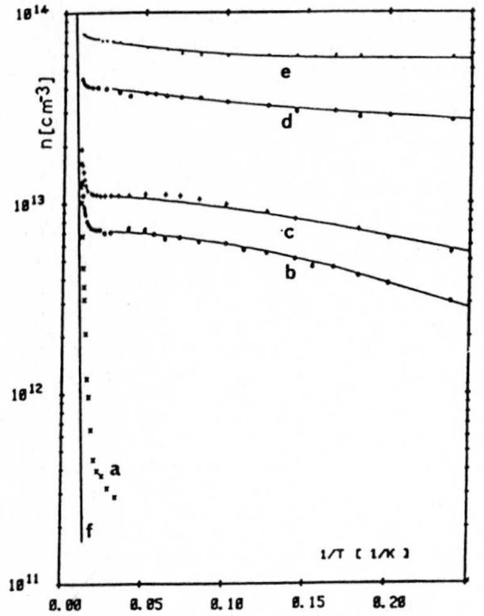

Fig.4. Electron concentration versus 1/T

Consider two different semiconductors (1) and (2), the material (2) having a wider gap E_g than the material (1) ($Eg_2 > Eg_1$). In an intuitive way we expect the conduction band discontinuity to be given by the difference between the electron affinities of the two materials. Although this simple idea is not exact [10] it gives a physical idea of the origin of the formation of the band discontinuity.

A detailed analysis of the heterojunction containing both depletion and accumulation regions has been carried out by Cserveny [11] . The energy band profile of an n-n type heterojunction at equilibrium is shown in Fig.5 .

Assuming that all the donors (Nd) are ionized, Cserveny has obtained an approximate relationship between the total charge stored in the accumulation layer and the part of the diffusion voltage Vd_2 appearing in semiconductor 2 :

$$n_s = \sqrt{2\ \varepsilon_2\ e\ Nd_2\ Vd_2}\ . \qquad \text{III-1}$$

Although this expression is not valid for degenerate conditions (i.e. the usual case in the heterojunction we study), it has the merit to show in analytical way that the free carrier density in the accumulation layer is directly connected to the characteristics of semiconductor 2. Consequently one can expect some changes in the electrical properties of the 2D electron gas when the parameters of semiconductor 2 are modified when pressure is applied.

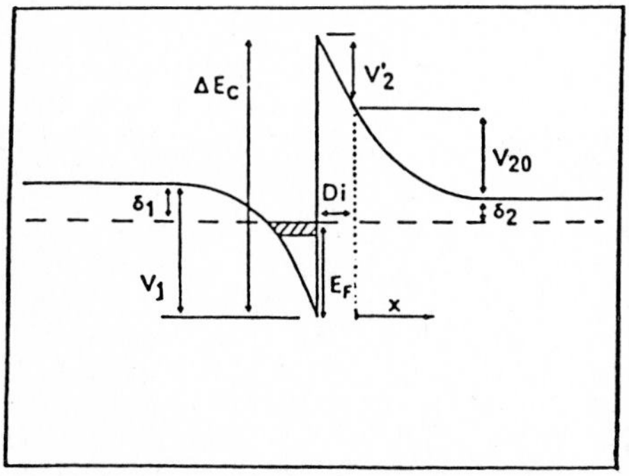

Fig. 5. Energy band diagram in the case of MOCVD samples

a) Effect of pressure on the carrier density of $Ga_{0.7}Al_{0.3}As/GaAs$ heterojunctions. Two types of heterojunctions have been studied, which were grown by MOCVD or MBE techniques with Si as a doping in GaAlAs. Most of the samples have a bridge shape but some of them are Van der Pauw samples. In all the experiments the pressure is applied at room temperature and the carrier concentration is measured by Hall effect in the whole range of temperature or by Shubnikov-de Haas experiments near the helium temperature.

The conduction in the bulk GaAlAs can be neglected in the high temperature range as long as the thickness of the GaAlAs layer does not exceed 500 A°. At helium temperature the conductivity is always governed by the two-dimensional electron gas even for GaAlAs layer thickness as high as 1800 A°.

Experimental results

The main feature observed when the pressure is applied is a linear decrease of the density ns , Fig. 6 .
To explain this behaviour we assume that the silicon impurity level in GaAlAs is strongly dependent on the pressure and associated with a satellite minimum of the conduction band.
Because of the existence of persistent conductivity in bulk GaAlAs, this level could be associated with the X minimum.

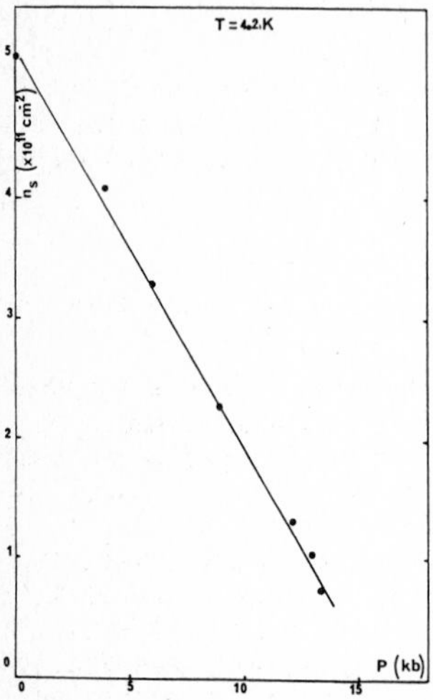

Fig.6. Experimental variations of ns with pressure

In order to verify this hypothesis we have proposed an abrupt heterojunction model in which GaAlAs is assumed to be partially compensated and the donors are not completely ionized even at room temperature.
The following analysis remains valid in the high temperature range - to avoid the consequences of the metastable behaviour of the donor level. In the low temperature range, the system is frozen out.

Theoretical analysis

The quantum well is treated in the triangular well approach - and a solution is found by requiring that the electron density equals the depleted charge in GaAlAs.
The spacer layer is considered to be perfectly compensated and therefore the electric field is constant in the spacer.
The electric field in the GaAlAs side is found by integrating the Poisson equation.

With the notations of Fig.5 , we found :

$$\left(\frac{dv_2}{dx}\right)^2_{x=0} = \frac{2\,q^2 N_d}{\varepsilon_2}\, v_{eff} \qquad \text{III-2}$$

$$v_{eff} = kT\left[\,Ln\,\frac{a+1}{a+\exp v_{20}/kT} + \frac{N_a}{N_d}\,\frac{v_{20}}{kT} + \frac{4\,N_c}{N_d}\,L_n\,\frac{1+\frac{1}{4}\exp -\delta_2/kT}{1+\frac{1}{4}\exp -\frac{(\delta_2+v_{20})}{kT}}\right]$$

where $a = 2\exp\frac{\varepsilon_d - \delta_2}{kT}$ $\qquad v_{20} = v_2\ (x=0)$.

The familiar looking relation is obtained :

$$ns_1 = \frac{\varepsilon_2}{q^2}\left(\frac{d\,v_2}{dx}\right)_{x=0} = \left(\frac{2\,\varepsilon_2\,N_d\,v_{eff}}{q^2}\right)^{1/2} \qquad \text{III-3}$$

in which v_{eff} is the effective band binding.

In order to fit dn_s/dp two parameters must be adjusted
i) the compensation ratio
ii) the pressure dependence of the impurity level β (we take $\varepsilon d\ (o) = 60$ meV).

A good agreement is obtained with $N_a/N_d = 0.32$ and $\beta = 11$ meV/Kbar.

b) Effect of pressure on the quantum properties of $Ga_{.7}Al_{.3}As/GaAs$. It is obvious that the decrease of n_s due to the application of pressure leads to a change in the quantum properties of the heterojunctions in the low temperature range.

The Shubnikov-de Haas experiments reported on Fig.7 show clearly the shift of the peaks of the magnetoresistance towards lower magnetic fields as the pressure is increased.

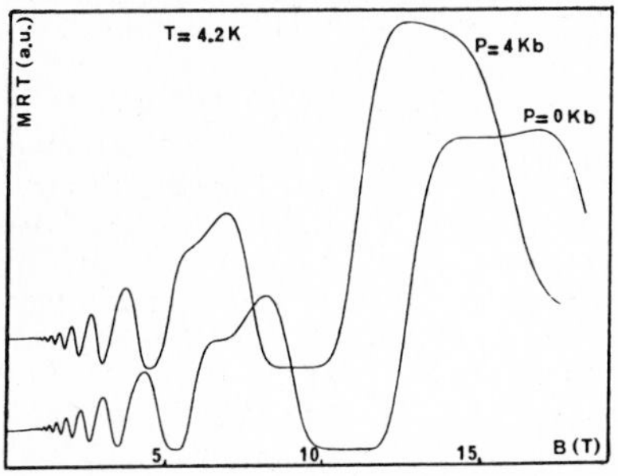

Fig.7. Transverse magnetoresistance (MRT) of GaAlAs/GaAs heterojunction

The observation of the ultra quantum limit in rather low magnetic fields becomes possible even for structures with high carrier concentrations under zero pressure. This is clearly shown on Fig.8 , where the Hall voltage versus B, for the same sample, is reported for several pressures. It would be certain y interesting to review this kind of experiment in very low temperature range to look at the anomalous quantum Hall effect (fractional quantum numbers).

c) Other consequences of the pressure-induced modulation

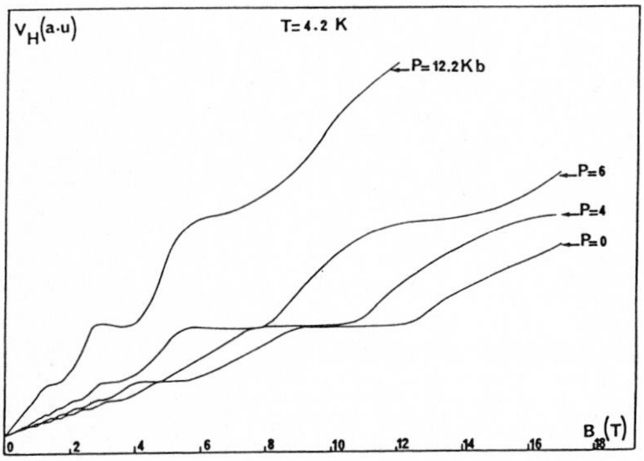

Fig.8. Quantum Hall effect of the same sample

One of the most interesting problems in 2D systems is the knowledge of the mobility dependence with carrier density n_s. The usual way to obtain a variation of n_s is to work on several samples with different spacer thickness or doping levels. It is, however, well known that mobility is strongly dependent on residual impurity concentration level, which is a parameter mostly difficult to control. Therefore, a comparison between theoretical models giving the mobility and the experimental results obtained on different samples is not realistic. On the other hand, the use of hydrostatic pressure offers the possibility to study the mobility dependence versus n_s on the same sample, Fig.9 . Comparison between theory [12] and experiments still has to be done.

Another possibility is to study the metal-non-metal transition in connection with the change in the carrier density ; this is shown in Fig.9 . Such a transition may be induced by magnetic field when the concentration is low enough and the field high enough to reach a situation where only the last Landau level is populated and the Mott criterion is attained (i.e. when the cyclotron orbit is sufficiently small to avoid any overlap of the wave functions). The free carrier density in Fig.10 is deduced from the conductivity tensor, assuming that the localized electrons do not participate in the Hall effect:

$$\left(n_s = \frac{1}{q} \frac{R_H B^2}{\rho_\perp^2 + R_H^2 B^2}\right) . \quad [2]$$

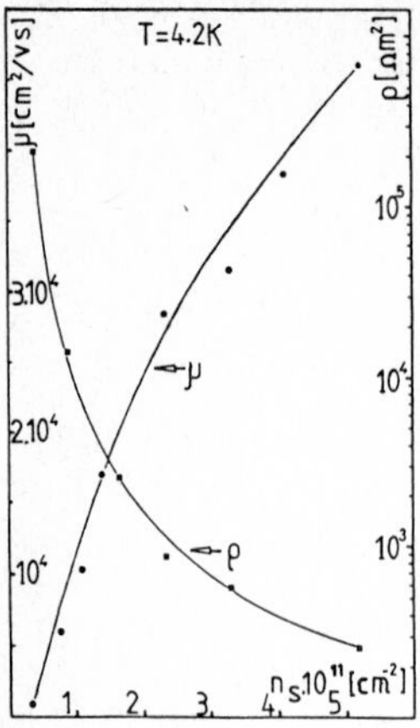

Fig.9. Mobility and resistivity versus n_s

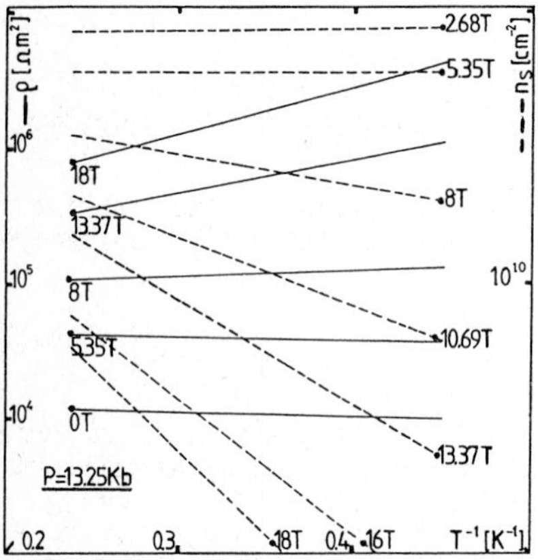

Fig.10. ρ and n_s versus 1/T for various magnetic fields

4. MAGNETOPHONON RESONANCE EFFECT UNDER HYDROSTATIC PRESSURE IN A 2D ELECTRONIC SYSTEM

The recent developments and applications of 2D electronic systems have reawakened particular interest in the use of the well-known magnetophonon resonance effect in bulk III-V semiconductors [13] to study the electronic properties of these systems [14]. The detailed analysis of the experimental results has enabled the study of the two-dimensional behaviour of the electrons confined at the interface of the heterojunction over a wide range of temperature up to 300K. The magnetophonon effect has thus been used to obtain very significant information on the electron-phonon interactions and to give an access to the electron effective mass. We want to show here what can be expected from magnetophonon measurements under hydrostatic pressure. We will first review the magnetophonon effect and compare the situation in bulk semiconductors and two-dimensional systems and we will pay some attention to the non-parabolic correction in both cases. We will then look at the effect of pressure and at some experimental results.

The magnetophonon resonance effect is observed through an oscillatory behaviour of the resistance with magnetic field. Since this is the effect of a resonant scattering of electrons between Landau levels by longitudinal optical phonons, two essential but competitive conditions are required :

- the Landau quantization condition $\mu B \gg 1$ where μ is the electron mobility and B is the magnetic field. This implies relatively low temperature.
- a sufficiently large optical phonon generation is necessary which means that temperature must not be too low.

In the three-dimensional case, these conditions imply the use of high purity and non-degenerate material where mobility reaches the highest values. The situation in a two-dimensional system is drastically different because high density goes with high Fermi wave vector k_F and therefore high mobility [15]. Then magnetophonon effect can be observed in the degenerate case of a two-dimensional system at high temperature. The scattering occurs now between Landau levels situated on both sides of the Fermi level.
The resonance condition is :

$$\omega_{Lo} = N\,\omega_c = N\,\frac{e\,B_N}{m^*} \qquad \text{(IV-1)}$$

where ω_{Lo} is the phonon frequency, generally precisely measured by Raman scattering, ω_c is the cyclotron frequency, N the resonance index and m* the effective mass. In the three-dimensional case, R.A. Stradling[13] derives the band edge effective mass from the Landau level energy given by E.D. Palik et al.[16] :

$$m^*_o = m^* \left[1 + k_2 \ (1 + (2L+1)/N) \ \frac{\hbar \omega_{Lo}}{E_G} \right] \qquad \text{(IV-2)}$$

where L is the quantum number of the initial Landau level, E_G the energy band gap and k_2 is a function of E_G/Δ_0 and m^*_o for non-parabolic correction. Δ_o is the spin-orbit splitting term. k_2 is about - 0.85 in GaInAs.

In the two-dimensional system, the wave vector along the z axis is quantized at a fixed value k_{z0}. We therefore have another expression for m^*_o assuming in Palik equations :

$$\frac{\hbar^2 k^2_{z0}}{2 m^*_o} = \varepsilon_0 \ (1 + \varepsilon_0/E_G) \qquad \text{(IV-3)}$$

where ε_0 is the energy separation between the lowest subband E_0 and the conduction band edge at the mean abscissa z_0 given by F. Stern[17] in the simple model of the triangular quantum well ($z_0 = 2\ E_0/3\ eF$). Then we get :

$$m^*_o = m^* \left[1 + 2\ k_2 \ \frac{\varepsilon_0(1+\varepsilon_0/E_G)}{E_G} + k_2 \ (1 + (2L+1)/N \ \frac{h\omega_{Lo}}{E_G} \right]. \qquad \text{(IV-4)}$$

The band edge mass obtained this way in a GaInAs-AlInAs heterostructure is in quite good agreement with the one measured in bulk GaInAs|18| at low temperature, respectively 0.0395, and 0.040 m_o.

This experimental effective mass value can be now compared to the one calculated with a model involving five levels derived by C. Hermann[19] from the Kane k.p model :

$$\frac{m_e}{m^*_o} = 1 + \frac{P^2}{3} \left[\frac{2}{E_G} + \frac{1}{E_G+\Delta_0} \right] - \frac{P'^2}{3} \left[\frac{2}{E\Gamma_7 c-E_0} + \frac{1}{E\Gamma_8 c-E_0} \right] + c. \qquad \text{(IV-5)}$$

In ternary alloys, a good agreement is achieved at zero pressure as one considers a decrease of the momentum matrix element P^2 due to alloy disorder of about 10 % [20] . In bulk material E_0 and m^*_o are found to vary linearly with pressure [21] . Therefore, m^*_o is linear with energy band gap as predicted by simplest Kane model in a first approximation. The variations of other parameters (Δ_0, P^2,ω_{Lo}) under pressure in the bulk have been measured or can be estimated [22] . In AlInAs-GaInAs heterostructure we find a linear variation of the effective mass m* with pressure[23] (to be published).

Figure 11 shows typical recordings of the oscillatory transverse magnetoresistance at 140K at different pressure values. The band edge effective mass m^*_o derived with equation (IV-4) is well described by the C. Hermann model (IV-5) over all the range of pressure from 0 to 16 kbars.

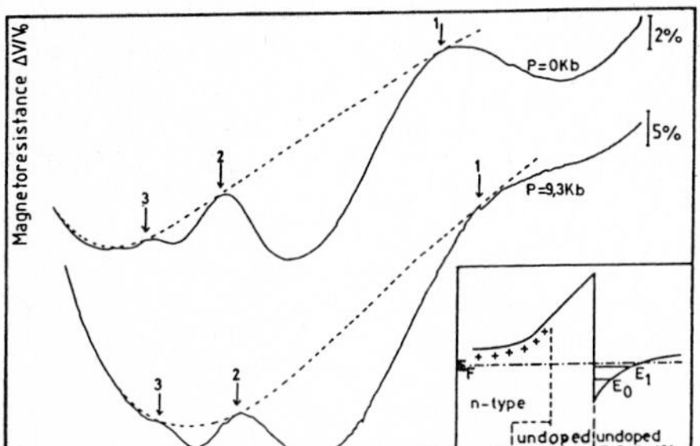

Fig.11. Transverse magnetophonon resonance ΔV/Vo at an AlInAs-GaInAs heterostructure interface at 140K

It is noteworthy that in the case of the AlInAs/GaInAs heterostructure the variation of the surface charge density n_s with pressure is relatively weak [23] (to be published) ; the GaAlAs/GaAs is better suited for the verification of the non-parabolic correction in two-dimensional systems because of the large variation of the Fermi level position which can be obtained.

So in two-dimensional systems like in bulk semiconductors the effective mass is in first approximation simply related to the energy band gap by a linear relation. We point out that this behaviour could not be described by a C. Hermann model without considering the specific non-parabolic correction in the quantum well.

5. CONCLUSION

We have made a description of what one can expect from application of hydrostatic pressure on typical 2D systems (GaAlAs/GaAs and GaInAs/AlInAs). We have shown that this method of investigation presents a large interest in studying these structures. Further developments can be expected on these 2Dsystems as well as on other type of heterojunctions.

Acknowledgments

We want to thank the Laboratories Thompson-CSF,LEP,and Bell for providing the samples. Most of the experiments have been made at the SNCI in Grenoble. We would like to acknowledge this support. This work is supported by the MIR and the CNRS.

References

1. J.M. Mercy, C. Bousquet, J.L. Robert, A. Raymond, G. Gregoris, J. Beerens, J.C. Portal, P.M. Fijlink, P. Delescluse, T. Chevrier, T. Linh : 5° Conference Electronic Properties of 2D systems, Oxford, 1983.
2. J.L. Robert, A. Raymond, R.L. Aulombard, C. Bousquet : Phil. Mag. B.42, 1003-1980.
3. A. Raymond, J.L. Robert, D.S. Kyriakos, M. Royer : 16th Int. Conf. Physics of Semiconductors, Montpellier, 1982.Physica 117-118B No.Hol.Pu.Com.
4. S. Forowski : E.P. Society, Lausanne, 1983.
5. J.R. Sandercock : Solid State Com. 7. 721, 1969.
6. H. Miyazawa, H. Ikoma : J. Phys. Soc. Japan, 23, 290, 1967.
7. S. Porowski, L. Konczewicz, A. Raymond, R.L. Aulombard, J.L. Robert, M. Baj : Application of high magnetic fields in SC Physics. Vol. 177. Springer Verlag, 1982.
8. A. Kadri, R.L. Aulombard, C. Bousquet, A. Raymond, J.L. Robert, S. Porowski, L. Konczenvicz : 16th Int. Conf. Physics of Semiconductors, Montpellier, 1982. Physica 117-118B, North-Holland Publishing Company.
9. A.K. Saxena : J. Phys. C. 13, 4323-4334, 1980.
10. R.S. Bauer, P. Zucher, H.N. Sang : Appl. Phys. Letters, n° 43, p. 663, 1983.
11. S.I. Cserveny, Int. T. Electronics, Vol. 2F n°1,6560, 1968.
12. G. Fishman, D. Calecki : 16th Int. Conf. Physics of Semiconductors, Montpellier, 1982. Physica 117-118B, North-Holland Publishing Company.

    G. Bastard : 5° Conference Electronics Properties of 2D Systems, Oxford, 1983.

13. R.A. Stradling and R.A. Wood : J. Phys. C. (Proc. Phys. Soc). Ser. 2, Vol. 1, 1711, 1968.

14. D.C. Tsui, T.H. Englert, A.Y. Cho and A.C. Gossard : Phys. Rev. Lett. 44, 341, 1980.
T.H. Englert, D.C. Tsui, J.C. Portal, J. Beerens and A.C. Gossard : Solid State Commun. 44, 1301, 1982.
G. Kido, N. Miura, M. Ohno, H. Sakaki : J. Phys. Soc. Japan, 51, 2168, 1982.
J.C. Portal, J. Cisowski, R.J. Nicholas, M.A. Brummell, N. Razechi and M.A. Poisson : J. Phys. C ; Solid State Phys. 16, L.573, 1983.
M.A. Brummell, R.J. Nicholas, J.C. Portal, K.Y. Cheng and A.Y. Cho : J. Phys. C. Solid State Phys. 16, L579, 1983.

15. H.L. Stormer, A.C. Gossard, W. Wiegmann and K. Baldwin : Appl. Phys. Let. 39, 912, 1981.

16. E.D. Palik, G.S. Picus, Steitler and R.F. Wallis : Phys. Rev. 122, 475, 1961.

17. T. Ando, A.B. Fowler and F. Stern : Rev. of Modern Physics 54, 466, 1982.

18. C.K. Sarka, R.J. Nicholas, Ph. D.T. Thesis Oxford, Unpublished.

19. C. Hermann and C. Weisbuch : Phys. Rev. B.15, 823, 1977.

20. J.A. Van Vechten and T.K. Bergstresser : Phys. Rev. B1, 3351, 1970.
E.D. Siggia : Phys. Rev. B.10, 5147, 1974.

21. G.C. Pitt, J. Lees, R.A. Hoult and R.A. Stradling : J. Phys. C. Solid State Phys. 6, 3282, 1983.

22. P.J. Melz and I.B. Ortenburger : Phys. Rev. B.3, 3257, 1971.
G. Martinez, Handbook on Semiconductors, edited by T.S. Moss, Vol. 2 edited by M. Balkanski, North-Holland Publis. Comp. p. 181, 1980
S.S. Mitra, O. Brasman, W.B. Daniels and R.K. Crawford : Phys. Rev. 186, 942, 1969.

23. G. Gregoris, Thesis, in preparation.

Fractional Quantization of the Hall Effect

Arthur C. Gossard
AT&T Bell Laboratories
Murray Hill, NJ 07974

Fractional quantization of the Hall effect and the two-dimensional structures in which it occurs are described. The fractional quantization to date has been observed in high mobility modulation-doped $Al_xGa_{1-x}As/GaAs$ heterostructures at high magnetic fields and low temperatures where all electrons lie in the lowest Landau level. At certain fractional Landau level occupations $\nu = p/q$ (q = 3, 5, 7; p = 1, 2, 3 . . .) minima and Hall resistance plateaus develop. The value of the Hall resistance, ρ_{xy}, approaches $h/\nu e^2$ at the plateaus. The observed phenomena and their temperature dependence suggest the occurrence of a series of correlated electron states at fractional occupation of the lowest Landau level. Both two-dimensional electron and hole systems exhibit fractional quantization.

1. Introduction

When systems of two-dimensional electrons are placed in high magnetic fields, magneto-quantum oscillations in their electrical properties may be observed which are associated with occupation of successive electron Landau levels. This has led to the observation of the quantized Hall effect at integral occupations of Landau levels in silicon MOS channels [1] and at compound semiconductor interfaces [2].

When the two-dimensional electron gas is subjected to still more intense normal magnetic field and to sufficiently low temperature, all electrons may be expected to fall into the lowest Landau level and spin state. In this limit, the possibility exists that the electrons will order under the influence of their mutual interactions and their low residual kinetic energy. Electrons on the surface of liquid helium have clearly exhibited ordering [3], while 2D electrons in Si MOSFETs have shown more ambiguous behavior [4]. The most remarkable ordering phenomena have been seen in GaAs, however, where an apparent succession of correlated electron states has been found at fractional occupations, ν, of the lowest Landau level [5,6].

The correlated states are manifested in the previously unobserved and unexpected fractional quantization of the Hall effect. It will be the purpose of this paper to describe the fractional quantum Hall effect and the materials and conditions under which it is found. The effects have thus far been seen for high-mobility electrons confined at modulation-doped interfaces between GaAs and $Al_xGa_{1-x}As$. The new states yield minima in electrical resistance and plateaus in the Hall resistance for current flow along the two-dimensional layers. Whereas the integral quantum Hall effect occurs because of minima in the density of electron states at energies between Landau levels, the fractional quantization is interpreted in terms of new gaps in the spectrum of electron energy levels appearing within the lowest Landau level under the influence of electron-electron interactions.

2. Samples for Accessing the Extreme Quantum Limit

Access to the extreme quantum limit, $\nu = nh/eB < 1$, requires that the Landau level spacing, $\hbar\omega_c = eB\hbar/m^*$, exceed the zero-field Fermi energy, $E_F = \pi n\hbar^2/2m^*$ or equivalently that the magnetic length or cyclotron radius $\sqrt{\hbar/eB}$ be less than the average interparticle spacing, $n^{-1/2}$, where n is the surface density of electrons. In order to observe electron ordering, it is desirable that interactions between electrons be large and that electron interactions with impurities be small. This requires the minimization of uncertainty broadening of the electron levels and inhomogeneous broadening caused by electron scattering and potential fluctuations.

For the purposes of achieving the above conditions, the modulation-doped $GaAs/Al_xGa_{1-x}As$ interface presents a number of advantages relative to a Si MOSFET. The GaAs system has a lower electron effective mass, longer scattering times and fewer charged defects in the confining layer relative to the Si/SiO_x interface of a Si MOSFET. The structures in which the fractional Hall effect has been observed to date have been grown by multiple-layer molecular beam epitaxy [7]. A typical modulation-doped structure used in the experiments is shown in Fig. 1. The active layer which contains the two-dimensional electron gas is epitaxial undoped GaAs of ≈ 1 μm thickness grown on a semi-insulating GaAs substrate crystal. Charge is introduced to the active layer from an $Al_xGa_{1-x}As$ layer, part of which is doped with Si donor atoms. A region of several hundred Å in the barrier immediately adjacent to the GaAs channel is left undoped in order to achieve separation between the donor atoms and the electrons in the GaAs. The electrons are confined at the $Al_xGa_{1-x}As$ interface by the Coulomb field of the ionized donors. The effective potential perpendicular to the interface is nearly triangular in shape, and the electrons occupy quantum bound states in the triangular potential well. They are not constrained with respect to motion parallel to the interface, however,

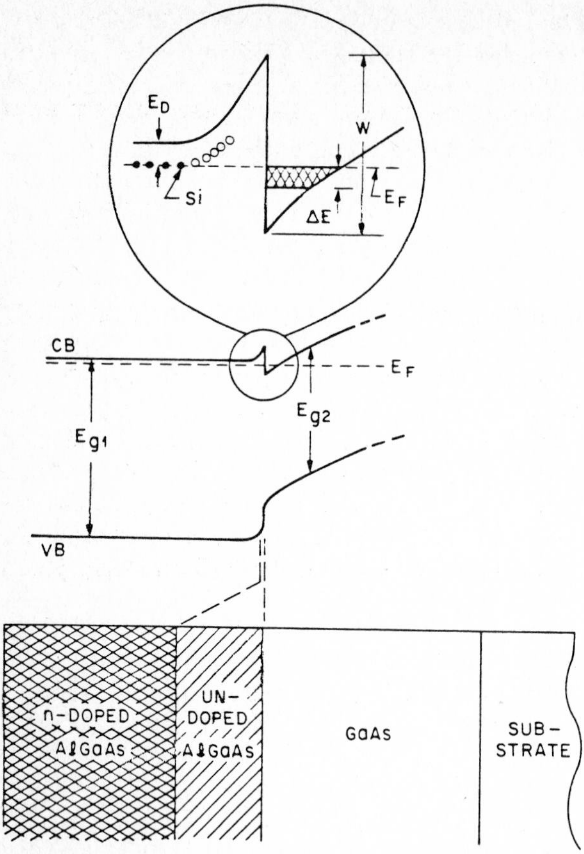

Fig. 1 Modulation-doped n-type GaAs$-$Al$_x$Ga$_{1-x}$As heterostructure and its energy band structure. CB and VB refer to conduction band and valence band edges; E_{g1} and E_{g2} are the energy gaps of Al$_x$Ga$_{1-x}$As and GaAs. ΔE is the zero-magnetic-field filling of the two-dimensional electron gas lowest quantum subband, and E_F is the Fermi energy. W is the step height between the GaAs conduction band and the Al$_x$Ga$_{1-x}$As conduction band at the interface

and move freely in that direction with minimal scattering from residual impurities in the channel and dopant impurities in the barrier beyond the undoped spacer layer. For the case where only the lowest quantum bound state is occupied, they form an excellent representation of the idealized two-dimensional electron gas. A region of several hundred Å of doped Al$_x$Ga$_{1-x}$As nearer to the surface and further from the GaAs channel is depleted of electrons by the surface potential so that there are not free electrons within the Al$_x$Ga$_{1-x}$As layer. The density of carriers in the GaAs channel needed to access the extreme quantum limit in 2OT fields is $\leq 2 \times 10^{11}$ cm^{-2}. Low-field electron mobilities up to $\approx$500,000 cm^2/Vs at $T \lesssim 4.2K$ have been obtained in the modulation-doped GaAs heterostructures in this range of carrier densities, corresponding to carrier scattering times of 2 x 10^{-11} sec [5]. Experimental measurements are made on photolithographically defined Hall bridges using microampere currents and alloyed Ohmic current and potential contacts.

3. Experimental Observation of Fractional Quantization

Experimental results for the resistivity and Hall resistance versus magnetic field in the fractional quantum Hall effect regime are shown for several samples in Figs. 2 and 3. Minima develop in the diagonal (in-plane) resistivity ρ_{xx} at magnetic fields corresponding both to integral filling and to certain fractional fillings of Landau levels. The Hall resistivity ρ_{xy} develops plateaus at the same integral and fractional filling factors. The classical value of the Hall resistance ρ_{xy} for n_A carriers per unit area is $\rho_{xy} = B/n_A e$. B/n_A is the magnetic flux per electron which is the flux quantum divided by the Landau level filling ν, i.e., $B/n_A = h/e^2\nu$ at filling factor ν. In two dimensions, the Hall resistance R_{xy} is equal to the Hall resistivity ρ_{xy}.

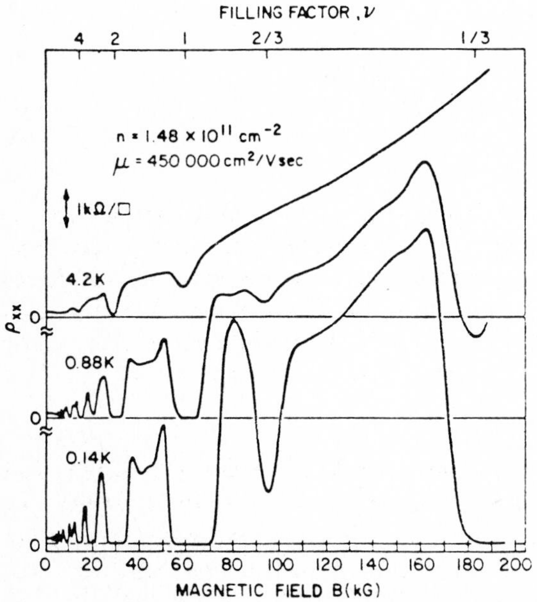

Fig. 2 Magnetic field dependence of the diagonal resistivity ρ_{xx} of $GaAs-Al_xGa_{1-x}As$ modulation-doped heterostructure at T = 4.2, 0.88, and 0.14K. Electron density n = 1.48 x 10^{11} electrons /cm^2 and electron mobility μ = 450,000 cm^2/Vs (Ref. [8])

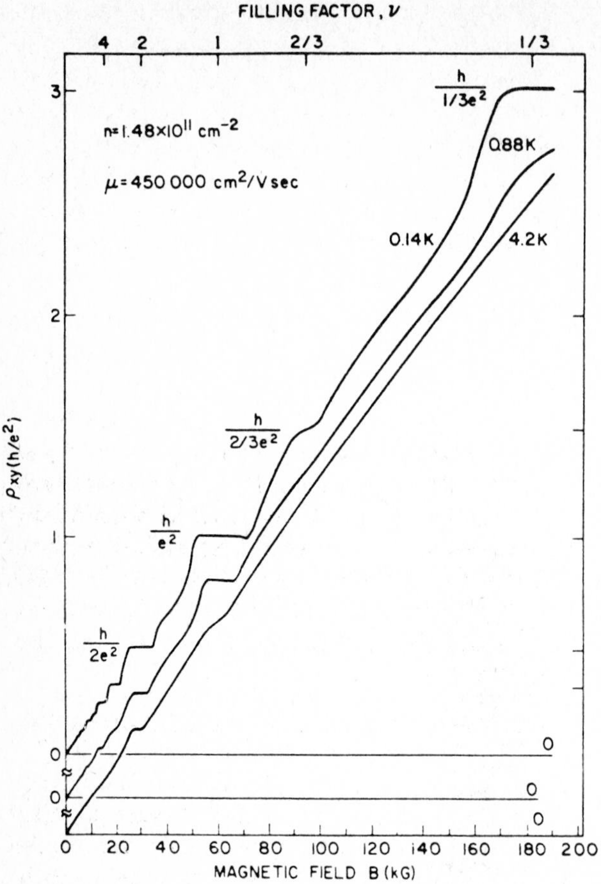

Fig. 3 Magnetic field dependence of the Hall resistance for the same structure as Fig. 2 at T = 4.2, 0.88, and 0.14K. Hall resistance ρ_{xy} and Hall resistivity ρ_{xy} are identical in two dimensions (Ref. [8])

The widths and depths of the resistivity minima and the widths and flatness of the Hall resistance plateaus are strongly dependent on the filling factors, ν, of the Landau levels and on the individual sample characteristics. The positions of the minima and plateaus and the Hall resistance values at the plateaus do not vary from sample to sample. The value of ρ_{xy} is $h/\nu e^2$ at filling factor ν. It has been measured at $\nu = 1/3$ to equal $h/\nu e^2$ to within one part in 10^4 [8]. The fractional quantization of the Hall resistance and appearance of minima in electrical resistance at fractional occupation of the lowest Landau level had been quite unexpected before their initial experimental observation. In analogy with the interpretation of the integral quantum Hall effect, these features suggest singularities in the electron state density and electron localization at the Fermi level at the fractional occupation values of the Landau level. The absence of any known one-electron effect which would remove the degeneracy of the lowest Landau level at these occupation fractions, and the known presence of electron-electron interactions of magnitude $n^{1/2}e^2/\epsilon \approx 3$ meV (30 K) at $n = 10^{11}$ cm^{-2} suggest electron-electron interactions as the driving mechanism for the new states. Down to $T \approx 0.4$ K, the diagonal resistivity is proportional to exp $(-\Delta/kT)$ with $\Delta = 0.28$ meV at $\nu = 1/3$ and $\Delta = 0.033$ meV at $\nu = 2/3$, indicating the presence of an energy gap in the states at fractional filling [8,9].

Increased current is also observed to suppress the resistivity minima and the Hall plateaus. The minima and plateaus are not observed in samples with lower mobility, in contrast to the case of the integer quantum Hall effect where the same lower mobility samples have wider plateaus and resistance minima [10]. The relative roles of potential fluctuations and inhomogenous broadening and of carrier scattering and dynamic broadening of levels are not yet fully determined in the integral and fractional regimes. It is probable that when the broadening approaches or exceeds the respective gaps of the ordered states, the ordering and concomitant plateaus and minima are suppressed. The gap between Landau levels exceeds the gaps of the fractionally quantized states. Hence, the fractionally quantized state is suppressed before the integrally quantized state.

In addition to quantization at quantum numbers 1/3 and 2/3, quantization has also been observed at a number of other fractions [6]. This is seen in Fig. 4, where experimental traces for ρ_{xy} and ρ_{xx} are shown. Resistivity minima are found to develop at ν = 4/3, 5/3, 2/5, 3/5, 4/5 and 2/7, suggesting that fractional quantization exists in multiple series, with each series based on the inverse of an odd integer. Quantization at $\nu = 1/5$ has also been recently reported [11].

Theoretical calculations suggest that the electron state at odd fractional filling of the lowest Landau level is an ordered state which is not an electron crystal but is an incompressible quantum fluid, separated by an energy gap from excitations which are fractionally charged quasiparticles [9]. The theoretical understanding of the fractional quantization is discussed elsewhere in this Winterschool by LAUGHLIN [9] and will not be treated in detail here.

4. Two–Dimensional Hole Gas

Two-dimensional hole gas samples with enhanced mobilities have also recently been fabricated and investigated for quantized Hall behavior. Trace d in Fig. 4 represents data for a hole sample. The hole-gas samples are prepared by MBE of GaAs in a manner analogous to the electron gas samples [12]. The sample structure and the energy band configuration of the hole gas samples is shown in

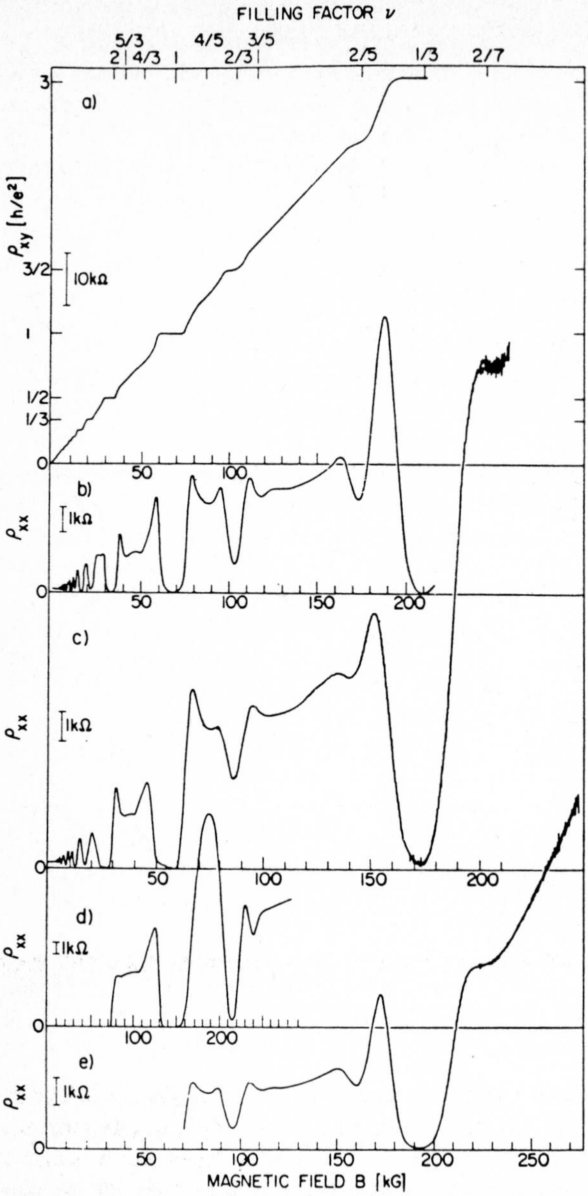

Fig. 4 Experimental magnetic field dependence of ρ_{xy} and ρ_{xx} for four samples at $T \approx 0.55K$. a) and b) are for a structure with $n = 1.7 \times 10^{11}\ cm^{-2}$ and $\mu = 500{,}000\ cm^2/Vs$. c) is for a sample with $n = 1.4 \times 10^{11}\ cm^{-2}$ and $\mu = 510{,}000\ cm^2/Vs$. d) is for a two-dimensional hole gas structure with $n = 3.5 \times 10^{11}$ holes/cm^2 and $\mu_{hole} = 36{,}000\ cm^2/Vs$. e) is for an electron-gas sample with $n = 1.5 \times 10^{11}\ cm^{-2}$ and $\mu = 400{,}000\ cm^2/Vs$. The magnetic field scales are normalized to align fields with equal filling factor ν. Complete parameters are given in Table I of Ref. [6]

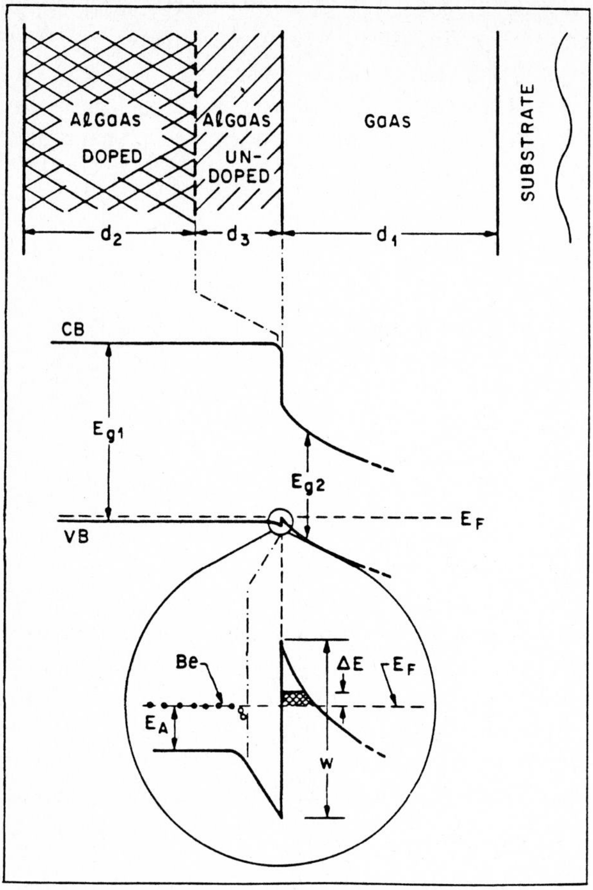

Fig. 5 p-type modulation-doped GaAs–$Al_xGa_{1-x}As$ heterostructure and band structure

Fig. 5. The acceptor dopant beryllium is introduced in the $Al_xGa_{1-x}As$ barrier layer in lieu of the donors used for electron-gas samples. $Al_xGa_{1-x}As$ acts as a confining barrier for holes which are generated in the GaAs by transfer of valence band electrons to lower-lying states on the acceptor in the barrier. Hole mobilities to over 50,000 cm^2/Vs have been produced. This is more than 100 times greater than the highest mobilities in uniformly doped p-type layers of equivalent carrier concentration. Hole scattering lifetimes are of the same magnitude as electron scattering times for electrons at 300,000 cm^2/Vs mobility, by virtue of the heavier mass of the hole relative to the electron. Both integral and fractional quantization of the Hall effect have been observed for the hole samples [6]. The Hall quantization features are comparably strong to those in high electron mobility samples, suggesting that carrier lifetime and not the absolute magnitude of the mobility is the more relevant factor in determining the existence of the quantization. The values of the Hall resistance in the hole samples at the integral Hall plateaus were compared with a precision of one part in 10^4 to the value in electron-gas samples. At this level of precision the quantized resistance values for holes and electrons are equal.

5. Conclusion

The main outlines of the experimental observations of the fractional quantization of the Hall effect are beginning to be filled in. The fascinating question of the complete extent of the series of fractional states remains to be answered as well as the question of the ultimate crystallization of the electron system. Other interesting questions concern the transitions to the new states, the effects on other physical properties produced by the states, and the spatial and geometrical features of the fractional quantization.

The author wishes to acknowledge his principal collaborators, H. L. Störmer, D. C. Tsui and W. Wiegmann. He also acknowledges the important contributions of A. M. Chang, J. C. M. Hwang, K. Baldwin, Z. Schlesinger, S. J. Allen, Jr., A. Pinczuk, T. Haavasoja and M. Paalanen, the continued support of M. B. Panish and V. Narayanamurti, and the communication of results from R. B. Laughlin, B. I. Halperin, P. W. Anderson, P. M. Platzman and E. E. Mendez. Many of the high-field results were obtained at the National Magnet Laboratory.

1. K. von Klitzing, G. Dorda, M. Pepper: Phys. Rev. Lett. *45*, 494 (1980).
2. D. C. Tsui, A. C. Gossard: Appl. Phys. Lett. *37*, 550 (1981).
3. C. C. Grimes, G. Adams: Phys. Rev. Lett. *42*, 795 (1979).
4. B. A. Wilson, S. J. Allen, Jr., D. C. Tsui: Phys. Rev. Lett. *44*, 479 (1980).
5. D. C. Tsui, H. L. Störmer, A. C. Gossard: Phys. Rev. Lett. *48*, 1559 (1982).
6. H. L. Störmer, A. Chang, D. C. Tsui, J. C. M. Hwang, A. C. Gossard, W. Wiegmann: Phys. Rev. Lett. *50*, 1953 (1983).
7. A. C. Gossard: "Treatise on Materials Science and Technology, Volume 24, Preparation and Properties of Thin Films," ed. K. N. Tu and R. Rosenberg, p. 13 (Academic Press, New York, 1982).

8. D. C. Tsui, H. L. Störmer, J. C. M. Hwang, J. S. Brooks, M. J. Naughton: Phys. Rev. B. *28*, 2274 (1983).

9. R. B. Laughlin: Phys. Rev. Lett. *50*, 1395 (1983) and this volume.

10. M. A. Paalanen, D. C. Tsui, A. C. Gossard, J. C. M. Hwang: Solid State Comm. (in press).

11. E. E. Mendez, M. Heiblum, L. L. Chang, L. Esaki: Phys. Rev. B *28*, 4886 (1983).

12. H. L. Störmer, A. C. Gossard, W. Wiegmann, R. Blondel, K. Baldwin: Appl. Phys. Lett. *44*, 139 (1984).

The Gauge Argument for Accurate Quantization of the Hall Conductance

R.B. Laughlin

Lawrence Livermore National Laboratory, P.O. Box 808
Livermore, CA 94550, USA

In this paper I review my theory [1] of the Ordinary Quantum Hall Effect which identifies the quantum measured in these experiments as the electron charge. This concept is important for understanding why the effect is so accurate, and also for understanding why fractionally charged quasiparticles need be invoked to explain the Fractional Quantum Hall Effect. The vehicle for making this identification is a thought experiment which I shall motivate and describe in detail.

I begin by reviewing the facts we wish to explain. In Fig. 1, I have reproduced von Klitzing's original data [2]. The experiment is illustrated in the figure. A current I is forced to flow between the source and drain of a MOSFET subjected to a magnetic field of strength H_0 normal to its surface. Both the voltage drop in the direction of current flow and the Hall voltage observed at low temperature are plotted against fermi level (gate voltage). Over a range of fermi levels one sees the voltage drop in the direction of current flow going to zero and the Hall voltage plateauing to a constant value. Von Klitzing originally reported that this value obeyed the equation

$$\frac{I}{V_H} = n \frac{e^2}{h} \tag{1}$$

to one part in a million. More recent measurements by the U.S. Bureau of

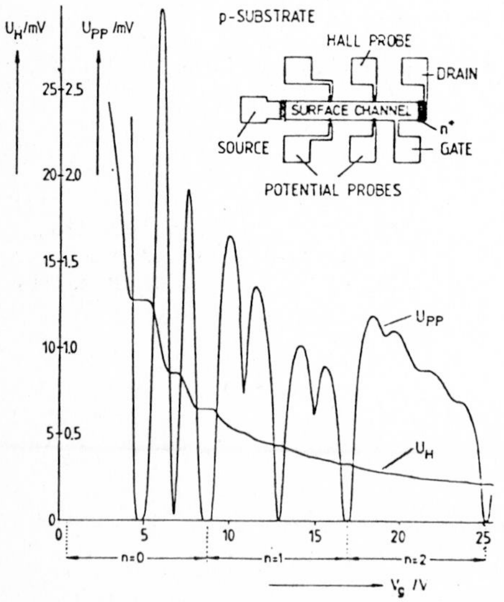

Figure 1: Original Quantum Hall measurements. Hall voltage and voltage drop along current flow direction versus gate voltage.

Standards in collaboration with Bell Laboratories have decreased the error to 10^{-7}. Thus the facts we wish to explain are the following:

1. The system exhibits a zero resistance state.
2. It does this over a range of carrier densities.
3. The Hall conductance is constant over this range.
4. The constant value of the Hall conductance obeys Eqn. 1.

I shall now describe the elementary quantum mechanics of this system and review the <u>wrong</u> explanation of the effect in terms of quantization of the charge <u>density</u>. This explanation is wrong fundamentally because no quantum of charge density exists, but more practically because the plateau phenomenon as observed cannot occur in the absence of dirt in the sample. The idealized Hamiltonian for the system is

$$H = \frac{1}{2m} \left| \frac{\hbar}{i} \vec{\nabla} - \frac{e}{c} \vec{A} \right|^2 + V \quad , \tag{2}$$

except at the sample edges where it is complicated and unknown. The electrons are confined to the x-y plane. I take the vector potential to be in Landau gauge

$$A = H_0 \, y \, \hat{x} \quad . \tag{3}$$

V is an external potential which I presently take to represent only impurities and lattice imperfections. This problem has a fundamental <u>energy</u> unit, the cyclotron frequency,

$$\hbar\omega_c = \hbar \left(\frac{eH_0}{mc}\right) \quad , \tag{4}$$

and a fundamental <u>length</u> unit, the magnetic length, which is just the length associated with an harmonic oscillator with frequency $\hbar\omega_c$:

$$a_0 = \sqrt{\frac{\hbar}{m\omega_c}} = \sqrt{\frac{\hbar c}{eH_0}} \quad . \tag{5}$$

In terms of these natural units, the eigenvectors and eigenvalues of H when V is zero, i.e.,when there is no dirt, are given by

$$|k,n\rangle = (\sqrt{\pi}\, 2^n n!)^{-1/2} \, e^{ikx} \, e^{1/2(y-k)^2} \left(\frac{\partial}{\partial y}\right)^n e^{-(y-k)^2} \quad , \tag{6}$$

and

$$H|k,n\rangle = (n + 1/2)\, |n,k\rangle \quad . \tag{7}$$

The nature of these wavefunctions is indicated in Fig. 2. Each is a track of width a_0 in which the electron travels with momentum k. In addition, the center of the track is displaced from the center of the sample by an amount

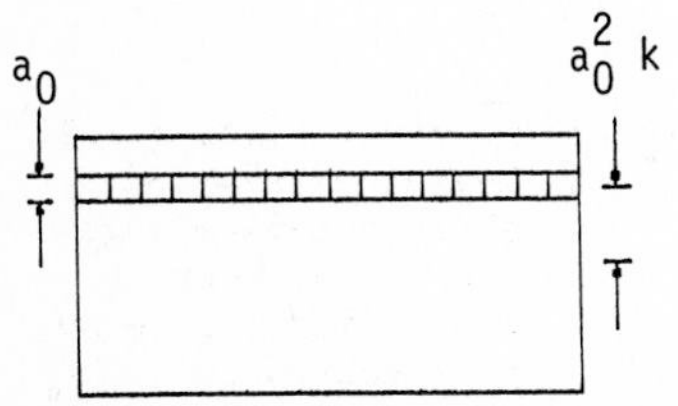

Figure 2: Illustration of Landau gauge wavefunction $|k,n\rangle$.

$a_0^2 k$. The manifold of states with the same n is the nth Landau level. The number of states N in a Landau level is

$$N = \frac{L_x L_y}{2\pi a_0^2} \quad , \tag{8}$$

where L_x and L_y are the sample dimensions. This result is obtained from the conditions $k_y = j\,(2\pi)/L_x$ with j and integer and $-L_y/2 < a_0^2 k < L_y/2$. It defines the "quantum" of surface charge density:

$$\sigma_1 = \frac{1}{2\pi a_0^2} \quad . \tag{9}$$

The charge density is an integral multiple of this value whenever an integral number of Landau levels is filled with electrons. Quantization of the surface charge density will lead to quantization of the Hall conductance, insofar as the classical expression for the Hall conductance

$$\frac{I}{V_H} = \frac{\sigma ec}{H_0} \tag{10}$$

is valid. The zero resistance state can also be rationalized as occurring when an integral number of Landau levels is filled, as this situation presents the electrons with an energy gap across which they must scatter in order to dissipate energy. The density of states of this idealized system is illustrated in Fig. 3.

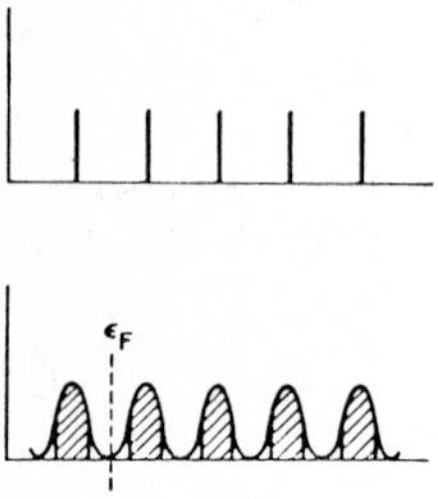

Figure 3: Densities of states versus energy. Top: V=0. Bottom: V≠0. The shaded regions consist of extended states.

Also illustrated in Fig. 3 is the density of states of the system when the dirt potential V is not zero. Rather than being a series of δ functions, it is a smooth continuum which, according as our understanding of localization phenomena, consists of two distinct regions. Near $(n + 1/2)\,\hbar\omega_c$ there is a narrow band of extended states, those which are contiguous from one side of the sample to the other and which are capable of carrying electric current. These are separated by tail of localized states, those not contiguous from one side of the sample to the other and not capable of carrying electric current. The Quantum Hall Effect occurs whenever the fermi level lies in a localized state band. This, however, presents us with a problem. The surface charge density is no longer quantized, since the density of states is now continuous. Furthermore, the number of states actually carrying current, the extended ones, is sample dependent, since V varies from sample to sample. It is thus impossible to understand accurate quantization of the Hall conductance in real samples in terms of quantization of the surface charge density because the surface charge density is not quantized.

Faced with this problem, we explore ways to circumvent it. One idea is to presume that V is zero, for some unknown reason. This will not do, however, because without localized states in the gap there is no reason for the fermi level to tarry in the gap. The addition of a single electron to an otherwise full Landau level would by definition raise the fermi level across the gap. The experiment says that the effect occurs over a range of fermi levels. Another idea is to presume that V is "weak", so that it is possible to describe the system with a perturbation theory that neglects Landau level mixing by V. This approximation is tacitly made in many of the papers of Ando [3]. It leads simply, but incorrectly, to the conclusion that the Hall conductance is quantized. The conclusion is incorrect because the perturbation theory at the same time yields the physically absurd result that localized states carry current. This was pointed out in an elegant paper by R.E. Prange. I shall summarize its content [4].

Let us assume that V has no matrix elements between Landau levels, that is

$$\langle k,n|V|k',n'\rangle = 0 \quad ; \; n \neq n' \quad . \tag{11}$$

Let us also assume that a uniform electric field E_0 is present in the y direction. This does not affect the form of the eigenstates when V is zero. The eigenstates of the Hamiltonian when V is not zero

$$H|\ell,n\rangle = E_{\ell,n}|\ell,n\rangle \tag{12}$$

are then linear combinations of eigenstates of the Hamiltonian when V is zero, but only lying within a given Landau level:

$$|\ell,n\rangle = \sum_k U_{\ell,k}|k,n\rangle \quad . \tag{13}$$

The transformation matrix is unitary. The expectation value of any operator O across all the new states derived from a Landau level is thus equal to its expectation value across the Landau level:

$$\sum_\ell \langle \ell,n|O|\ell,n\rangle = \sum_k \langle k,n|O|k,n\rangle \quad . \tag{14}$$

In particular, if O is the velocity operator v_x, given by

$$v_x = \frac{1}{m}\left(\frac{\hbar}{i}\frac{\partial}{\partial x} - \frac{e}{c}A_x\right) \quad , \tag{15}$$

then the total current carried by the new states equals that carried by the Landau level when V is zero. This "proves" that the Hall conductance is unaffected by V. However, the matrix elements of v_x satisfy

$$\langle k,n|v_x|k',n\rangle = \delta_{kk'}\, v_D \quad , \tag{16}$$

where $v_D = c(E_0/H_0)$ is the classical drift velocity. This implies that

$$\langle \ell,n|v_x|\ell,n\rangle = v_D \quad . \tag{17}$$

In other words, all the new states also have velocity v_D, even though we know some of them must be localized. This is ridiculous. It implies that

Eqn. (11) is never a valid approximation. Prange subsequently solved, without resorting to this approximation, the problem of a single δ-function impurity in an otherwise dirt-free environment. He found one localized state to be bound down out of each Landau level. He calculated the current carried by this localized state and found, consistent with one's intuition, that the localized state did not carry current. However, he also found that the remaining N-1 extended states carried too much current, and that the sum of the excess exactly canceled the current not carried by the localized state. This is miraculous! It is as though the extended states know when one of their colleagues dies!

It is clear that this clairvoyance of the extended states is not a peculiarity of the model, but is behavior that occurs for any V, and that there is a deep and fundamental reason why it occurs. I show now that accurate quantization of the Hall conductance is the logical consequence of two things:

1. Gauge invariance of the interaction of light with matter.
2. The existence of a mobility gap.

By a mobility gap, I mean a region of the density of states that is localized. I shall demand that the fermi level lie in it.

I begin by observing that the second condition automatically implies a zero resistance state, because the parallel conductance σ_{xx} is known from the Kubo formula to be a fermi surface property. Since all the states at the fermi surface are localized by assumption, σ_{xx} is zero. Thus it remains to show that Eqn (1) is valid.

I now observe that Eqn (2) is the exact Hamiltonian of the system, if one sums the kinetic energy term over electrons and reinterprets V to mean the sum of all electron-ion and electron-electron interactions. For any V, the velocity operator v_x is given formally by the expression

$$v_x = -\frac{c}{e}\frac{\partial H}{\partial A_x} \quad . \tag{18}$$

Eqn (18) suggests a way of calculating the current carried by the ground state without calculating any wavefunctions. Rather than diagonalizing the Hamiltonian and then calculating the current carried by each state, one might calculate the total energy of the system, and differentiate in the end by the x component of the vector potential. One would take care in doing this not to perform a gauge transformation, since that would be unphysical.

I have realized this idea with a thought experiment in which the ribbon of two-dimensional metal is bent into a loop, as shown in Fig. 4. A magnetic field H_0 pierces the surface of this ribbon. A current I flows around it. There is a voltage drop V between one edge and the other which we wish to relate to I. Since there is no resistance, energy is conserved, and we can write a version of Faraday's Law of Induction relating I to the adiabatic derivative of the total energy of the system U with respect to some magnetic flux ϕ threading the loop:

$$I = c\frac{\partial U}{\partial \phi} \quad . \tag{19}$$

I emphasize that this flux is different from the flux associated with H_0. This derivative may be accomplished by differentiating instead with respect to a uniform vector potential A pointing around the loop, in the manner

$$I = \frac{c}{L} \frac{\partial U}{\partial A} \quad , \tag{20}$$

where L is the loop circumference. The connection with Eqn (18) should be apparent. This derivative is nonzero only by virtue of the fact that at least some of the electrons reside in extended states, for if a state ψ is localized, the addition of a vector potential increment δA to it looks like a gauge transformation. It responds to this addition in the manner

$$\psi \rightarrow \psi \, e^{i (e \, \delta A/c) \, x} \quad , \tag{21}$$

where x is the coordinate around the loop. This leaves its energy invariant:

$$E_{\psi} \rightarrow E_{\psi} \quad . \tag{22}$$

Thus the state does not contribute to the energy derivative. Localized states do not carry current. On the other hand, if ψ is contiguous about the loop, then its phase is locked in, and Eqns (21) and (22) do not apply. The state responds in a different manner. It can be seen by examining Eqn (3) that adding a vector potential increment in the x direction just redefines the origin of y. Thus in the absence of dirt, a wavefunction of the form of Eqn (6) responds by moving in the y direction, and this changes its energy because an electric field is present. It is possible to add up the energy changes of all the states to evaluate Eqn (19). An easier way, however, is to replace the adiabatic derivative in Eqn (19) with an adiabatic differential, in the manner

$$I = c \, \frac{\Delta U}{\Delta \phi} \quad , \tag{23}$$

where $\Delta\phi = hc/e$ is a flux quantum. Once an entire flux quantum threads the loop, the electrons cannot tell any flux is present. The states must map identically into themselves, up to an unimportant phase. When V is zero, the states may be seen to map under this operation in the manner

$$|k,n\rangle \rightarrow |k + \frac{2\pi}{L_x}, n\rangle \quad . \tag{24}$$

This is illustrated in Fig. 5. Each state maps into its neighbor, as in a shift register. The net result is the transfer of exactly one state from one edge of the ribbon to the other. The current is then c times the number of electrons transferred times the energy gained per electron divided by a flux quantum, or

$$I = c \, \frac{(n) \, (eV_H)}{(hc/e)} = n \, \frac{e^2}{h} \quad . \tag{25}$$

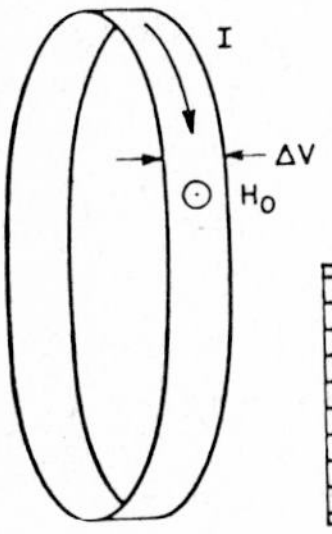

Figure 4: Loop of thought experiment.

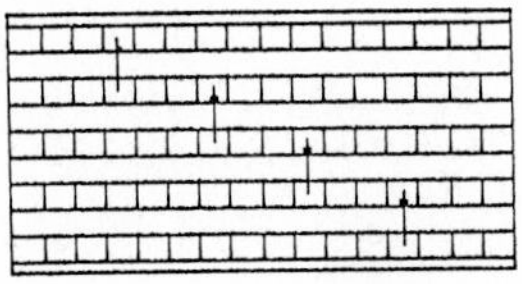

Figure 5: Illustration of shift-register motion of eigenstates during adiabatic addition of $\Delta\phi$.

Thus far, I have shown only what one knew already, i.e., that the Hall current is quantized when there is no dirt and the Landau level is full. However, this reasoning also works when dirt is present. As in the dirt-free case, gauge invariance is an exact symmetry of the system, so that the loop with $\Delta\phi$ threading is physically indistinguishable from the loop without $\Delta\phi$. Also, there is a gap in the problem, although it is not a real gap but a mobility gap. The localized states in this gap map into themselves under this operation and thus are in essence "not there". The extended states cannot move across this gap because the operation is adiabatic and because extended and localized states cannot coexist at the same energy. Accordingly, the only excitations which can result from this operation are the charge-transfer variety that I described in the dirt-free case. Thus, almost trivially the current carried by the ground state satisfies Eqn (1), for some integer, whenever the fermi level lies in a mobility gap.

The above result may be summarized succinctly in the following way. The Quantum Hall Effect does not measure any quantum of surface charge density, because there is no such quantum. It measures instead the number of electrons transferred in a thought experiment. It measures e. That is why it is so accurate.

The last issue with which I shall deal is the question of edges. Real samples are approximately 10^4 a_0 wide, yet the experiment is accurate to at least parts in 10^7. Why do the edges not matter? To understand edges, it is necessary to realize that they are normal metals with dissipation. Were this not the case, the experiment would be impossible, because the voltage drop V, a chemical potential difference, would be undefined for lack of local thermal equilibrium. If, for example, one were actually to perform my thought experiment in a loop without resistive loss in the contacts, one would adiabatically accelerate the electrons in the contacts during the course of the experiment. There would be energy associated with the diamagnetic currents in the "superconducting" edges reckoned into the derivative in Eqn (19). In a real sample, the mobility gap collapses in a complicated and as yet unknown way near the edge, so as to make the edge a normal metal with a well-defined chemical potential. The detailed nature of the edge is unimportant, so long as the net result in my thought experiment is the transfer of an electron from the local fermi level at one edge to the local fermi level at the other, without dissipating energy. For then the change to the system's energy is the chemical potential difference, which is the definition of e V. The extent to which the system always does this in an energy conserving way is an outstanding question. However, insofar as a macroscopic conductivity is appropriate for describing the normal metal at the edge, the flux can always be added sufficiently slowly to ensure an arbitrarily small heat production. An appropriate analogy would be one's ability to penetrate a normal metal with a magnetic field in a completely reversible way, so long as the penetration is slow.

This work was performed under the auspices of the U.S. Department of Energy by the Lawrence Livermore National Laboratory under Contract No. W-7405-Eng-48.

[1]R.B. LAUGHLIN: Phys. Rev. B23, 5632 (1981).
[2]K. von KLITZING, G. DORDA and M. PEPPER: Phys. Rev. Lett. 45, 494 (1980).
[3]H. AOKI and T. ANDO: Solid State Comm. 38, 1079 (1981); references may also be found in Surf. Sci. 113, 27 (1981).
[4]R.E. PRANGE: Phys. Rev. B23, 4802 (1981).

Fractional Quantization of the Hall Effect

R.B. Laughlin

Lawrence Livermore National Laboratory, P.O. Box 808
Livermore, CA 94550, USA

The Fractional Quantum Hall Effect is caused by the condensation of a two-dimensional electron gas in a strong magnetic field into a new type of macroscopic ground state, the elementary excitations of which are fermions of charge 1/m, where m is an odd integer.

1 Preliminary Considerations

We consider a two-dimensional metal in the x-y plane subject to a magnetic field H_0 in the z direction. The many-body Hamiltonian is

$$H = \sum_j \left[\frac{1}{2m} \left| \frac{\hbar}{i} \vec{\nabla} - \frac{e}{c} \vec{A} \right|^2 + V(z_j) \right] + \sum_{j<k} \frac{e^2}{|z_j - z_k|} \quad , \tag{1}$$

where $z_j = x_j - iy_j$ is a complex number locating the jth electron, $V(z_j)$ is the potential generated by a uniform neutralizing background of density σ

$$V(z) = -\sigma e^2 \int \frac{d^2 z'}{|z - z'|} \quad , \tag{2}$$

and $\vec{A} = \frac{H_0}{2} (y\hat{x} - x\hat{y})$ is the symmetric gauge vector potential. We restrict our attention to the lowest Landau level, for which the single-body wave-functions are

$$|n\rangle = \frac{1}{\sqrt{2^{n+1} \pi n!}} \, z^n \, e^{-\frac{1}{4}|z|^2} \quad , \tag{3}$$

with the magnetic length $a_0 = (\hbar c / eH_0)^{1/2}$ set to 1. These states are degenerate at energy $\hbar\omega_c / 2$, with $\omega_c = eH_0/mc$ the cyclotron frequency. We assume $\hbar\omega_c > e^2/a_0$.

2 Ground State

By analogy with liquid helium, we propose a variational wavefunction for this system of the Jastrow form

$$\psi = \left[\prod_{j<k} f(z_j - z_k) \right] e^{-\frac{1}{4}\sum_\ell |z_\ell|^2} \quad , \tag{4}$$

as such wavefunctions are efficient in keeping the particle apart. Restriction to the lowest Landau level requires f to be a polynomial, the Pauli principle requires f to be odd, and conservation of angular momentum by H requires f to be homogeneous. Thus the only allowed wavefunctions of the Jastrow form are

$$|m\rangle \equiv \psi_m = \prod_{j<k} (z_j - z_k)^m \; e^{-\frac{1}{4}\sum_\ell |z_\ell|^2} \quad , \tag{5}$$

with m an odd integer. The nature of this state is understood by interpreting its square as the probability distribution function of a classical plasma, in the manner

$$|\psi_m|^2 = e^{-\beta\Phi} \quad , \tag{6}$$

with $\beta = 1/m$ and

$$\Phi = -2m^2 \sum_{j<k} \ln|z_j - z_k| + \frac{m}{2}\sum_\ell |z_\ell|^2 \quad . \tag{7}$$

Φ describes particles of "charge" m repelling one another logarithmically and being attracted logarithmically to a uniform background of "charge" density $\sigma_1 = 1/2\pi$. Local neutrality of this "charge" requires that the electrons be spread out to a density $\sigma_m = \sigma_1/m$. The fractional quantum Hall effect occurs when $\sigma = \sigma_m$.

We calculate $\langle m|m\rangle$ and $\langle m|H|m\rangle$ using the hypernetted chain approximation for the radial distribution function g(r) of the plasma. If we let $x = r/\sqrt{2m}$ and define Fourier transforms in the manner

$$\hat{h}(k) = \int_0^\infty h(x)\, J_0(kx)\, x dx \quad , \tag{8}$$

where J_0 is an ordinary Bessel function of the first kind, then the equations we solve are [1,2]

$$g(x) = \exp\{\, h(x) - c_s(x) - 2mK_0(Qx) \,\} \quad , \tag{9}$$

where K_0 is a modified Bessel function of the second kind, Q is an arbitrary cutoff parameter, and

$$\hat{h}(k) = \hat{c}(k) + 2\hat{c}(k)\hat{h}(k) \quad , \quad \text{with} \tag{10}$$

$$\hat{c}_s(k) = \hat{c}(k) + \frac{2mQ^2}{k^2(k^2+Q^2)} \quad , \tag{11}$$

and $h(x) = g(x) - 1$. The numerical solution to these equations for m=3 is displayed in Figs. 1 and 2. The absence of structure in g(x) beyond x=4 reflects the liquid nature of the state. In terms of g(x), the total energy per electron is

$$U_{total} \equiv \frac{\langle m|H|m\rangle}{\langle m|m\rangle}/N - \frac{1}{2}\hbar\omega_c = \frac{1}{\sqrt{2m}}\int_0^\infty h(x)\, dx \quad , \tag{12}$$

in units of e^2/a_0. N is the number of electrons. We have fited a sequence of such calculations to the semiempirical formula

$$U_{total}(m) = \frac{0.814}{\sqrt{m}}\left(\frac{0.23}{m^{0.64}} - 1\right) \quad . \tag{13}$$

The cohesive energy per electron, defined by

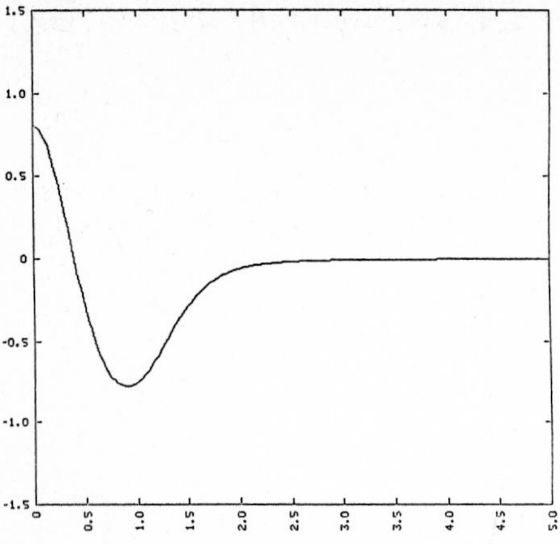

Figure 1: $c_s(x)$ versus x for m=3 and Q=2

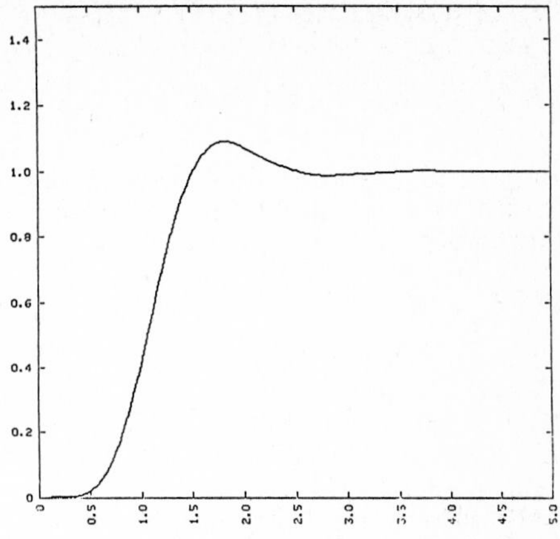

Figure 2: g(x) versus x for m=3

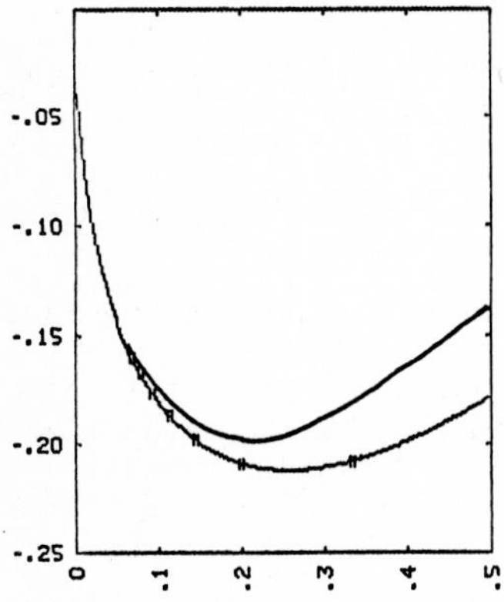

Figure 3: Cohesive energy per electron in units of e^2/a_0 versus filling factor $\nu = 1/m$. Top curve is charge density wave value from [3]. Bottom curve is (13)

$$U_{coh} = U_{total} - \sqrt{\frac{\pi}{8}}\ \frac{1}{m} \quad , \tag{14}$$

is compared with that calculated by YOSHIOKA and FUKUYAMA [3] for a charge density wave in Fig. 3. The normalization integral $\langle m|m\rangle$ is the plasma partition function, and is given by

$$\frac{1}{N}\,\ell n(\ \langle m|m\rangle\ ) = mN\left(\frac{1}{2}\ell n(2mN) - \frac{3}{4}\right) + \ell n(2mN) - \frac{m}{2}\ell n(2m) - 2mf(2m) + O\left[\frac{\ell n(N)}{N}\right] \quad , \tag{15}$$

where f is a slowly varying function of order 1 fited from Monte Carlo experiments [4] to the formula

$$f(\Gamma) = A + \frac{B}{\Gamma^{\alpha}} + \frac{C}{\Gamma^{\gamma}} + \frac{D}{\Gamma} \quad , \tag{16}$$

with $\Gamma = 2m$, valid in the range of interest. The parameters are listed in

Table 1. The function f is the excess free energy of the plasma, while the remaining terms are "electrostatic" in nature, except for $\ln(2mN)$, which is just the log of the volume.

Table 1

A = -0.3755	D = -1.2862
B = 1.6922	α = 0.74
C = 0.1494	γ = 1.70

3 Quasiparticles

The elementary excitations of ψ_m are made with a thought experiment in which the exact ground state is pierced at location z_0 with an infinitely thin magnetic solenoid through which is passed adiabatically a flux quantum hc/e . The solenoid may then be removed by a gauge transformation, leaving behind an exact excited state of the many-body Hamiltonian. Operators which approximate the effect of this procedure are

$$S_{z_0} = \prod_i (a_i^\dagger - z_0) \quad , \tag{17}$$

and its Hermitean adjoint $S_{z_0}^\dagger$, where a_j is the ladder operator

$$a_j = \frac{x_j + iy_j}{2} + \left(\frac{\partial}{\partial x_j} + i \frac{\partial}{\partial y_j} \right) \quad . \tag{18}$$

That they do so may be seen from the fact that the thought experiment maps the single-body states (3) in the manner $|n\rangle \to |n\pm1\rangle$, whereas

$$a|n\rangle = \sqrt{2n}\ |n-1\rangle \tag{19}$$

and

$$a^\dagger|n\rangle = \sqrt{2(n+1)}\ |n+1\rangle \quad . \tag{20}$$

The operator a annihilates $|0\rangle$, consistent with the thought experiment's mapping it to the next Landau level. Note that S_{z_0} and $S_{z_0}^\dagger$ are exact for non-interacting electrons when they are described by ψ_m. ψ_m is a single Slater determinant of the single-body functions $|n\rangle$.

We calculate quasiparticle properties with the hypernetted chain. For the quasihole wavefunction

$$S_{z_0}|m\rangle \equiv \psi_m^{+z_0} = e^{-\frac{1}{4}\sum_\ell |z_\ell|^2} \prod_i (z_i - z_0) \prod_{j<k} (z_j - z_k)^m \quad , \tag{21}$$

we write $|\psi_m^{+z_0}|^2 = e^{-\beta\Phi'}$, with $\beta = 1/m$ and

$$\Phi' = \Phi - 2m \sum_i \ln|z_i - z_0| \quad . \tag{22}$$

This is a plasma with two components, N particles of "charge" m and one particle of "charge" 1. The two-component hypernetted chain equations are

$$g_{ij}(x) = \exp\{ -\beta v_{ij}(x) + h_{ij}(x) - c_{ij}(x) \} \quad , \tag{23}$$

and

$$\hat{h}_{ij}(k) = \hat{c}_{ij}(k) + 2 \sum_{\ell} \hat{h}_{i\ell}(k)\, \rho_\ell\, \hat{c}_{\ell j}(k) \quad , \tag{24}$$

where the indices run over the two kinds of particle. With x defined as before, the densities are $\rho_1 = 1$ and $\rho_2 = 1/N$. To solve the problem, we use perturbation theory in ρ_2. The zero-order solution to g_{11} is given by (9) through (11) . For $g_{12}(x)^2$ we have

$$\hat{h}_{12}(k) = \{ 1 + 2\hat{h}_{11}(k) \} \hat{c}_{12}(k) \quad , \tag{25}$$

$$\hat{c}_{12_s}(k) = \hat{c}_{12}(k) + \frac{2Q^2}{k^2(k^2+Q^2)} \quad , \tag{26}$$

and

$$g_{12}(x) = \exp\{ h_{12}(x) - c_{12_s}(x) - 2K_0(Qx) \} \quad . \tag{27}$$

The numerical solution of these equations for m=3 is shown in Figs. 4 and 5. Note that the divergence of (26) as $k \to 0$ requires the total excess charge accumulated around z_0 to be exactly -1/m of an electron. Using $g_{12}(x)$, we construct the change to $g_{11}(x)$ resulting from the presence of the quasihole.

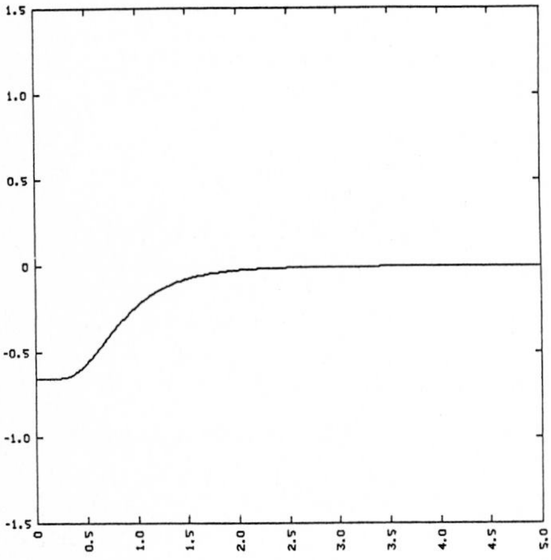

Figure 4: $c_{12_s}(x)$ versus x for m=3 and Q=2

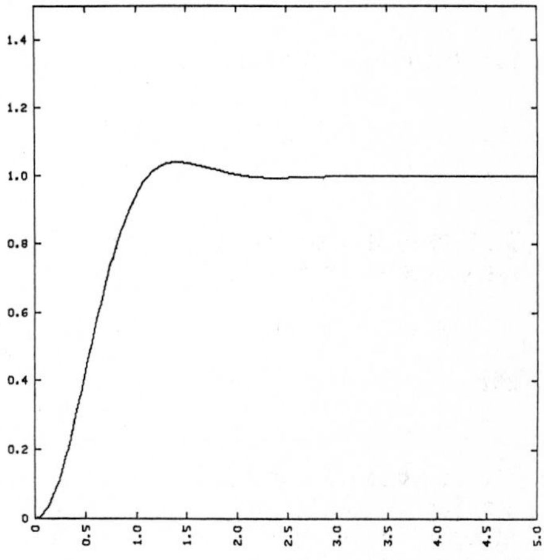

Figure 5: $g_{12}(x)$ versus x for m=3

We have

$$\delta\hat{h}_{11}(k) \simeq \{1 + 2\hat{h}_{11}(k)\}^2\, \delta\hat{c}_{11}(k) + \frac{2}{N}\, \hat{h}_{12}(k) \quad , \tag{28}$$

and

$$\delta c_{11}(x) = \left(\frac{h_{11}(x)}{1 + h_{11}(x)} \right) \delta h_{11}(x) \quad . \tag{29}$$

The solution $N\delta h_{11}(x)$ to these equations for m=3 is plotted in Fig. 6. The energy to make a quasihole can be calculated from it in the manner

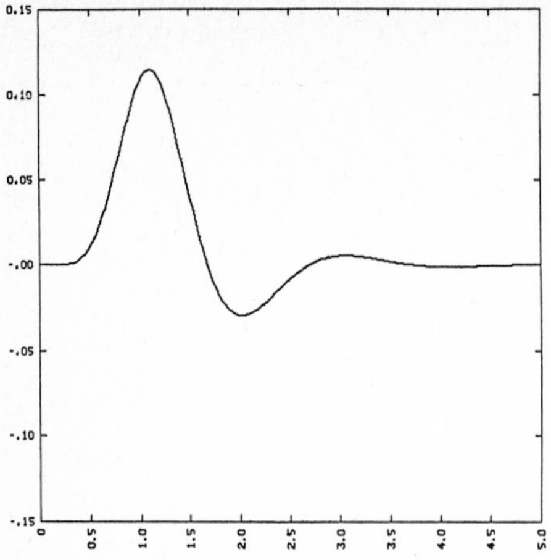

Figure 6: $\delta h_{11}(x)$ versus x for quasihole at m=3

$$\Delta_{\text{Quasihole}} = \frac{N}{\sqrt{2m}} \int_0^\infty \delta h_{11}(x)\, dx \quad , \tag{30}$$

in units of e^2/a_0. We obtain 0.026, which is considerably lower than the "Debye" estimate of 0.062.

A similar procedure may be used for the quasielectron. We have

$$S^\dagger_{z_0} |m\rangle \equiv \psi_m^{-z_0} = e^{-\frac{1}{4}\sum_\ell |z_\ell|^2} \prod_i (2\frac{\partial}{\partial z_i} - z_0^*) \prod_{j<k} (z_j - z_k)^m \quad . \tag{31}$$

Normalizing this wavefunction and calculating its charge density involve integrating over spatial variables, which allows us to integrate by parts and then consider a situation similar to (21) and (22) but with [1]

$$\Phi' = \Phi - 2m \sum_i \ell n\{ |z_i - z_0|^2 - 2 \} \quad . \tag{32}$$

For this problem, we obtain an "integrated by parts" $\tilde{g}_{12}(x)$ and $\tilde{c}_{12_s}(x)$ satisfying (25) and (26), but with

$$\tilde{g}_{12}(x) = \left(\frac{x^2 - 2}{x^2} \right) \exp\{ \tilde{h}_{12}(x) - \tilde{c}_{12}(x) - 2K_0(Qx) \} \quad . \tag{33}$$

The numerical solution of these equations with m=3 is shown in Figs. 7 and 8. As with the quasihole, the Ornstein-Zernicke relation (25) forces the total charge accumulated around z_0 to be -1/m electrons. However, the *actual* $g_{12}(x)$, given by

$$g_{12}(x) = \left[\frac{1}{2m} \left(\frac{\partial^2}{\partial x^2} + \frac{1}{x}\frac{\partial}{\partial x}\right) + 2x\frac{\partial}{\partial x} + 2mx^2 + 2 \right] \left(\frac{\tilde{g}_{12}(x)}{2mx^2 - 2} \right) \tag{34}$$

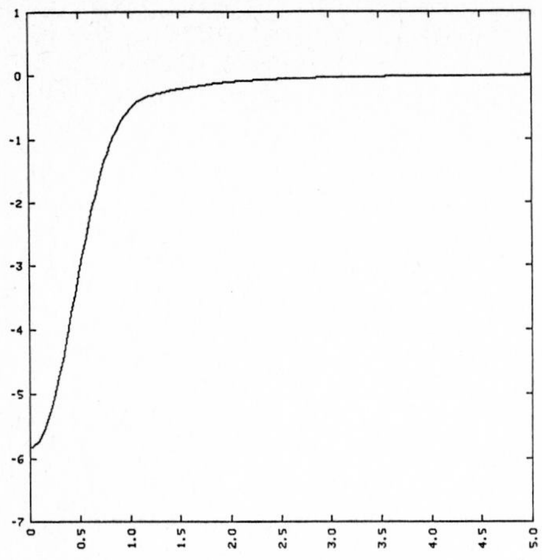

Figure 7: $\tilde{c}_{12_s}(x)$ versus x for m=3 and Q=2

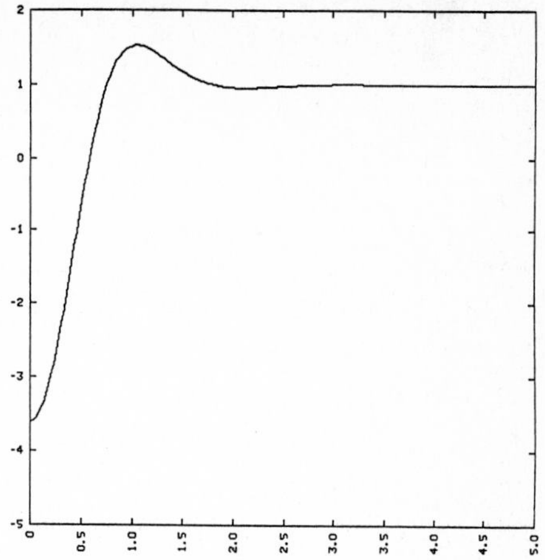

Figure 8: $\tilde{g}_{12}(x)$ versus x for m=3

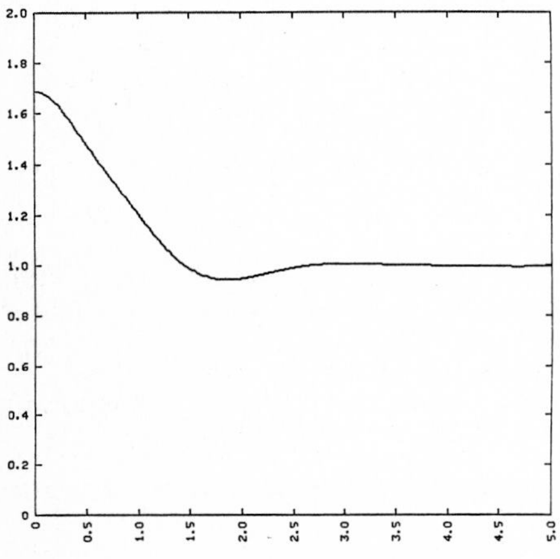

Figure 9: $g_{12}(x)$ versus x for quasielectron at m=3

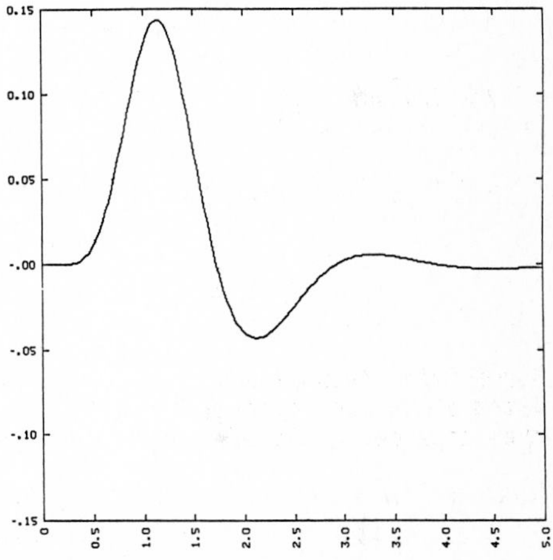

Figure 10: $\delta h_{11}(x)$ versus x for quasielectron at m=3

correctly accumulates +1/m of an electron. $g_{12}(x)$ is shown in Fig. 9. To calculate the quasielectron creation energy, we employ the somewhat uncontrolled approximation of assuming the existence of a "pseudopotential" which when used as $v_{12}(x)$ in (23) and (24) reproduces $g_{12}(x)$. To the extent such a potential is physical, we can calculate $\delta h_{11}(x)$ using (28) and (29), and then calculate the quasielectron creation energy using (30). In Fig. 10, we show the $\delta h_{11}(x)$ obtained using this procedure. Note the similarity to Fig. 6. The quasielectron creation energy we obtain using this $\delta h_{11}(x)$ is 0.030 in units of e^2/a_0.

Operators S_k and $S_k^\dagger$ creating a quasiparticle in an angular momentum state analogous to the single-body state $|n>$ in (3) are the elementary symmetric polynomials [5], defined by the expression

$$S_{z_0} = \sum_k S_k z_0^k \quad . \tag{35}$$

We have explicitly

$$S_0 = z_1 z_2 z_3 \ldots z_N \quad , \tag{36}$$

$$S_1 = -\sum_j z_1 z_2 \ldots \hat{z}_j \ldots z_N \quad , \tag{37}$$

$$\vdots \qquad\qquad \vdots$$

$$S_{N-1} = (-1)^{N-1} (z_1 + \ldots + z_N) \quad , \tag{38}$$

where $\hat{z}_j$ means omit this factor from the product. When m=1, the state $S_k|m>$ is a full Landau level, but for a hole in $|k>$, that is, a hole with orbit radius $\sqrt{2k+2}$. We now show that the quasiparticle behaves *kinematically* as though it has charge e/m: the orbit radius of $S_k|m>$ or $S_k^\dagger|m>$ is exactly $\sqrt{2mk+2}$.

We first observe that since there are no thermodynamic forces on plasma particles, provided they feel the neutralizing background potential, we have

$$<m|S_{z_0}^\dagger S_{z_0}|m> = e^{\frac{1}{2m}|z_0|^2} <m|S_0^\dagger S_0|m> \quad . \tag{39}$$

However, we also have

$$<m|S_{z_0}^\dagger S_{z_0}| > = \sum_{k,k'} (z_0^*)^{k'} (z_0)^k <m|S_{k'}^\dagger S_k|m> \quad , \tag{40}$$

so that

$$<m|S_{k'}^\dagger S_k|m> = \frac{\delta_{kk'}}{(2m)^k k!} <m|S_0^\dagger S_0|m> \quad , \tag{41}$$

and similarly for the adjoint. We next observe that from translational invariance of the plasma, matrix elements of the charge density operator $\rho(z)$ may be computed from the relation

$$\begin{aligned} <m|S_{z_0}^\dagger \rho(z) S_{z_0}|m> &= \sum_{k,k'} (z_0^*)^{k'} (z_0)^k <m|S_{k'}^\dagger \rho(z) S_k|m> \\ &= \frac{<m|S_0^\dagger S_0|m>}{2\pi m} e^{\frac{1}{2m}|z_0|^2} g_{12}(|z-z_0|) \quad . \end{aligned} \tag{42}$$

Thus

$$\frac{<m|S_k^\dagger \rho(z) S_k|m>}{<m|S_k^\dagger S_k|m>} = \frac{1}{2\pi m} \left(1 + \frac{(2m)^k}{k!} (\frac{\partial}{\partial z_0^*} \frac{\partial}{\partial z_0})^k \{ e^{\frac{1}{2m}|\ |^2} \times h_{12}(|z-z_0|) \} \Big|_{z_0=0} \right) \quad . \tag{43}$$

Since $h_{12}(x)$ is short-ranged, the charge density is $(2\pi m)^{-1}$ almost everywhere. Also, since from the charge-neutrality sum rule

$$\frac{1}{2\pi m} \int h_{12}(|z|)\, d^2z = -\frac{1}{m} \quad , \tag{44}$$

we have

$$\int \left(\frac{\langle m|S_k^\dagger \rho(z) S_k|m\rangle}{\langle m|S_k^\dagger S_k|m\rangle} - \frac{1}{2\pi m} \right) d^2z = -\frac{1}{m} \quad . \tag{45}$$

Similarly, the constant-screening sum rule [2]

$$\frac{1}{2\pi m} \int h_{12}(|z|)\, |z|^2\, d^2z = -\frac{2}{m} \tag{46}$$

implies that

$$\int \left(\frac{\langle m|S_k^\dagger \rho(z) S_k|m\rangle}{\langle m|S_k^\dagger S_k|m\rangle} - \frac{1}{2\pi m} \right) |z|^2\, d^2z$$

$$= -\frac{2}{m} + \frac{1}{2\pi m} \left(\frac{(2m)^k}{k!} \left(\frac{\partial}{\partial z_0^*} \frac{\partial}{\partial z_0}\right)^k \left\{ e^{\frac{1}{2m}|z_0|^2} |z_0|^2 \right\} \Big|_{z_0=0} \right.$$

$$= -\frac{1}{m} \left(2(km+1) \right) \tag{47}$$

and similarly for quasielectrons.

4 Acknowledgements

This work was performed under the auspices of the U.S. Department of Energy at Lawrence Livermore National Laboratory under Contract No. W-7405-Eng-48.

5 References

[1] R.B. LAUGHLIN: Phys. Rev. Lett. 50, 1395 (1983); Proceedings of the Fifth International Conference on Electronic Properties of Two-Dimensional Systems, Oxford, England, Published in Surface Science.
[2] J.P. HANSEN and D. LEVESQUE: J. Phys. C14, L603 (1981).
[3] D. YOSHIOKA and H. FUKUYAMA: J. Phys. Soc. Jpn. 47, 394 (1979).
[4] J.M. CAILLOL, D. LEVESQUE, J.J. WEIS, and J.P. HANSEN: J. Stat. Phys. 28, 325 (1982).
[5] S. LANG: Algebra (Addison-Wesley, Reading, Mass., 1965), p. 132.

Quantum Hall Effect – Questions and Answers

Questions by J. Hajdu

Institut für Theoretische Physik, Universität zu Köln
D-5000 Köln 41, Fed. Rep. of Germany

Answers by R.B. Laughlin

Lawrence Livermore, National Laboratory, P.O. Box 808
Livermore, CA 95440, USA

1. Dr. Laughlin, your gauge theory of the quantum Hall effect (QHE) is based on the equilibrium current formula

$$J_y = -\frac{1}{Ar}\left(\frac{\partial F}{\partial A_y}\right)_{T,\zeta} = -\frac{e}{Ar}\langle v_y \rangle_{eq} \qquad (1)$$

where F is the free energy and Ar is the area of the system, and v_y is the single-electron velocity. How do we know that the isothermal current (1) is the one which is measured in QHE experiments? This question arises because (1) is in general not equal to the current of an isolated system as given by the Kubo formula. In the case of an electron-impurity system which is infinite in the direction along the electric field (no edges), the two currents coincide if the Fermi energy is situated in any gap (zero density of states). Is it possible to demonstrate that the two types of current are equal as long as the Fermi energy lies in a mobility gap (σ_{xx} = 0)? The situation seems to be even more difficult to survey if the system is finite in the direction of the electric field and, therefore, in (1), edge current contributions have to be taken into account. Or is this a trivial matter?

The reason Eqn (1) is incompatible with the Kubo formula is that it is incorrect. The correct expression is

$$I = c\,\frac{\partial U}{\partial \phi} \quad .$$

The formula

$$I = c\,\frac{\partial A}{\partial \phi}$$

is valid only in the limit $T\to 0$, when A=U. The use of a free energy in this manner was an unfortunate short-cut made in my original paper on this subject (Phys. Rev. B23, 5632 (1981)). I had needed a very terse way to introduce chemical potentials. Derivation of this expression is trivial. One has

$$\vec{\nabla} \times \vec{E} = -\frac{1}{c}\dot{B}$$

$$\dot{U} = I \times \int \vec{E}\cdot d\vec{s} = -\frac{1}{c}\dot{\phi} \times I$$

$$I = -c\;\dot{U}/\dot{\phi}\,.$$

As I remark in my contribution to these Proceedings, the reason for invoking thermodynamics at all is to remove unphysical diamagnetism in the metallic contacts. The key point is that dissipation in the contacts can cause thermal equilibrium to be achieved locally without generating heat.

2. Essential for the gauge theory is the assumption of a double-connected belt (or annular) geometry. In this case gauge transformation can change the state of the system. The diamagnetic edge currents consist of two disconnected loops. In a real rectangular system, however, gauge invariance is maintained and there is only one edge current loop. What principle legitimates the transfer of results from the belt model to the single-connected reality?

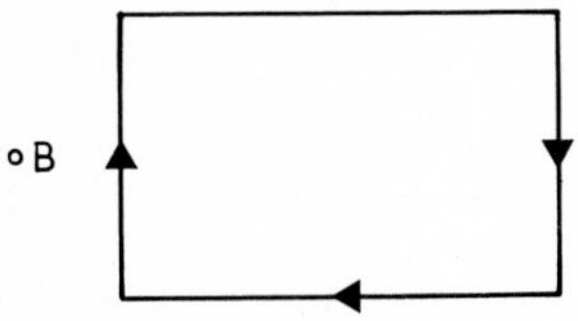

Reality

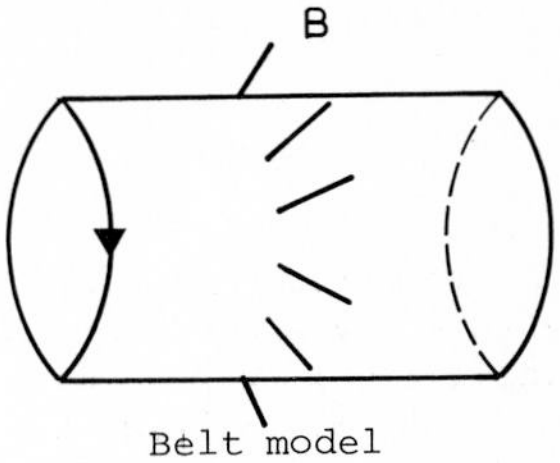

Belt model

The error incurred in approximating the strip geometry of a real experiment with a band have not been quantitatively assessed by anyone, including myself. However, if the conductivity is a bulk property, as are most conductivities, then "end" effects should be physically similar to "edge" effects, which are known experimentally to be small. Of course, it is conceivable that this conductivity is not a bulk property, as would be the case if all the current were flowing in the edges. I do not believe this is the case, firstly because the edges display ordinary metallic conductivity, and secondly because Hall currents are routinely seen in sample interiors.

3. Is the "gauge argument" a heuristic principle or a theorem which can be proved by the usual means of solid state theory? Why is it allowed to replace the gauge derivative $\partial F/\partial Ay$ by the quotient $\Delta F/\Delta Ay = Ly\Delta F/\phi_o$ ($\phi_o = h/e$)?

Not to be glib, but it seems to me that it is a theorem which already has been proved by the usual means of Solid State Physics. If you are asking whether a more complicated derivation exists, I would have to suppose so. It is allowed to replace the derivative with a differential only when the energy increases monotonically with flux through the loop. This is clearly not the case when the loop is small, so a necessary condition is that the system be macroscopic. Nonmonotonic increase of energy with flux just means physically that the conductance is flux dependent. Someone ought to calculate this Bohm-Ahronov component to see how big it is.

4. Consider a sequence of samples with different mobility.

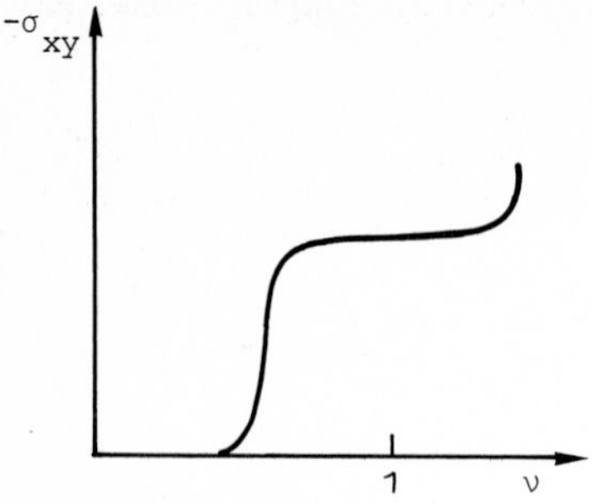

Quite good sample

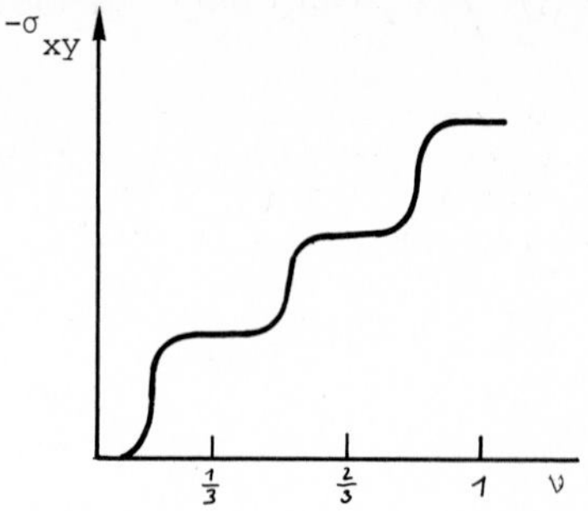

Good sample

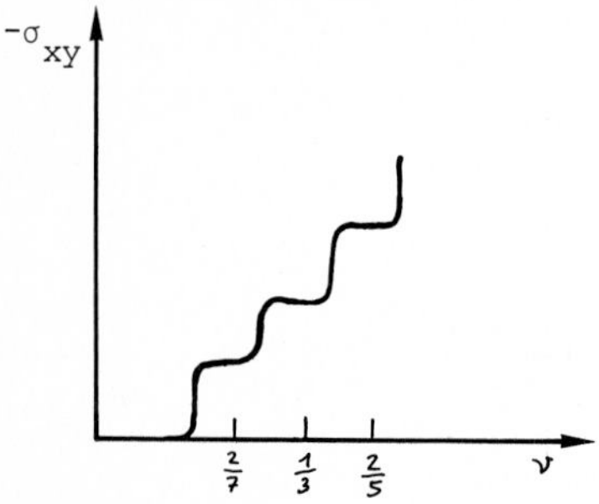

Better sample

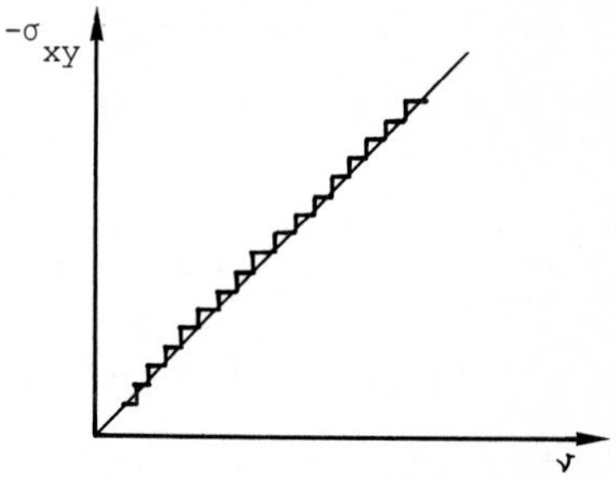

Nearly ideal sample (?)

What limits the width of the, say 1/3 plateau? Extrinsic reasons (localization) or intrinsic: "domains of dominance" of neighboring plateaux? At still higher mobilities: intrinsic behavior?, approach to free sliding? How? Any end to poliferation of plateaux?

Surely both "domains of dominance" and dirt are implicated in the plateau width. We still have no quantitative understanding of the interplay between these two factors. I cannot presently tell you how the system goes from broad plateaux to no plateaux as the disorder is turned off. That is a good problem! However, I can tell you that many fractions do not exist because they are unstable to crystallization. If not too many fractions are missing, the pictures you have drawn could be qualitatively correct.

5. Would you kindly comment on the nature of the fractionally charged excited states? In your description these are current-carrying quasiparticles whereas Anderson views them as equivalent ground states (having equal energy density in the thermodynamic limit). This corresponds to m-fold discrete symmetry breaking and provides an alternative explanation of the 2/m,... plateaux.

There is insufficient space to describe the quasiparticles in detail. Let it suffice to say that they behave almost in every way like electrons of charge e/m confined to their lowest Landau level. I think Anderson is wrong. I do not believe states differing in energy by the gap Δ should be considered degenerate.

6. Concerning the statistics of excitations different authors come to different conclusions: Laughlin: Fermi, Haldane: Bose, Halperin: any. Are they all justifiable?

I unfortunately do not understand my competitors' theories well enough to criticize them. The three of us are getting together this summer at Aspen for a "shoot-out at the O.K. Corral", during which I presume the truth will emerge. I am suspicious that ours are semantic differences. Here is a simple example of how that can happen. If Ψ is a fermion wavefunction in the lowest Landau level, then it can be written

$$\Psi(z_1,\ldots,z_N) = A(z_1,\ldots,z_N) \; e^{-1/4 \sum_\ell |z_\ell|^2} ,$$

where A is an antisymmetric polynomial. However, A can always be factored in the manner

$$A = P \prod_{j<k} (z_j - z_k) ,$$

where P is a symmetric polynomial. Thus to every fermion wavefunction Ψ there is a boson wavefunction Φ given by

$$\Phi(z_1,\ldots,z_N) = P(z_1,\ldots,z_N) \; e^{-1/4 \sum_\ell |z_\ell|^2} ,$$

and vice versa. Do we have fermions or bosons?

7. In your theory both particle and hole excitation exist. How about the formation of excitons and an excitonic condensate?

This subject would fill a third manuscript. Let me just say that I have just completed some work on this matter: yes excitons do exist, yes they do have an impact on the Fractional Quantum Hall Effect, and yes there has been talk of exciton-mediated phase transitions.

Patrik Fazekas kindly contributed to the formulation of the questions.

Index on Contributors

Applications of the Monte Carlo Method

in Statistical Physics

Editor: **K. Binder**

1984. 90 figures. XIV, 311 pages.
(Topics in Current Physics, Volume 36)
ISBN 3-540-12764-X

Contents: *K. Binder, D. Stauffer:* A Simple Introduction to Monte Carlo Simulation and Some Specialized Topics. – *D. Levesque, J. J. Weis, J. P. Hansen:* Recent Developments in the Simulation of Classical Fluids. – *D. P. Landau:* Monte Carlo Studies of Critical and Multicritical Phenomena. – *K. E. Schmidt, M. H. Kalos:* Few- and Many-Fermion Problems. – *A. Baumgärtner:* Simulations of Polymer Models. – *K. W. Kehr, K. Binder:* Simulation of Diffusion in Lattice Gases and Related Kinetic Phenomena. – *Y. Saito, H. Müller-Krumbhaar:* Roughening and Melting in Two Dimensions. – *K. Binder, D. Stauffer:* Monte Carlo of "Random" Systems. – *C. Rebbi:* Monte Carlo Calculations in Lattice Gauge Theories. – Additional References with Titles. – Subject Index.

Monte Carlo Methods

in Statistical Physics

Editor: **K. Binder**

1979. 91 figures, 10 tables. XV, 376 pages.
(Topics in Current Physics, Volume 7)
ISBN 3-540-09018-5

Contents: *K. Binder:* Introduction: Theory and "Technical" Aspects of Monte Carlo Simulations. – *D. Levesque, J. J. Weis, J. P. Hansen:* Simulation of Classical Fluids. – *D. P. Landau:* Phase Diagrams of Mixtures and Magnetic Systems. – *D. M. Ceperley, M. H. Kalos:* Quantum Many-Body Problems. – *H. Müller-Krumbhaar:* Simulation of Small Systems. – *K. Binder, M. H. Kalos:* Monte Carlo Studies of Relaxation Phenomena: Kinetics of Phase Changes and Critical Slowing Down. – *H. Müller-Krumbhaar:* Monte Carlo Simulation of Crystal Growth. – *K. Binder, D. Stauffer:* Monte Carlo Studies of Systems with Disorder. – *D. P. Landau:* Applications in Surface Physics.

Springer-Verlag
Berlin
Heidelberg
New York
Tokyo

Light Scattering in Solids I

Introductory Concepts

Editor: **M. Cardona**

2nd corrected and updated edition. 1983.
111 figures. XV, 363 pages.
(Topics in Applied Physics, Volume 8)
ISBN 3-540-11913-2

Contents: *M. Cardona:* Introduction. – *A. Pinczuk, E. Burstein:* Fundamentals of Inelastic Light Scattering in Semiconductors and Insulators. – *R. M. Martin, L. M. Falicov:* Resonant Raman Scattering. – *M. V. Klein:* Electronic Raman Scattering. – *M. H. Brodsky:* Raman Scattering in Amorphous Semiconductors. – *A. S. Pine:* Brillouin Scattering in Semiconductors. – *Y.-R. Shen:* Stimulated Raman Scattering. – Overview. – Additional References with Titles. – Subject Index. – Contents of Light Scattering in Solids II, III and IV.

Light Scattering in Solids II

Basic Concepts and Instrumentation

Editors: **M. Cardona, G. Güntherodt**

1982. 88 figures. XIII, 251 pages.
(Topics in Applied Physics, Volume 50)
ISBN 3-540-11380-0

Contents: *M. Cardona, G. Güntherodt:* Introduction. – *M. Cardona:* Resonance Phenomena. – *R. K. Chang, M. B. Long:* Optical Multichannel Detection. – *H. Vogt:* Coherent and Hyper-Raman Techniques. – Subject Index.

Springer-Verlag
Berlin
Heidelberg
New York
Tokyo

Light Scattering in Solids III

Recent Results

Editors: **M. Cardona, G. Güntherodt**

1982. 128 figures. XI, 281 pages.
(Topics in Applied Physics, Volume 51)
ISBN 3-540-11513-7

Contents: *M. Cardona, G. Güntherodt:* Introduction. – *M. S. Dresselhaus, G. Dresselhaus:* Light Scattering in Graphite Intercalation Compounds. – *D. J. Lockwood:* Light Scattering from Electronic and Magnetic Excitations in Transition-Metal Halides. – *W. Hayes:* Light Scattering by Superionic Conductors. – *M. V. Klein:* Raman Studies of Phonon Anomalies in Transition-Metal Compounds. – *J. R. Sandercock:* Trends in Brillouin Scattering: Studies of Opaque Materials, Supported Films, and Central Modes. – *C. Weisbuch, R. G. Ulbrich:* Resonant Light Scattering Mediated by Excitonic Polaritons in Semiconductors. – Subject Index.

Light Scattering in Solids IV

Electronic Scattering, Spin Effects, SERS, and Morphic Effects

Editors: **M. Cardona, G. Güntherodt**
1984. 322 figures. XIV, 542 pages.
(Topics in Applied Physics, Volume 54)
ISBN 3-540-11942-6

Contents: *M. Cardona, G. Güntherodt:* Introduction. – *G. Abstreiter, M. Cardona, A. Pinczuk:* Light Scattering by Free Carrier Excitations in Semiconductors. – *S. Geschwind, R. Romestain:* High Resolution Spin-Flip Raman Scattering in CdS. – *G. Güntherodt, R. Zeyher:* Spin-Dependent Raman Scattering in Magnetic Semiconductors. – *G. Güntherodt, R. Merlin:* Raman Scattering in Rare-Earth Chalcogenides. – *A. Otto:* Surface-Endhanced Raman Scattering: "Classical" and "Chemical" Origins. – *K. Arya, R. Zeyher:* Theory of Surface-Enhanced Raman Scattering. – *B. A. Weinstein, R. Zallen:* Pressure-Raman Effects in Covalent and Molecular Solids. – Errata for Light Scattering in Solids II (TAP 50). – Subject Index.